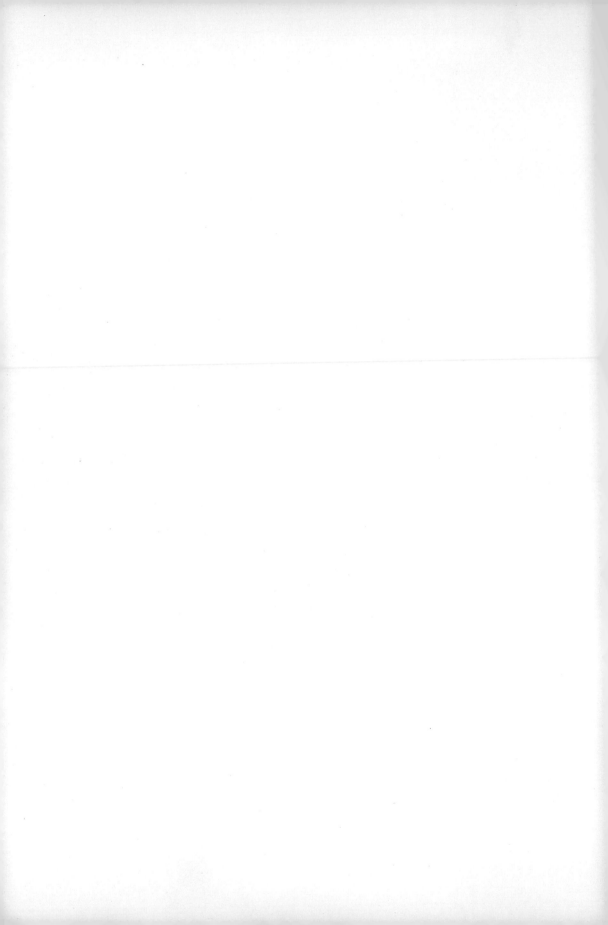

Lipids and Essential Oils as Antimicrobial Agents

Lipids and Essential Oils as Antimicrobial Agents

Editor

HALLDOR THORMAR

Faculty of Life and Environmental Sciences, University of Iceland, Reykjavik, Iceland

A John Wiley and Sons, Ltd., Publication

This edition first published 2011
© 2011 John Wiley & Sons, Ltd

Registered office
John Wiley & Sons Ltd, The Atrium, Southern Gate, Chichester, West Sussex, PO19 8SQ, United Kingdom

For details of our global editorial offices, for customer services and for information about how to apply for permission to reuse the copyright material in this book please see our website at www.wiley.com.

Library of Congress Cataloging-in-Publication Data

Lipids and essential oils as antimicrobial agents / editor, Halldor Thormar.
    p. ; cm.
  Includes bibliographical references and index.
  ISBN 978-0-470-74178-8 (cloth) – ISBN 978-0-470-97661-6 (ebook) – ISBN 978-0-470-97662-3 (obook)
 1. Anti-infective agents.   2. Lipids–Therapeutic use.   3. Essences and essential oils–Therapeutic use.   I. Thormar, Halldor.
  [DNLM: 1. Anti-Infective Agents.   2. Lipids–pharmacology.   3. Oils, Volatile–pharmacology.   QV 250]
  RM267.L57 2011
  615′.792–dc22

                                                                                    2010037014

A catalogue record for this book is available from the British Library.

Print ISBN: 9780470741788
eBook ISBN: 9780470976616
oBook ISBN: 9780470976623
ePub ISBN: 9780470976678

Typeset in 10/12pt Times by Aptara Inc., New Delhi, India.
Printed and bound in Singapore by Markono Print Media Pte Ltd.

Contents

List of Contributors

Akram Astani Department of Infectious Diseases, Virology, University of Heidelberg, Heidelberg, Germany

Gudmundur Bergsson Institute of Molecular Medicine, Trinity College Dublin, Dublin, Ireland

Carol L. Bratt Dows Institute for Dental Research, College of Dentistry, The University of Iowa, Iowa City, IA, USA

Kim A. Brogden Dows Institute for Dental Research, and The Department of Periodontics, College of Dentistry, The University of Iowa, Iowa City, IA, USA

Christine F. Carson Discipline of Microbiology and Immunology (M502), School of Biomedical, Biomolecular and Chemical Sciences, The University of Western Australia, Crawley, Western Australia, Australia

Deborah V. Dawson Dows Institute for Dental Research, and The Department of Preventive and Community Dentistry, College of Dentistry, The University of Iowa, Iowa City, IA, USA

David Drake Dows Institute for Dental Research, and The Department of Endodontics, College of Dentistry, The University of Iowa, Iowa City, IA, USA

Katherine A. Hammer Discipline of Microbiology and Immunology (M502), School of Biomedical, Biomolecular and Chemical Sciences, The University of Western Australia, Crawley, Western Australia, Australia

Hilmar Hilmarsson Faculty of Life and Environmental Sciences, University of Iceland, Reykjavik, Iceland

Charles E. Isaacs Department of Developmental Biochemistry, New York State Institute for Basic Research in Developmental Disabilities, NY, USA

Thórdís Kristmundsdóttir Faculty of Pharmaceutical Sciences, University of Iceland, Reykjavik, Iceland

Peter J. Quinn Department of Biochemistry, King's College London, London, United Kingdom

Jürgen Reichling Department of Biology, Institute of Pharmacy and Molecular Biotechnology, University of Heidelberg, Heidelberg, Germany

Paul Schnitzler Department of Infectious Diseases, Virology, University of Heidelberg, Heidelberg, Germany

Skúli Skúlason Faculty of Pharmaceutical Sciences, University of Iceland, Reykjavik, Iceland

Halldor Thormar Faculty of Life and Environmental Sciences, University of Iceland, Reykjavik, Iceland

Phil Wertz Dows Institute for Dental Research, and The Department of Oral Pathology, Radiology & Medicine, College of Dentistry, The University of Iowa, Iowa City, IA, USA

Introduction

There has recently been a renewed interest in the antimicrobial effects of natural compounds which were commonly used as health remedies in the Western world until the advent of antibiotic drugs in the 1940s and 50s. After the emergence of antibiotics many previously fatal infections and infectious diseases were brought under control and millions of lives were saved. Due to the dramatic effect of the new synthetic drugs, some health professionals even believed that the threat to mankind of pathogenic microorganisms had finally been eliminated.

The great success of chemotherapy, using synthetic antibiotics against bacterial and fungal infections and nucleoside analogues against viral infections, discouraged researchers and the pharmaceutical industry from making serious efforts to develop drugs containing simple natural compounds. However, this may now be changing, with the increasing problem of drug-resistant bacterial and viral strains, partly caused by drug overuse. Because the development of new drugs has not in all cases kept up with the emergence of new resistant strains of pathogens, such strains cause thousands of deaths annually, many in hospitals. Also, most synthetic drugs have more or less severe side effects, which affect a considerable number of patients. In spite of these drawbacks, the health benefits of antibiotics to humans and their domestic animals can hardly be overestimated.

It has become apparent to many medical microbiologists and health professionals that besides synthetic drugs, which inhibit the replication of pathogenic microorganisms in a specific way, there may be a place for less specific antimicrobial compounds, microbicides, which kill the pathogens on contact. Microbicides could act in concert with specific antibiotics, launching a two-pronged attack on the invading pathogens. Direct killing, in addition to growth inhibition of pathogens, might make the formation of antibiotic-resistant strains less likely. Due to their broad antimicrobial actions, resistance to microbicides is rarely observed.

The success of antibiotic drugs is due to the fact that our knowledge of their actions is based on a solid scientific ground. Their actions are in most cases predictable and their side effects known, because they have undergone a thorough scientific scrutiny, for both safety and activity, before being applied to the general public. In contrast, the use of natural health remedies was for a long time mainly based on anecdotal evidence and on accumulated experience of their beneficial effects obtained over centuries. The knowledge was mostly empirical. Recently, and mostly during the past few decades, the antimicrobial actions of the natural compounds, which originate in both the animal and the plant kingdom, have been studied by modern scientific methods similar to those applied in the study of synthetic drugs. This research has confirmed and extended the prior empirical knowledge of their antimicrobial activity.

In this book, scientific studies of the antimicrobial actions of two groups of naturally occurring organic materials are reviewed, namely lipids and essential oils. Lipids are diverse constituents of plants and animals which are insoluble in water but soluble in nonpolar organic solvents such as ethanol and ether. The main types of lipid are fats and oils, phospholipids, waxes and steroids. These lipids have various functions in the body. Animal fats and vegetable oils are triglycerides composed of fatty acids and glycerol and are a source of energy. Phospholipid molecules contain two fatty acids and are a major component of cell membranes. The hydrolytic products of triglycerides and phospholipids, particularly the fatty acids, have antimicrobial activities. In addition to being natural compounds they have the advantage of being both environmentally safe and generally harmless to the body in concentrations which kill pathogenic microbes. They are nonallergenic and are fully metabolized in the body. Lipids, particularly triglycerides, are abundant in nature and are an inexpensive source of antimicrobial products.

The first part of this book deals with various aspects of lipids as antimicrobial agents, beginning with an examination of the chemical nature of lipids in Chapter 1. The history of lipids as antimicrobial agents, from the discovery of the antibacterial activity of natural soaps in the 1880s until 1960, is told in Chapter 2, followed by a discussion in Chapter 3 of more recent studies of the antibacterial, antiviral and antifungal actions of lipids. After chapters on antimicrobial lipids in mother's milk (Chapter 4) and the skin (Chapter 5), Chapter 6 looks at the role of lipids in the natural host defence against pathogenic microorganisms, discussed in the context of other factors of the innate immune system. Recent studies strongly support earlier observations that natural fatty acids on the surface of the skin and mucous membranes contribute significantly to the host defence against infections by pathogenic microbes. Triglycerides in breast milk are hydrolysed by lipases in the gastrointestinal tract of infants, where they release free fatty acids, which seem to have an important protective effect against enteric pathogens. It has been suggested that the natural protective function of lipids could be enhanced by prophylactive or therapeutic application of drugs containing lipids as active ingredients. Chapter 7 reviews the application of lipids in pharmaceuticals, cosmetics and health foods and their possible use in therapeutics. A broad overview is given not only of antimicrobial activity but also of other health-related functions of lipids. Finally, Chapter 8 discusses antimicrobial lipids as disinfectants, antiseptics and sanitizers, for example in the food industry and in the home. The advent of antibiotics and inexpensive, synthetic detergents in the middle of the twentieth century caused a decline of interest in common hygiene outside of the hospital setting, for example in the home and in public places such as schools. The problem of drug-resistant pathogens and of environmental hazards caused by some synthetic disinfectants has led to an awareness of the advantage of using natural and environmentally friendly disinfectants and sanitizers. Thus, antibacterial lipids could, for example, be used to reduce the risk of contamination by foodborne pathogens in the kitchen and in food-preparing and food-processing facilities.

Essential oils of flowering plants are secondary metabolites which are a part of the defence system of the plants, defending them against herbivorous animals and microorganisms. They have been used as health remedies for centuries but until recently little scientific research was carried out to establish the antimicrobial effect of essential oils and their chemical components. In the past few decades, a vast amount of scientific data has been collected on this subject and the second part of the book gives an overview of the current knowledge of the antimicrobial and biological functions of essential oils. Chapter 9 reviews the chemistry and bioactivity of essential oils and their use in food and cosmetic products. Chapter 10 describes the antiviral activity of essential oils and their effect in treatment of herpes simplex. Finally, Chapter 11 gives a comprehensive overview of the antibacterial and antifungal effects of essential oils, their use in pharmaceutical formulations and their clinical efficacy and toxicity in humans and animals.

Although focussing on the antimicrobial action of lipids and essential oils, the book also describes various other health-related aspects of these natural products. It thus gives comprehensive and detailed information on the biological effects of lipids and essential oils based on the results of scientific research. Each chapter stands by itself and need not be read in the context of the others. Therefore, the chapters do not have to be read in sequence starting at the beginning of the book. There is thus a certain degree of overlap of data between chapters, but not redundancy, since the same data are viewed from different points of view and in different contexts by the different authors. Although written by scientists and aimed primarily at health professionals, the book is written in language which should be understandable to readers in general. It should be of interest to anyone concerned about health issues and particularly to those who are conscious of the benefits of health food and natural products.

Halldor Thormar
Reykjavik, June 2010

1

Membranes as Targets of Antimicrobial Lipids

Peter J. Quinn

Department of Biochemistry, King's College London, London, United Kingdom

Lipids and Essential Oils as Antimicrobial Agents   Halldor Thormar
© 2011 John Wiley & Sons, Ltd

1.1   Introduction

Certain lipids are known to inhibit growth or kill microbes. A wide variety of lipids have been tested and they vary widely in chemical structure and efficacy. Because of the lack of a systematic relationship between lipid structure and antimicrobial activity an explanation of their mode of action is problematic. The two possible molecular mechanisms to account for the antimicrobial action of lipids are 1) a specific interaction with sites within the microorganism that influences biochemical functions and loss of viability or 2) a general nonspecific interaction that perturbs the structure of the microorganism, thereby inhibiting normal physiological functions.

Lipids are a diverse class of compounds that defy definition by simple chemical characteristics but can be broadly categorized by their solubility in solvents of relatively low polarity. In this way they can be readily distinguished from the other constituents of living cells, such as nucleic acids, carbohydrates and proteins. Indeed, this is the operational basis for the extraction and purification of lipids from biological tissues. The conventional methods of lipid extraction employ solvent mixtures with a relatively polar solvent, initially to loosen up the tissue and dislodge lipids from their interaction with other cellular constituents and culminate in isolation of the lipids in a two-phase system in which the polar molecules partition into an aqueous phase [1, 2].

Within the general class of lipids, subdivisions are recognized on the basis of their relative solubility in solvents of low polarity, or to put it another way, solubility in solvents of different polarities. Again, no systematic chemical criteria can be adopted to account for solubility of lipids in solvents of different polarities. However, polarity of solvents can be defined by objective criteria [3, 4] and solubility of lipids in solvents can be measured by a variety of biophysical parameters.

This chapter will provide an account of lipid solubility in solvents of different polarities and explain the general chemical principles that govern this property. The relevance of lipid solubility to antimicrobial action will be discussed in the context of the role of lipids in the structure of cell membranes of living organisms and how these structures are disrupted by antimicrobial lipids.

1.2   Oil and Water Don't Mix!

We are all familiar with the old adage that oil does not mix with water. Equally, we know intuitively that if a drop of ink is placed in a beaker of water the ink will diffuse out from the concentrated drop, eventually distributing randomly throughout the aqueous phase. Both of these situations have expressions in the Law of Thermodynamics, which states that all systems move to their state of lowest free energy, and in the cases we are considering, to a more random and chaotic state. The formulation is:

$$\Delta G = \Delta H - T \Delta S \qquad (1.1)$$

where ΔG is the change in free energy of the system, ΔH is the change in heat, T is the absolute temperature and ΔS is the change in entropy. The negative sign on the $T \Delta S$ component signifies a spontaneous change from a more ordered to a disordered state.

At first sight there seems to be a contradiction in the two examples given above. The two-phase system of oil and water appears to be a perfectly ordered system, yet clearly this is the equilibrium position. To understand why this is the lowest free energy of the system we need to consider the consequences on the order of the molecules if we attempt to place oil into an aqueous environment. Oils are largely composed of hydrocarbon, and hydrocarbons are nonpolar, since they have electron distributions about the constituent atoms that are relatively even. By contrast, water is highly polar, with electrons attracted to the oxygen atom generating a surfeit of negative charge (a δ-negative charge) and creating a deficiency of negative charges (δ-positive charges) associated with the two hydrogen atoms. This is the basis of molecular polarity. When hydrocarbon is exposed to water the molecules of water in contact with the hydrocarbon lose their freedom to interact with like polar water molecules and consequently become ordered. It is this molecular order that results in a decrease in entropy and a consequent increase in free energy of the system.

Lipids of biological origin are not composed purely of hydrocarbon, and constituent atoms like oxygen and nitrogen bring about an asymmetric distribution of electrons within the lipid molecule. This provides opportunities for polar interactions with water, thereby increasing the entropy of the system. We next consider the origin of polarity of biological lipids and the common chemical strategies used in nature to achieve a polar character.

1.3   Polar Lipids

Living cells do not synthesize pure hydrocarbons directly, although compounds such as methane and ethane are byproducts of lipid metabolism of some microorganisms. In general, therefore, the lipids of living cells contain electrophilic atoms like oxygen, nitrogen, sulfur, phosphate and so on that confer a polar character on the lipid. The presence of polar groups on lipids renders them less soluble in nonpolar solvents and induces them to assemble into characteristic aggregates or dispersions in water rather than forming a two-phase oil and water system.

1.3.1   The Amphiphilic Character of Polar Lipids

The presence of a polar group in a lipid has considerable influence on the properties of the molecule apart from its solubility in water. This is particularly the case with complex lipid molecules in which domains of nonpolar hydrocarbon within the molecule are separate from polar moieties. This arrangement confers on the lipid the physical properties of an amphiphile (Greek: *amphi*, on both sides; *philos*, loving), in which the hydrocarbon portion achieves a lowest free energy in a nonpolar environment and the polar group can be hydrated.

Amongst the weakest amphiphiles in biological tissues are the fats and oils, which largely make up the nutritional reserves of triacylglycerides. Such lipids generally form two-phase systems in water because the carbonyl oxygens of the acyl ester bonds linking the fatty acids to the glycerol backbone are not sufficiently polar to balance the remaining hydrocarbon character of the lipid. This is reflected in the locations of oils and fats within cells in phase-separated droplets bounded by a membrane and sequestered from contact with water.

As the balance of hydrophilic affinity created by the hydrocarbon component increases, so does its tendency to interact with water. Polar lipid molecules are therefore able to bridge a hydrophilic and a hydrophobic environment and are said to be surface-active or posses the properties of a surfactant.

1.3.2   Hydrophobic Constituents of Lipids

The biosynthesis of lipids is conducted primarily by two metabolic pathways: the pathway for synthesis of sterols and related prenyl compounds, and fatty acid synthase. The products of both pathways provide lipids that are common constituents of cell membranes.

Sterols are all derived from isoprene substrate and perform essential structural and signalling roles in eukaryotes. In vertebrates cholesterol is the major structural sterol, whereas in plants a variety of sterols are found, including stigmasterol, sitosterol and camposterol, with other minor sterols. The sterol found mainly in microorganisms is ergosterol. Sterols are weakly polar, with an amphiphilic balance favouring hydrophobic interactions. The dominant feature of their structure is the polycyclic ring and the associated hydrocarbon side chain. The polar affinity resides in a hydroxyl group located on the sterol ring. Other important prenyl-derived compounds are the ubiquinols and plastoquinols, which are typified by relatively long hydrocarbon chains attached to a fully substituted benzoquinone ring system conferring a weakly polar character to the lipid.

The products of fatty acid synthase are fatty acids with a hydrocarbon carbon chain length of 16 carbons – palmitic acid. Palmitic acid can be elongated by the sequential addition of two-carbon units up to aliphatic carbon chain lengths of C24. Oxidative processes are also able to desaturate fatty acids at specific locations in the hydrocarbon chain, resulting in the insertion of *cis*-double bonds. The first double bond is inserted in the centre of the palmitate chain at position $\omega 9$ (the nomenclature counts from the terminal methyl carbon because biochemically this represents the sequence of desaturation of the alkyl chain) and subsequent desaturation occurs at $\omega 6$ and then $\omega 3$. Only plants are able to perform the last two desaturations, so that linoleic and linolenic acids are essential fatty acids for animals and represent vitamin F [5]. Other modifications of the aliphatic chain include the formation of branched chains by the attachment of methyl side-chain groups. Oxidative reactions of arachidonic acid represent the precursors of important signalling compounds, comprising the prostaglandins, thromboxanes and leucotrienes.

Fatty acids represent basic building blocks of complex polar lipids that are major constituents of biological membranes. The two principle classes of complex lipid that incorporate fatty acids are the glycerolipids and sphingolipids. The backbone of the glycerolipids is the anomeric polyol glycerol. Long-chain fatty acids are esterified to the C1 and C2 positions of the glycerol to form a diglyceride. Fatty acids are attached to the long-chain base, sphingosine, by an N-acyl linkage to form the basic building block of the sphingolipids, ceramide. The chemical structures of these hydrophobic components of complex polar lipids are illustrated in Figure 1.1.

Another group of long-chain hydrocarbons that are widely found in both eukaryotes and prokaryotes are waxes. These relatively nonpolar lipids constitute components of the so-called neutral lipid fraction and are composed of long-chain n-alkanolic acids and n-alken-1-ols of an even number of carbon atoms ranging in length from C12 to C32. Carbon chains of between C20 and C24 are commonly found in the acyl portion, whilst C24 and

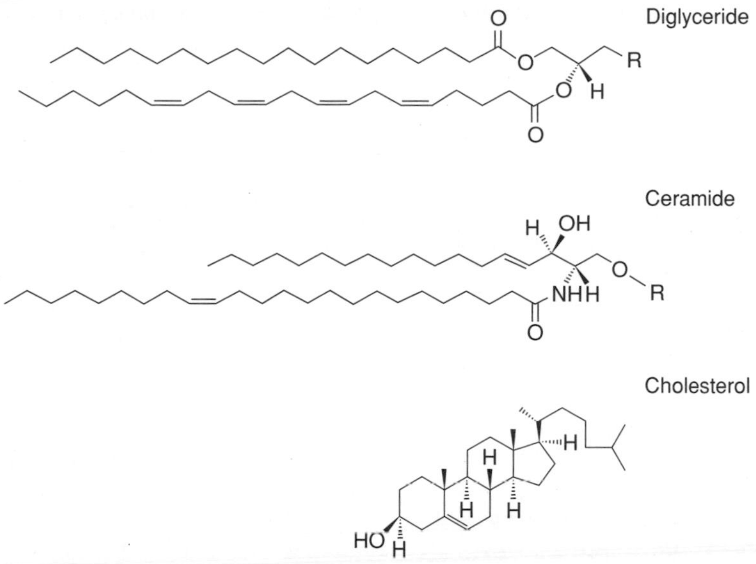

Figure 1.1 *The chemical structures of the hydrophobic components of complex polar lipids.*

C28 predominate in the alcohol component. A certain amount of unesterified hydrocarbon material may also be associated with wax esters as well as fatty aldehydes. Waxes can fulfil a protective role and the hydrophobic character of their structure can act as a reservoir that traps lipids with potential antimicrobial activities. Other lipids classified with the neutral lipids are carotenoids, sterol esters and glycerides.

1.3.3 Polar Groups of Complex Lipids

Polar groups that are responsible for the amphiphilic properties of complex cellular lipids are generally attached to hydroxyl groups of the diacylglycerol or ceramide. The type of polar group is used to designate the class of complex lipid. The structures of polar groups that represent the basis of classification of complex lipids are shown in Figure 1.2.

Arguably the most abundant polar lipids in nature are the glycosylglycerides, in which the diacyl glycerol is acylated to hexose sugars. While many different types of hexose may be attached glysidically to the C3 position of the glycerol, galactose is the most common in plant and algal systems. Mannose and glucose are more frequently encountered in bacterial species. Monogalactosyldiacylglycerols and digalactosyldiacylglycerols are the major lipids of the chloroplast thylakaoid membranes and their polar character is due to hydration of the sugar groups. The extent of hydration is increased in another lipid, 6-sulphoquinovosydiacylglycerol, by the presence of a sulphonate group on the sugar, and this lipid has been identified as a constituent of the membranes of all photosynthetic plants, algae and bacteria so far examined.

The most ubiquitous complex polar lipids are the phospholipids. The basic structure is phosphatidic acid, in which a phosphate group is esterified to the C3 of the diacylglycerol. Although this phospholipid is present only in minor proportions in tissue lipid extracts it is

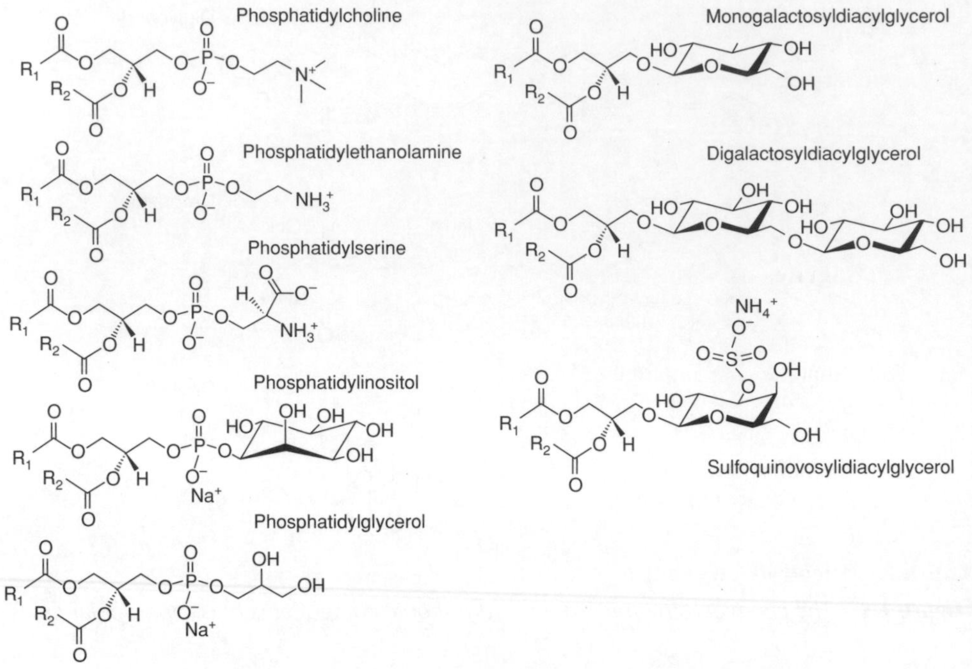

Figure 1.2   *The structures of polar groups of membrane lipids.*

a pivotal intermediate in the biosynthesis of the major phospholipid classes. Thus, different groups attached to the phosphatidic acid define the particular class of phospholipid. The major classes of phospholipids are based on choline, ethanolamine, serine, glycerol and inositol substituents on the phosphate group of phosphatidic acid. The amphiphilic property of the phospholipids is due not only to hydration of the negatively-charged phosphate but to hydroxyls, carboxyls and amino groups of the polar moieties.

A major class of sphingolipid is based on attachment of a phosphocholine group to ceramide to form sphingomyelin. The remaining sphingolipid classes, however, are not phospholipids but rely on the sugar residues attached to the hydroxyl residue of ceramide for their amphiphilic properties. The attachment of a single hexose such as glucose or galactose to the ceramide backbone constitutes the class of cerebrosides. More complex glycosphingolipids are the gangliosides and globosides. These polar lipids contain branched-chain oligosaccharides with as many as seven neutral and amino sugars attached to the ceramide. The presence of such large polar groups renders these classes considerably less soluble in nonpolar solvents.

In addition to the polar lipids normally found associated with living cells, a range of surfactant compounds are known to be biosynthesized by various microorganisms. Such biosurfactants have unique properties such as biodegradability and production under relatively benign environmental conditions and can be produced from substrates consisting of vegetable and even industrial waste materials. An important example of biosurfactants is the mannosylerythritol lipids produced by the yeast strain of the genus *Pseudozyma* [6, 7].

The amphiphilic properties of these glycolipid biosurfactants are due to sugar residues, which are often coupled to acetate groups and acylated with one or more hydrocarbon chains of varying length. Amongst the features of useful detergent-like properties [8,9], the glycolipid biosurfactants have been shown to promote differentiation [10] and apoptosis [11] of immortalized tumour cell lines in tissue culture as well as to bind antibodies [12,13] and lectins [14].

1.4   Properties of Surfactants

As described in the preceding sections, the dominant feature of structural lipids found in living cells is that they are amphiphilic. However, the balance of hydrophobic and hydrophilic affinities within the lipids varies greatly depending on the extent of the hydrocarbon domain of the molecule and the affinity of the polar group for water. This balance influences the ability of the polar lipids to act as surfactants bridging the interface between hydrophobic and hydrophilic environments.

1.4.1   Critical Micelle Concentration

A useful parameter to compare the relative surface activities of surfactants is the critical micelle concentration. The parameter is defined as the concentration of polar lipid in free solution that is in equilibrium with aggregates of surfactant. Where the solvent is water, the concentration of polar lipid in free solution decreases as the hydrophobic affinity within the molecule relative to the polar affinity increases.

Surfactants will tend to concentrate at interfacial regions between water and nonpolar solvents or air as this will represent the lowest free energy of the system. There are two aspects to this reduction in free energy. The first has been described above and results from the removal of hydrocarbon from contact with water. The second is to lower the free energy of the interface between the two media; that is, a reduction in surface tension. When all the interface is occupied by surfactant molecules a further increase in surfactant concentration will result in the formation of aggregates of the surfactant as again the association of surfactant molecules is configured so that hydrocarbon is sequestered away from water and the polar groups of the surfactant are exposed on the outside so as to form a stable aggregate.

A convenient method for measuring critical micelle concentration of a surfactant is to monitor the surface tension of water at an air interface with increasing concentration of surfactant in the subphase. This is illustrated in Figure 1.3. A comparison of the surfactant activity of common surfactants with the polar lipids of biological membranes shows that membrane lipids are only weakly surface-active. The critical micelle concentration for most membrane lipids is in the range of nM concentration. This means that membrane lipids are overwhelmingly present in aqueous systems in the form of aggregates. Biosurfactants, on the other hand, have considerably higher critical micelle concentrations. Mannosylerythritol lipids have critical micelle concentrations in the μM concentration range at interfacial surface tensions of about 30 mN/m [15]. This compares with the critical micelle concentration of typical domestic washing-up liquids, which are in the mM concentration range.

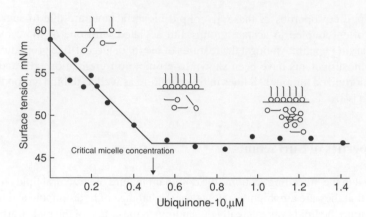

Figure 1.3   *Measurement of the critical micelle concentration of a surfactant by relating the decrease of the surface tension of formamide at an air interface as a function of surfactant concentration of the subphase. (Reprinted from [21]. With kind permission from Springer Science+Business Media.)*

1.4.2   Aggregation of Surface-Active Molecules

Amphiphiles self-assemble in solvents into a variety of aggregates with normal or reversed curvature. The structures have been classified according to their morphology and range from dispersed micellar to cubic to hexagonal to lamellar phases [16]. The rich polymorphism displayed by surfactants arises from the wide differences in amphiphilic balance within the molecules. The polymorphism of these structures can be modulated by changing the polar interaction with the solvent or by altering the van der Waals interactions between the hydrocarbon residues of the surfactant. Thus surfactants exhibit lyotropic mesomorphism in response to the changing hydration of the polar head group. Likewise, thermotropic mesomorphism can be demonstrated as a consequence of changes in packing order of the hydrocarbon domain at different temperatures.

This is illustrated by the simple binary mixture of glycerol monooleate. The temperature–composition phase diagram in the temperature range 20 to 105 °C and water content 0 to 100% (w/w) was first published 25 years ago [17]. The phase diagram was subsequently extended to temperatures below 20 °C, where complicated behaviour involving metastable phases was reported [18]. Nonionic detergents are known to have different effects on the phase behaviour of monoolein. For example, at full hydration at 20 °C alkyl glucosides in mole fractions up to 0.25 in monoolein form a cubic phase of space group Pn3m determined from X-ray scattering methods [19]. In the presence of higher mole fractions of detergent a liquid–crystal bilayer is the preferred phase, with an intermediate cubic phase of space group Ia3d observed in mole fractions of 0.4 nonionic detergents.

Somewhat different effects on the phase behaviour of monoolein/water systems have been reported for terpinen-4-ol, the active surfactant of tea-tree oil, as evidenced from optical and NMR studies [20]. The influence of the presence of 5 wt% terpinen-4-ol on the lyotropic phase behaviour of monoolein is shown in Figure 1.4. It can be seen that in low water contents (<5 wt%, D_2O) a micelle phase is formed by the monoolein, and this is not greatly altered by the presence of terpinen-4-ol. With increasing water content the

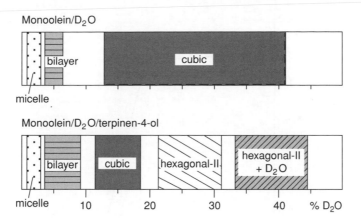

Figure 1.4 *The influence of the presence of 5 wt% terpinen-4-ol on the lyotropic phase behaviour of monoolein. (Reprinted with permission from [20]. Copyright 2002 American Chemical Society.)*

lamellar liquid–crystal phase is stabilized up to about 10 wt% water before forming the cubic phase. The major effect of the terpinen-4-ol is to destabilize the cubic phases and induce a hexagonal phase of type I. This effect is due to the partitioning of the terpinen-4-ol into the aqueous interfacial domain, where it serves to expand the area of lipid–water interface.

The general conclusion from studies of surfactants in aqueous systems is that the structure formed by the surfactant in concentrations above the critical micelle concentration depends primarily on the amphiphilic balance within the molecule. This can be illustrated in the case of a common membrane lipid, phosphatidylcholine. This phospholipid forms only lamellar phases in water over a wide range of temperatures. Removal of one of the fatty acyl residues from the glycerol produces lysophosphatidycholine. This nomenclature is in deference to the fact that the lipid will cause haemolysis of erythrocytes by a detergent action on the cell membrane. The dramatic shift in amphiphilic balance resulting from removal of one fatty acyl residue converts a weak surfactant into a relatively strong detergent. This is reflected in the type of aggregate formed by lysophosphatidycholine in water: hexagonal-I, which consists of tubes of lipid with acyl chains oriented into the interior and the alignment of the tubes in a hexagonal packing array.

1.4.3   The Influence of Solvent

A component of amphiphilic balance of surfactant molecules is the interaction of their polar groups with the solvent. The effect of the polarity of the solvent can be illustrated by the solubility of ubiquinone-10 in ethanol–water mixtures [21]. Figure 1.5 shows the concentration of ubiquinone-10 in the supernatant of centrifuged samples, judged by absorption at $\lambda = 275$, as a function of the volume of water added to an ethanol solution of the lipid. There is a marked decrease in solubility of ubiquinone-10 when the proportion of water in the solvent mixture exceeds about 12% by volume. Examination of the aggregated lipid by wide-angle X-ray scattering methods indicates that the benzoquinone ring of the molecule is not solvated in the aggregated form. Nevertheless, the demonstration shows that

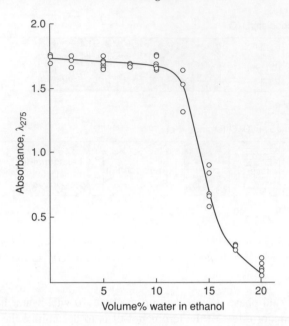

Figure 1.5 *The effect of polarity of the solvent on the solubility of ubiquinone-10 in ethanol–water mixtures. (Reprinted from [21]. With kind permission from Springer Science+Business Media.)*

increasing solvent polarity causes aggregation of the weak surfactant as the hydrocarbon component is forced to exclude increasingly polar solvent.

Modulation of the electrostatic charge of ionic surfactants greatly influences the phase behaviour of amphiphiles. The phospholipids of cell membranes contain charged phosphate groups and some classes contain additional amino or carboxyl groups. Counter-ions in the aqueous medium are known to play an important role in the phase behaviour of the lipids in the membrane bilayer matrix. An example of this is the physiological process of domain structure and fusion between bilayer membranes, which involves the creation of a nonlamellar transition state within the phospholipid bilayer. Calcium is a potent membrane fusogen which, by binding to the phosphate groups, reduces charge repulsion between adjacent bilayers, and by promoting aggregation is able to bridge the bilayers and promote fusion through the creation of defects in the lamellar structures [22].

A component of the charge neutralization of the phospholipid polar group by interaction with calcium is an alteration of the structure of the water layer that hydrates the interface. The thermodynamics of lipid hydration has been considered in thermal studies of multibilayer liposomes of membrane polar lipids that each take up between 10 and 30 molecules of water depending on the nature of the polar group and the state of the alkyl chains [23]. Measurements of the partial molar free energies and enthalpies of these swelling systems indicate that the first four water molecules hydrate the lipids in a favourable enthalpic interaction but the partial free energies and enthalpies of additional water molecules have opposite signs. There is also evidence that additional hydration involves changes in the thermal excitation of the lipid degrees of freedom.

1.5   Cell Membranes

As indicated in Section 1.3, there is a wide range of weakly polar lipids associated with cell membranes of living organisms. The other major component of membranes is the protein; because most incorporate sugar residues attached to the polypeptide chain, these are mainly glycoproteins. Since cell membranes represent important targets for antimicrobial lipids, we next consider the structure of cell membranes and the factors that govern their stability.

The contemporary view of the structure of biological membranes is formalized in the fluid-mosaic model proposed nearly 40 years ago [24]. The polar lipids are said to be arranged in a fluid bilayer structure which acts as a matrix for the orientation and organization of the different intrinsic and extrinsic proteins. While this view has proved to be reasonably durable, much greater detail has emerged about the disposition of the components on either side of the bilayer and within lateral domains of the bilayer. Furthermore, the notion that the lipid bilayer matrix is a fluid structure has been modified by the realization that various degrees of order in the lipids are created by the presence of sterols when they interact with sphingolipids [25].

1.5.1   Membrane Lipids

The types of polar lipid found in biological membranes were seen in Section 1.3 to include phospholipids, glycolipids and sterols. Apart from sterols and other neutral lipids, each of the major classes of polar lipid consists of a family of molecular species defined by the type of hydrocarbon chain associated with the lipid backbone. In the case of the diacylglycerophospholipids, the fatty acids acylated at the *sn*-1 and *sn*-2 position of the glycerol differ in length, degree of *cis*-unsaturation and position of attachment to the glycerol. In some members of the family the fatty acids may be branched-chain or they may be attached by ether or vinyl ether rather than ester bonds to the glycerol. The importance of this complexity can be understood by the differences that the hydrocarbon component of the lipid confers on the physical properties of the lipid.

One of the most important features governing the morphological characteristics of phospholipid assemblies in aqueous systems is the presence of *cis*-unsaturated bonds. These bonds, as opposed to double bonds in the *trans*-configuration, introduce a kink into the chain that prevents the close parallel alignment required to maximize van der Waals cohesive bonds between them. As a consequence, the temperature at which the chains are transformed from a solid to a liquid is greatly reduced. In the context of a lipid bilayer the transition between ordered and liquid–crystal lamellar phases as well as between lamellar and nonlamellar phases is determined by the number and location of *cis*-unsaturated bonds in the fatty acid residues associated with the phospholipid. Characteristics of the phases adopted by membrane lipids in aqueous systems are illustrated in Figure 1.6. A review of the behaviour of membrane lipids and their arrangement in the structure has been published recently [26].

With the advent of sophisticated mass-spectrometric methods for analysis of the lipid composition of biological membranes [27] it has been recognized that each morphologically distinct membrane contains a complex and distinctive lipid composition. In many membranes, hundreds of molecular species of lipids have been characterized. Moreover, the constituent lipids of each membrane appear to be preserved within relatively narrow

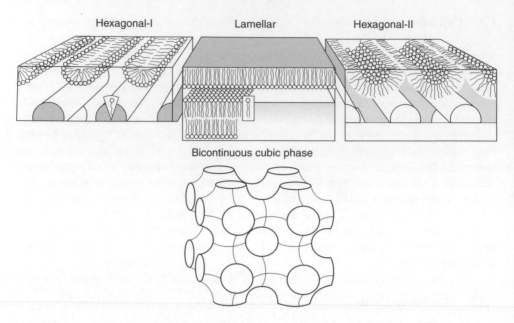

Figure 1.6   *Characteristics of the phases adopted by membrane lipids in aqueous systems.*

limits by biochemical mechanisms that are, as yet, not completely understood. The processes of membrane lipid turnover and homeostasis raise questions as to the reasons for the diversity of membrane lipid compositions. Why, for example, do the cells of higher organisms maintain such a complex lipid matrix despite being protected from variations in their environment by the homeostatic regulation of surrounding extracellular fluids or serum and the composition of the cytosol? We do not know whether the maintenance of narrow ranges (homeostatic regulation) of any particular molecular species of lipid in the membrane is critical to the cell or the functions performed by the constituent membranes and how much redundancy is incorporated into the system. There are also uncertainties as to whether lipid complexity is predicated by the assortment of proteins in particular membranes or whether lipids have their own rules for assembling which govern, in turn, the way proteins are 'passively' inserted into the matrix.

Lipid homeostasis for a complex mixture represents a cost in terms of metabolic expense and gene diversity, which encodes all enzymes responsible for catalysing the biosynthetic and degradation pathways. Multiple cross-regulation would be expected to achieve the characteristic composition of membrane. The maintenance of a complex lipid composition requires regulated pathways to repair the alterations (literally 'homeostasis') induced by the 'activated enzymes' perturbing the cell membranes.

Methods of establishing the physical consequences of the diversity of lipid composition have not kept pace with the biochemical characterization of this diversity in particular membranes. The main reason for this is that methods employed to characterize the conformation and structure of the membrane lipid matrix are averaging techniques, such as spectroscopy, that are unable to distinguish subtle differences in local environment created by domains dependent on lipid complexity. As a result, much information is indirect and based on the construction and examination of models to simulate the lipid matrix of biological membranes. Frequently such models diverge considerably from the real world because of the low energy barriers that separate conformational states in the complex mixtures which represent biological membranes that are less likely to occur in defined lipid mixtures. Furthermore, in terms of representation, some molecular species may be present in relatively minor proportions and would not be expected to greatly influence the phase structure of the membrane lipid matrix. Others, such as cholesterol and sphingomyelin in the plasma membrane, may dominate the lipid composition of the membrane and exert a major impact on the structure and properties of the membrane. Clearly, characterizing the physical consequences and influences of individual molecular species of membrane lipid on the structure and properties of the bilayer matrix remains a considerable challenge.

1.5.2 Lipid Domains in Membranes

The segregation of the lipids of cell membranes into separate domains is now known to underlie membrane functions like signal transduction, fusion and so on. [28, 29]. This realization has come about through the characterization of so-called liquid-ordered phases formed between choline phosphatides and cholesterol. Such phases are created by specific interactions between the molecules which segregate from domains of fluid lipids to form a platform or raft into which lipid-anchored membrane proteins are partitioned. The segregation of these proteins from the fluid phase of the membrane appears to be required for them to perform their function.

Cell membranes can be fractionated according to their solubility in mild detergents [30]. The detergent-insoluble fraction floats at low density upon gradient centrifugation. The membranes can easily be harvested, the residual detergent removed and the resulting membrane further purified or subfractionated. The membrane fractions isolated in this way are referred to as rafts and take the form of vesicles about 200 nm in diameter [31]. Membrane lipids act as platforms for the assembly of receptors on one side of the membrane and appropriate effecter proteins on the opposite side. This arrangement allows signals generated when a ligand binds to its receptor to be transduced across the membrane to the biochemical apparatus responsible for setting the physiological response in train. The polar lipids of the raft serve as filter devices, in order to include particular membrane proteins and exclude others. The lipids of the raft membrane are distinctive from those of the surrounding membrane and a specific interaction between the lipids is believed to be the mechanism underlying their segregation.

The characteristic feature of the lipid composition of membrane rafts is the predominance of phospholipids with saturated hydrocarbon chains and the high proportion of sterol. This is particularly evident in the molecular species of sphingomyelin in membrane rafts isolated from erythrocyte ghost membranes [32]. To investigate the factors that govern the partition of sphingomyelin into the rafts the molecular species of sphingomyelin recovered in the detergent-resistant membrane fractions were compared with those in the membrane ghosts from which they were derived. The fatty acids in amide ester linkage to the sphingosine isolated from human erythrocyte ghost membranes and corresponding raft fractions are shown in Table 1.1. Saturated molecular species of sphingomyelin dominate the raft membrane fraction at the expense of monoenoic fatty acids. This is achieved by an approximately threefold enrichment of C22 : 0 and C24 : 0 molecular species of sphingomyelin and the

Table 1.1   N-acyl-linked fatty acids of sphingomyelin in human erythrocyte ghosts and detergent-resistant membranes isolated from them by treatment with Triton X-100 and fractionation on a density gradient.

N-acyl fatty acid	Human erythrocyte sphingomyelin	
	Ghost membrane	Raft membrane
C16 : 0	41.45 ± 5.68	8.63 ± 0.74
C18 : 0	6.25 ± 0.74	6.36 ± 0.66
C18 : 1	1.15 ± 0.21	1.49 ± 0.21
C18 : 2	0.10 ± 0.01	0.18 ± 0.13
C20 : 0	0.70 ± 0.01	2.22 ± 0.03
C22 : 0	5.78 ± 0.48	14.72 ± 2.02
C24 : 0	20.02 ± 2.35	62.10 ± 5.68
C24 : 1	24.02 ± 1.87	1.47 ± 0.19
C26 : 0	0.33 ± 0.04	2.78 ± 0.21
C26 : 1	0.19 ± 0.01	0.05 ± 0.01
Total	100	100
Saturated	74.54	96.81
Monounsaturated	25.36	3.01
Polyunsaturated	0.10	0.18

Data from [32].

almost complete exclusion of molecular species of sphingomyelin associated with C24 : 1 fatty acid.

The raft fractions isolated under mild conditions appear to represent individual domains within the membrane. Evidence for this is that detergent-resistant membranes retain their original asymmetry and can be subfractionated by immunoprecipitation methods into vesicles that contain unique sets of antigens. In the case of neuronal cells, separation of vesicles with prion protein from vesicles displaying Thy-1 can be achieved. A similar segregation of these antigens is observed in the intact cell membrane [30]. Lipid analysis of prion protein and Thy-1 vesicles show that the two raft membranes have different lipid compositions and in turn these are different from that of the detergent-resistant membrane fraction from which they were derived [33]. Thus prion protein vesicles contain significantly more unsaturated, longer-chain lipids than Thy-1 vesicles and have a fivefold greater content of hexosylceramide. These results lead to the conclusion that unsaturation and glycosylation of lipids are major sources of diversity of raft structure.

While sphingomyelin and cholesterol have tended to achieve prominence in raft lipid composition, more recent studies have indicated that glycerophospholipids and diacylglycerols are also constituents of raft fractions. One method that has been used to identify such components is the detection of lipids from radioactive precursors that are isolated in raft fractions [34]. The method involves feeding cells with radiolabelled glycerol, fatty acids or water-soluble polar groups and identifying the complex lipid into which they are biosynthetically incorporated. Using this approach it was shown that raft fractions derived from human leukaemic T-cell line Jurkat had a considerably higher cholesterol content than the parent membrane and that polar lipids incorporating [^{3}H]-glycerol were present. These glycerophospholipids included choline, ethanolamine, serine and inositol phosphoglycerides, with a preponderance of phosphatidylcholine and phosphatidylserine. Incorporation of radiolabelled fatty acid precursors into the phospholipids showed preferential labelling of raft lipids with saturated fatty acids such as palmitic and stearic acids, rather than oleic, linoleic and arachidonic acids. These results are consistent with other studies indicating that the lipids isolated in the raft fractions contain predominantly saturated molecular species of membrane lipids [35, 36].

1.5.3 Membrane Proteins

Proteins are major constituents of membranes and vary in proportion to the polar lipids from 0.25 wt/wt lipid in central-nerve myelin to more than 3.5 wt/wt lipid in the inner mitochondrial and photosynthetic thylakaoid membranes. The association of the protein with the lipid bilayer matrix can occur via polar interactions, in which case the protein can be dislodged from the structure by modulating the ionic environment. These proteins are referred to as extrinsic or peripheral membrane proteins. Another group of proteins are the intrinsic proteins and these are in contact with the hydrocarbon chains of the lipids. They require detergents to extract them from the lipid bilayer matrix. A third group are lipid-anchored proteins, which are essentially water-soluble proteins that incorporate a covalently-bound lipid which anchors them to the bilayer. A typical anchor tethering such proteins to the external surface of the plasma membrane is glycerylphosphorylinositol, forming the (so-called) GPI-anchored proteins [37]. Lipid anchors associated with the proteins on the cytoplasmic surface of membranes are saturated fatty acids and isoprenyl groups [38].

Because there are no proteins or groups of proteins common to biological membranes it is agreed that proteins are not required to organize or direct the structure of membranes. This means that the structure of membranes is not a process that is directly under genetic control and the assembly of components is driven by entropic forces. Nevertheless, membrane proteins can exert an essential role in the differentiation of particular membrane structures. Examples of such structures include intercellular junctions and the grana stacks of chloroplast thylakoid membrane. Indeed, such structures serve to confine different proteins to domains within the same contiguous membrane.

The orientation of proteins within the bilayer matrix is vectoral. The unique arrangement of each protein with respect to the lipid bilayer is necessary to perform transport and other biochemical functions in a unidirectional manner. The structure of membranes is established during membrane biogenesis, which occurs by a process of expansion and differentiation of preexisting membrane. This involves incorporation of newly synthesized proteins and lipids, the post-translational modification of the proteins and metabolic retailoring of the lipids.

1.5.4 Membrane Stability

The thickness of biological membranes is in the order of 5 to 6 nm, while the surface area can be very large, considering for example the size of an ostrich egg. Soap bubbles are another example of stable thin structures held together by the forces holding the weak surfactants of membrane lipids together. These thickness-to-area-ratio arguments suggest a structure that is quite robust, but the cohesive forces keeping the molecules together are in fact relatively weak van der Waals cohesive forces. These forces arise from fluctuating dipoles created by the asymmetric electrostatic charge distribution about all atoms and the interacting force decreases according to the sixth power of the distance separating the atoms. Although they are relatively weak forces they add up over the surface of adjacent hydrocarbons to provide strong cohesion between lipid molecules. This, together with the entropic forces associated with exclusion of water considered in Section 1.2, is why biological membranes are inherently stable structures.

This stability may seem to contradict the dynamic properties of biological membranes. Many cellular processes, including membrane biogenesis, secretion, endocytosis and so on, require fusion between membranes. In order for membrane fusion to take place in an orderly manner, specific membrane proteins are involved to make sure that indiscriminate fusion is avoided. Secondly, the bilayer arrangement at the point of fusion must be disturbed and transient nonbilayer arrangements of the lipids are required to complete the fusion process. The question is, how can an essentially stable structure have a built-in mechanism to cater for a dynamic process like fusion? The answer is found in the diverse amphiphilic balance of the lipids that constitute the membrane bilayer matrix.

It has long been known that the total polar lipid extracts of biological membranes do not form bilayer structures when dispersed in aqueous media under physiological conditions [39]. Instead they form a range of structures from bilayer to hexagonal-II and cubic arrangements. We can conclude from this observation that a proportion of the lipids found in cell membranes are inherently unstable in a bilayer configuration under physiological conditions and that interaction with other membrane components is required for them to conform to a bilayer arrangement. The membrane lipids that, when dispersed in pure form in aqueous media, assume a nonbilayer structure (mainly hexagonal-II structure) are the phosphatidylethanolamines and monohexosyldiacylglycerols. Indeed, all biological membranes contain a proportion of membrane lipids that form nonbilayer structures. It can now be seen that dynamic processes like the fusion between membranes can be accomplished by controlled dissociation of nonbilayer-forming lipids from the constraints that normally preserve them in a bilayer configuration.

Most biological membranes have a much greater proportion of nonbilayer-forming lipids than would be required to participate in membrane fusion events, so other functions may be ascribed to these lipids. One such function is to seal the interface between intrinsic membrane proteins and the lipid bilayer matrix. Because the amphiphilic balance in nonbilayer-forming lipids favours a hydrophobic affinity they are ideally placed to occupy the voids created by the irregular polypeptide surface, thereby blocking the passage of water-soluble solutes across the membrane. Their function could be seen as sealing the intrinsic proteins into the lipid bilayer matrix. To put this in context, the inner mitochondrial and chloroplast thylakoid membranes both have a relatively high proportion of nonbilayer-forming lipids: cardiolipin in the case of mitochondrial membrane and monogalactosyldiacylglycerol in thylakoid membrane. These membranes have a high intrinsic membrane protein-to-lipid ratio, are required to be impermeable to protons and, despite coming into close proximity with adjacent membrane surfaces, do not fuse.

The consensus picture that emerges from consideration of these factors is that biological membranes are stable barriers to the passive diffusion of water-soluble solutes and yet are potentially unstable due to possession of nonbilayer-forming polar lipids that, if separated into pure phases, could destroy the properties of the membrane and ultimately lead to loss of viability of the cell. The critical factor that preserves the functional integrity of cell membranes is the maintenance of a balance of components within the structure that prevents uncontrolled dissociation of nonbilayer-forming lipids from their interaction with molecules constraining them into a bilayer configuration. Because environmental factors play a critical role in the tendency of nonbilayer-forming lipids to form nonbilayer structures there is a dichotomy between homoiothermic and poikilothermic organisms in how they handle this process.

1.5.4.1 Membrane Lipid Phase Behaviour

The phase that membrane lipids form when dispersed in physiological media at the growth temperature depends on both the type of hydrocarbon component and the hydration and charge of the polar group. In general, the molecular species of choline-, serine-, glycerol- and dihexosyl-based polar lipids extracted from biological membranes form fluid bilayer structures. The molecular species of ethanolamine- and monohexosyl-based polar lipids and cardiolipins form hexagonal-II structures. The factors that determine the particular phase that will form include the packing constraints determined by the structure of the lipid, and environmental factors like temperature, pressure, water activity and solutes [40].

Nonbilayer-forming lipids will form bilayer structures at temperatures below the growth temperature. The temperature at which these lipids undergo a bilayer to nonbilayer transition is increased by increasing the length of the hydrocarbon chains, decreasing the number of *cis*-unsaturated bonds and increasing the hydration of the head group [41, 42]. These thermotropic and lyotropic effects have considerable influence on the stability of biological membranes when shifted from environmental conditions that support growth of cells.

Consider the consequences of exposing cell membranes to low temperatures [43]. The lamellar gel to liquid–crystal phase transition temperature of nonbilayer-forming lipids is considerably higher than the same transition observed in bilayer-forming lipids with equiv- alent hydrocarbon-chain components. This means that in mixtures of molecular species of lipid that occur in cell membranes, cooling below the growth temperature first causes the nonbilayer lipids to be transformed into fluid bilayers and mix with the fluid bilayer-forming lipids. This is usually a reversible process in cell membranes because on rewarming the nonbilayer lipids reassociate with membrane components that constrain them into a bilayer configuration. However, further cooling results in the phase separation of the nonbilayer- forming lipids into lamellar gel phase under conditions in which the bilayer-forming lipids are in a fluid phase. The permeability-barrier properties and integrity of the membrane are compromised when the membrane is rewarmed because the phase-separated gel phase is able to form nonbilayer structures as it becomes concentrated during formation of the gel phase. The lipid–phase separation model is illustrated by the cartoon in Figure 1.7.

Equally damaging lipid–phase separations have been identified in biological membranes that have been subjected to heat stress. Thus chloroplast thylakoid membranes first destack when heated to temperatures higher than 35 °C, and this is associated with dissociation of the light-harvesting chlorophyll–protein complexes of photosystem II from the reac- tion centre and the phase separation of nonbilayer lipid structures at temperatures above 45 °C [44]. Similar effects have been noted as a result of salt stress, in which there is a loss

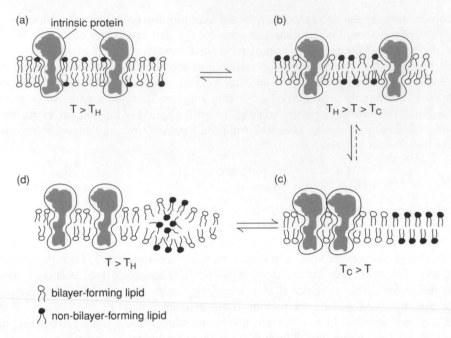

Figure 1.7   *The lipid–phase separation model of low-temperature damage to biological membranes.*

of photosynthetic functions that can be counteracted by an increase in the proportion of membrane lipids with polyunsaturated fatty acyl residues and the presence of compatible solutes [45].

1.5.4.2   *Membrane Lipid Homeostasis in Homoiothermic Organisms*

As has been noted in the above discussion, the lipid composition of the membranes of homoiothermic organisms is highly complex, consisting of hundreds of individual molecular species. Cells have many subcellular organelles which are distinguished not only by the particular proteins but by the mix of lipid molecular species. Clearly, because the identity of each morphologically distinct membrane is preserved within relatively narrow limits, biochemical mechanisms must be in place to sense the membrane composition and to replace lipids that have been oxidized or otherwise damaged.

Integral to this is the process of membrane biogenesis. Membranes are not synthesized *de novo* but are differentiated from existing membrane by insertion of newly synthesized lipids and proteins. With the notable exceptions of egg yolk, lung surfactant and the prolamellar bodies of protoplasts, cells do not synthesize more membrane lipids than they need to satisfy their growth requirements. The primary site of synthesis of membrane lipids is the endoplasmic reticulum; with some classes of lipid required for mitochondria and chloroplasts synthesized in these particular organelles. Membrane buds off from the endoplasmic reticulum to undergo differentiation through the *trans*-Golgi network, ultimately forming plasma membrane at the cell periphery. Lipids may also be exchanged between different membranes by a process mediated by soluble lipid-exchange proteins, of which many specific for particular lipids have been identified. Lipids are also constantly undergoing metabolic turnover: this may be by routes that involve reactions which do not result in products that lose their identity as lipids, for example partial turnover by acyl exchange between lipids. Alternatively, lipids may be degraded to water-soluble products and replaced by new lipids.

It can be seen that preserving the huge array of molecular species of lipid in each subcellular membrane of cells is a prodigious task. A significant proportion of the energy consumption of cells is devoted to the biosynthesis of membrane lipids to serve the process. Indeed, the greatest part of the oxygen uptake by most cells is consumed in the desaturation of fatty acids rather than in synthesis of ATP.

1.5.4.3   Membrane Lipid Adaption in Poikilothermic Organisms

When poikilothermic organisms are exposed to changed environmental conditions they respond by changing the lipid component of their cell membranes. This process of adaptation is based on biochemical sensors that detect the physical properties of the lipid bilayer and modulate biosynthetic pathways responsible for retailoring the membrane lipids [46]. The changes in membrane lipid composition resulting from a shift to lower temperatures vary depending on the particular organism but can include one or a combination of the following factors:

1. An increase in the number of unsaturated double bonds in the fatty acyl chains. This is accomplished by the action of fatty acid desaturases, which insert *cis*-unsaturated bonds at specific locations in the fatty acid.
2. A partial turnover of the membrane lipids *in situ*. This can involve the retailoring of molecular species of lipid by acyl-exchange reactions.
3. A change in the proportions of the major lipid classes. Some phospholipid classes are enzymaticlly interchangeable, so that the exchange results in generation of new molecular species within each phospholipid class.
4. A shortening of the fatty acyl chains. This can be brought about by shortening existing chains or by *de novo* synthesis of shorter chains.
5. A change in the type of fatty acid, such as incorporation of branched-chain fatty acids or an alteration in the proportions of *iso-* and *antiiso*-branched fatty acids.
6. An overall increase in the rate of lipid biosynthesis, which results in an increase in the lipid : protein ratio of the membrane.

Upward shifts in the growth temperature are generally associated with a reversal of the changes that are seen on exposure to low temperatures.

Because these adaptive changes involve biochemical reactions, the rate of adaptation is reflected in the extent of the environmental perturbation. Furthermore, strategies have evolved in some poikilothermic organisms, like higher plants, to attenuate the effects of temperature by changing the way the constituents of the membrane are organized. Plants, it must be emphasized, must be able to adapt to diurnal temperature ranges of many tens of degrees, which requires more than biochemical responses.

1.6   The Action of Antimicrobial Lipids on Cell Membranes

Since the physical properties of antimicrobial lipids are those of mild surfactants their primary target is the lipid bilayer matrix of cell membranes. There are no significant differences in the basic structures of membranes of pro- and eukaryotes with respect to the functions of the lipid components of these membranes but the proportion of the major lipid classes and the molecular species of lipid within each of these classes will differ from one membrane type to another. Thus there is currently little evidential basis for predicting how a particular antimicrobial lipid will act to selectively perturb one membrane as opposed to another. The basis of selectivity in the absence of this information is to target the antimicrobial lipid to the organism of interest and protect the host as far as possible by avoiding damaging exposure.

There is nevertheless a considerable amount of information about the action of surfactants on membrane lipids that may guide inquiry on the action of weak surfactants, in which antimicrobial lipids are included. The impetus for studies of detergent action on biological membranes comes from the recognition that selective detergent extraction of cell membranes provides a useful method of fractionation of functional domains of the structure. Detergent fractionation relies on the selective solubilization of membrane lipids, suggesting that antimicrobial lipids, like detergents currently employed in membrane fractionation, are targetted to particular lipids of the membrane bilayer matrix.

The detergents used in membrane fractionation are of the general type of nonionic surfactants, which comprise of an alkyl chain or other hydrocarbon moiety linked to a polyhydroxylated polar group. Not all are nonionic, as some zwitterionic surfactants, such as CHAPS, have similar properties to those of nonionic surfactants in solubilization of membrane lipids. Amongst the most widely used nonionic detergents for fractionating biological membranes is Triton X-100 [47]. This has a hydrophobic component consisting of octylphenol coupled with polyethyleneglycolether. It is a relatively strong surfactant with a critical micelle concentration of 0.2 mM at 25 °C. Other nonionic detergents, including octylglucoside, Lubrol and Brij96, have been examined for their selective action in solubilizing membrane lipids, and all have been found to be different from one another. From this it may be concluded that while the underlying mechanism of detergent solubilization of membrane lipids may be the same, the patterns of solubilization differ from one detergent to another. This suggests that while the mechanism of action of different antimicrobial lipids on the lipid bilayer matrix of target membranes may be identical they are likely to differ in their perturbation of particular membranes.

The example of Triton X-100 is instructive in providing a perspective on how a particular surfactant acts on different membrane lipids because of the considerable body of evidence available on its properties and behaviour. Bilayers of pure phosphatidylcholines are completely solubilized by Triton X-100 at effective detergent : lipid ratios of about 3.5 : 1 when the phospholipid is in the liquid-disordered state. Interestingly, phosphatidylcholines with unsaturated fatty acyl chains are more resistant to solubilization by Triton X-100 than are those with fully saturated chains. Solubilization of phosphatidylcholine with saturated chains in the gel phase appears to be dependent on the length of the acyl chains [48]. Homologues of phosphatidylcholines with diacyl chains C14 and C15 are readily solubilized by Triton X-100 at 20 °C with a detergent : lipid molar ratio of 5 : 1, whereas phospholipid homologues with C16 or longer chains remain completely insoluble under these conditions.

Fluid bilayers of sphingomyelin are more readily solubilized by Triton X-100 than are those of phosphatidylcholines under the same conditions, requiring detergent : lipid ratios of only about 1.5 : 1 to achieve complete solubilization. However, sphingomyelin in the gel phase is almost insoluble at 20 °C in the presence of Triton X-100.

The presence of cholesterol, but not other sterols like androsterol which lack the hydrocarbon tail, reduces the tendency of phospholipids to be solubilized by Triton X-100. This effect has been ascribed to the formation of liquid-ordered phases by interaction of the sterol with the phospholipid. It is noteworthy that androsterol is not able to form liquid-ordered phases with membrane phospholipids [49]. Liquid-ordered phases are more stable in binary mixtures of cholesterol and sphingomyelin. It has been shown that binary mixtures of sphingomyelin with 15 mole% cholesterol are insoluble in Triton X-100 at 50 °C with a detergent : lipid molar ratio of 5 : 1 and insoluble at 22 °C with 20 mole% cholesterol [50].

The selective solubilization of lipids in ternary mixtures of phosphatidyl-choline/sphingomyelin/cholesterol by Triton X-100 has been examined in mixtures where the proportions of sphingomyelin and cholesterol were maintained equimolar while the proportion of phosphatidylcholine varied [51]. The results indicated that regardless of the forces that maintain the liquid-ordered structure of sphingomyelin–cholesterol, the lipids became more susceptible to detergent solubilization as the proportion of phosphatidyl-choline in the mixture was increased. Such studies are complicated however by the fact that although the proportions of sphingomyelin and cholesterol in the mixture are equimolar, cholesterol also partitions into the phosphatidylcholine domain in proportion to the size of this domain. Also, caution is required in the interpretation of the results of such studies because the Triton X-100 itself may create ordered lipid domains from otherwise homogeneous fluid bilayers formed by lipid mixtures and some attempts have been made to account for this in explaining the selective solubility of fluid bilayer domains [52].

1.7 Conclusions

On the basis of these studies it is likely that antimicrobial lipids, which have a lower critical micelle concentration than Triton X-100 and related detergents, will interact with the less ordered domains of cell membranes. These domains are associated with the maintenance of the permeability-barrier properties and interpolation of intrinsic membrane proteins into the lipid bilayer matrix. Their mode of action on biological membranes in this case is to perturb the selective permeability properties of the target membrane, resulting in dissipation of solute gradients. The influence of such lipids on membrane protein functions cannot be discounted and indeed has been considered as the basis of a new approach to the modulation of cell behaviour [53]. Whether targeting of antimicrobial lipids to disordered lipid domains of prokaryotic membranes can be engineered to be selective is a challenge yet to be met.

Acknowledgements

The work was aided by a grant from the Human Frontier Science Programme (RGP0016/2005C).

References

1. Folch, J., Lees, M. and Sloane Stanley, G.H. (1957) A simple method for the isolation and purification of total lipids from animal tissues. *J. Biol. Chem.*, **226**, 497–509.
2. Dole, V.P. and Meinertz, H. (1960) Microdetermination of long-chain fatty acids in plasma and tissues. *J. Biol. Chem.*, **235**, 2595–2599.
3. Snyder, L.R. (1974) Classification of the solvent properties of common liquids. *J. Chromaotogr. A.*, **92**, 223–230.
4. Snyder, L.R. (1978) Classification of solvent properties of common liquids. *J. Chromatogr. Sci.*, **16**, 223–234.
5. Fulco, A.J. (1974) Metabolic alterations of fatty acids. *Ann. Rev. Biochem.*, **43**, 215–241.
6. Kitamoto, D., Isoda, H. and Nakahara, T. (2002) Functional and potential applications of glycolipid biosurfactants: from energy-saving materials to gene delivery carriers. *J. Biosc. Bioeng.*, **94**, 187–201.
7. Lang, S. (2002) Biological amphiphiles (microbial surfactants). *Curr. Opin. Colloid Interface Sci.*, **7**, 12–20.
8. Arutchelvi, J.I. Bhaduri, S., Uppara, P.V. and Doble, M. (2008) Mannosylerythritol lipids: a review. *J. Indust. Microb. Biotech.*, **35**, 1559–1570.
9. Imura, T., Ohta, N., Inoue, K. *et al.* (2006) Naturally engineered glycolipid surfactants leading to distinctive self-assembled structures. *Chem. Eur. J.*, **12**, 2434–2440.
10. Wakamatsu, Y., Zhao, X., Jin, C. *et al.* (2001) Mannosylerythritol lipid induces characteristics of neuronal differentiation in PC12 cells through an ERK-related signal cascade. *Eur. J. Biochem.*, **268**, 374–383.
11. Zhao, X., Wakamatsu, Y., Shibahara, M. *et al.* (1999) Mannosylerythritol lipid is a potent inducer of apoptosis and differentiation of mouse melanoma cells in culture. *Cancer Res.*, **59**, 482–486.
12. Imura, T., Ito, S., Azumi, R. *et al.* (2007) Monolayers assembled from a glycolipid biosurfactant from *Pseudozyma* (Candida) antarctica serve as a high-affinity ligand system for immunoglobulin G and M. *Biotechnol. Lett.*, **29**, 865–870.
13. Ito, S., Imura, T., Fukuoka, T. *et al.* (2007) Kinetic studies on the interactions between glycolipid biosurfactants assembled monolayers and various classes of immunoglobulins using surface plasmon resonance. *Colloids Surf. : Biosurfaces*, **58**, 165–171.
14. Konishi, M., Imura, T., Morita, T. *et al.* (2007) A yeast glycolipid biosurfactant, mannosylerythritol lipid, shows high binding affinity towards lectins on a self-assembled monolayer system. *Biotechnol. Lett.*, **29**, 473–480.
15. Morita, T., Konishi, M., Fukuoka, T. *et al.* (2008) Production of glycolipid biosurfactants, mannosyletythritol lipids, by *Pseudozyma siamensis* CBS 9960 and their interfacial properties. *J. Biosci. Bioeng.*, **105**, 493–502.
16. Larsson, K., Quinn, P.J., Sato, K. and Tyberg, F. (2006) *Lipids: Structure, Physical properties and Functionality*, The Oily Press, Bridgewater, UK.
17. Hyde, S.T., Ericsson, B., Andersson, S. and Larsson, K. (1984) A cubic structure consisting of a lipid bilayer forming and infinite periodic minimum surface of the gyroid type in the glycerolmonooleate-water system. *Z. Kristallogr.*, **168**, 213–219.
18. Qui, H. and Caffrey, M. (2000) The phase diagram of the monoolein/water system: metastability and equilibrium aspects. *Biomaterials*, **21**, 223–234.
19. Misquitta, Y. and Caffrey, M. (2003) Detergents stabilize the cubic phase of monoolein: Implications for membrane protein crystallization. *Biophys. J.*, **85**, 3084–3096.
20. Caboi, F., Murgia, S., Monduzzi, M. and Lazzari, P. (2002) NMR investigation on *Melaleuca alterniflolia* essential oil dispersed in the monoolein aqueous system: phase behaviour and dynamics. *Langmuir*, **18**, 7916–7922.
21. Ondarroa, M., Sharma, S.K. and Quinn, P.J. (1986) Solvation properties of ubiquinone-10 in solvents of different polarity. *Biosci. Rep.*, **6**, 783–796.
22. Estes, D.J., Lopez, S.R., Fuller, A.O. and Mayer, M. (2006) Triggering and visualizing the aggregation and fusion of lipid membranes in microfluidic chambers. *Biophys. J.*, **91**, 233–243.

23. Wennerstrom, H. and Sparr, E. (2003) Thermodynamics of membrane lipid hydration. *Pure Appl. Chem.*, **75**, 905–912.
24. Singer, S.J. and Nicolson, G.L. (1972) The fluid mosaic model of the structure of cell membranes. *Science*, **175**, 720–731.
25. Quinn, P.J. and Wolf, C. (2009) The liquid-ordered phase in membranes. *Biochim. Biophys. Acta*, **1788**, 33–46.
26. van Meer, G. Voelker, D.R. and Feigenson, G.W. (2008) Membrane lipids: where they are how they behave. *Nature Rev. Mol. Cell Biol.*, **9**, 112–124.
27. Wolf, C. and Quinn, P.J. (2008) Lipidomics: Practical aspects and applications. *Progr. Lipid Res.*, **47**, 15–36.
28. Harding, A.S. and Hancock, J.F. (2008) Using plasma membrane nanoclusters to build better signalling circuits. *Trends Cell Biol.*, **18**, 364–371
29. Shaikh, S.R. and Edidin, M.A. (2006) Membranes are not just rafts. *Chem. Phys. Lipids*, **144**, 103.
30. Chen, X., Jen, A., Warley, A. *et al.* (2009) Isolation at physiological temperature of detergent-resistant membranes with properties expected of lipid rafts: influence of buffer composition. *Biochem. J.*, **417**, 525–533.
31. Chen, X., Lawrence, M.J., Barlow, D.J. *et al.* (2009) The structure of detergent-resistant membrane vesicles from rat brain cells. *Biochim. Biophys. Acta*, **1788**, 477–483.
32. Koumanov, K.S., Tessier, C., Momchilova, A.B. *et al.* (2005) Comparative lipid analysis and structure of detergent-resistant membrane raft fractions isolated from human and ruminant erythrocytes. *Arch. Biochem. Biophys.*, **434**, 150–158.
33. Brugger, B., Graham, C., Leibrecht, I. *et al.* (2004) The membrane domains occupied by glycosylphosphatidylinositol-anchored prion protein and Thy-1 differ in lipid composition. *J. Biol. Chem.*, **279**, 7530–7536.
34. Rouquette-Jazdanian, A.K., Pelassy, C., Breittmayer, J.P., Cousin, J.L. and Aussel, C. (2002) Metabolic labelling of membrane microdomains/rafts in Jurkat cells indicates the presence of glycerophospholipids implicated in signal transduction by the CD3 T-cell receptor. *Biochem. J.*, **363**, 645–655.
35. Seminario, M.C. and Bunnell, S.C. (2008) Signal initiation in T-cell receptor microclusters. *Immunol. Rev.*, **221**, 90–106.
36. Dietrich, C., Bagatolli, L.A., Volovyk, Z.N. *et al.* (2001) Lipid rafts reconstituted in model membranes. *Biophys. J.*, **80**, 1417–1428.
37. Orlean, P. and Menon, A.K. (2007) GPI anchoring of protein in yeast and mammalian cells, or: how we learned to stop worrying and love glycolipids. *J. Lipid Res.*, **48**, 993–1011.
38. Resh, M.D. (2004) Membrane targeting of lipid modified signal transduction proteins. *Subcell. Biochem.*, **37**, 217–232.
39. Quinn, P.J. (1989) Membrane lipid phase behaviour and lipid-protein interactions. *Subcell. Biochem.*, **14**, 25–95.
40. Shearman, G.C., Ces, O., Templer, R.H. and Seddon, J.M. (2006) Inverse lyotropic phases of lipids and membrane curvature. *J. Phys. Condens. Matter*, **18**, S1105–S1124.
41. Seddon, J.M., Cevc, G., Kaye, R.D. and Marsh, D. (1984) X-ray diffraction study of the polymorphism of hydrated diacyl and dialkylphosphatidylethanolamines. *Biochemistry*, **23**, 2634–2644.
42. McIntosh, T.J. (1996) Hydration properties of lamellar and non-lamellar phases of phosphatidylcholine and phosphatidylethanolamine. *Chem. Phys. Lipids*, **81**, 117–131.
43. Quinn, P.J. (1985) A lipid-phase separation model of low-temperature damage to biological membranes. *Cryobiology*, **22**, 128–146.
44. Gournaris, K., Brain, A.P.R., Quinn, P.J. and Williams, W.P. (1984) Structural reorganisation of chloroplast thylakoid membranes in response to heat stress. *Biochim. Biophys. Acta*, **766**, 198–208.
45. Allakhverdiev, S.I. and Murata, N. (2008) Salt stress inhibits photosystems II and I in cyanobacteria. *Photosynth. Res.*, **98**, 529–539.
46. Guschina, I.A. and Harwood, J.L. (2006) Mechanisms of temperature adaptation in poikilotherms. *FEBS Lett.*, **580**, 5477–5483.

47. Goni, F.M., Alonso, A., Bagatolli, L.A. *et al.* (2008) Phase diagrams of lipid mixtures relative to the study of membrane rafts. *Biochim. Biophys. Acta*, **1781**, 665–684.
48. Patra, S.K., Alonso, A. and Goni, F.M. (1998) Detergent solubilization of phospholipid bilayers by surfactants. *Biochim. Biophys. Acta*, **1373**, 112–118.
49. Gao, W.Y., Chen, L., Wu, R.G. *et al.* (2008) Phase diagram of androsterol-dipalmitoylphosphatidylcholine mixtures dispersed in excess water. *J. Phys. Chem. B.*, **112**, 8375–8382.
50. Patra, S.K., Alonso, A., Arrondo, J.L.R. and Goni, F.M. (1999) Liposomes containing sphingomyelin and cholesterol: detergent solubilization and infrared spectroscopic studies. *J. Liposome Res.*, **9**, 247–260.
51. Sot, J., Collado, M.I., Arrondo, J.L.R. *et al.* (2002) Triton X-100-resistant bilayers: effect of lipid composition and relevance to the raft phenomenon. *Langmuir*, **18**, 2828–2835.
52. Keller, S., Tsamaloukas, A. and Heerklotz, H. (2005) A quantitative model describing the selective solubilization of membrane domains. *J. Am. Chem. Soc.*, **127**, 11469–11476.
53. Escriba, P.V., Gonzalez-Ros, J. M., Gonui, F.M. *et al.* (2008) Membranes: a meeting point for lipids, proteins and therapies. *J. Cell. Mol. Med.*, **12**, 829–875.

2

Antibacterial Effects of Lipids: Historical Review (1881 to 1960)

Halldor Thormar

Faculty of Life and Environmental Sciences, University of Iceland, Reykjavik, Iceland

2.1 Introduction

In 1881, Robert Koch [1] wrote a report on disinfection, which was based on a number of experiments, particularly on the effect of various chemical compounds on *Bacillus anthracis*. A few years earlier he had shown that this bacterium is the cause of anthrax, by eliciting the disease in experimental animals injected with the bacterium grown in pure culture. Thus, the anthrax bacillus was the first bacterium conclusively demonstrated to be the cause of a disease. This discovery made it possible to test the effects of various disinfectants *in vitro*, not only against the antrax bacillus but also against a number of other

Lipids and Essential Oils as Antimicrobial Agents   Halldor Thormar
© 2011 John Wiley & Sons, Ltd

bacterial species which in the following years were shown to be the causes of well-known diseases, such as cholera, typhoid fever, tuberculosis and diphtheria. On page 271 in his report on disinfection, Koch drew attention to an observation which suggested that certain fatty acids inhibited bacterial growth. However, he could not confirm this observation by further experiments, which showed that neither butyric acid nor oleic acid in 1 : 2000 dilution had an effect on the growth of anthrax bacillus. On the other hand, potassium soap in 1 : 5000 dilution had an inhibiting effect and in 1 : 1000 dilution (0.1%) completely stopped the growth of the anthrax bacillus. Since free potassium had a much smaller effect, Koch concluded that a different component of the soap he used, most likely some kind of fatty acid, caused the inhibition of bacterial growth. This was the first indication that lipids might have an antimicrobial effect and that ordinary soaps, which are potassium or sodium salts of fatty acids, could be used as disinfectants as well as for cleaning – a function that had been known for thousands of years. In the following years, Koch's discovery led to numerous studies on the antimicrobial effects of various types of soap on a variety of pathogenic bacteria, which at the end of the nineteenth century caused serious illnesses in the Western world, but are now a problem only in Third World countries.

The early studies of the effect of soaps on bacteria were largely of a practical nature; that is, they sought to determine to what extent soaps could be used for disinfection under various conditions. However, there was also an early interest in the question of which components of soaps were the active ingredients. After it had been established that the antibacterial activity resided in the fatty acid moiety, studies were done on the mechanism by which soaps either inhibit the growth of bacteria or kill them. There was even an interest in a possible therapeutic use of soaps or fatty acids and in whether or not fatty acids and their salts found in body tissues might play a role in natural defence against bacterial infection.

With the advent of antibiotics in the early 1940s the interest in fatty acids and other lipids as antibacterial agents decreased greatly, at least temporarily, but there was a new interest in their effect on viruses and particularly on fungi. Fungicidal dosage forms containing fatty acids or salts thereof were formulated for use against fungal infections, particularly of the skin and mucosal membranes.

2.2   Antibacterial Activity of Soaps

Following the report by Koch in 1881 on the antibacterial effect of potassium soap [1], a considerable effort was made to test the possible germicidal effect of different types of soap against pathogenic bacteria and thus their potential as general disinfectants. These early studies were reviewed by Reithoffer in 1896 [2], Serafini in 1898 [3] and Konrádi in 1902 [4]. According to these authors, the results of Koch on the anthrax bacillus were confirmed by Max Kuisl in 1885, but in his hands the potassium soap in 5% concentration had no inhibiting effect on typhoid or cholera bacteria. Furthermore, they quoted studies by Di Mattei from 1889, who tested sodium soaps against four different species of bacteria. In contrast to Kuisl, Di Mattei found that soaps killed cholera bacteria (*Vibrio cholerae*) in culture medium within minutes to hours, whereas it took longer to kill typhoid and anthrax bacilli, and *Staphylococcus pyogenes aureus* isolated from pus survived for at least 24 hours. In 1890, Behring [5] studied the effect of about 40 different commercial soaps on the anthrax bacillus and concluded that their activity depended on their free alkali contents.

In 1893, Jolles [6] tested the activity of five different soaps against the cholera bacterium in bouillon culture and found that all five soaps, in 8 to 10% solution, killed the bacteria within one to two minutes at 15 °C, as determined by plate culture of samples taken at different times. At decreasing soap concentration it took increasingly longer time to kill the bacteria, but the killing time decreased with increasing temperature. He found very little difference in the germicidal activities of the five soaps against the cholera bacilli. In a study published in 1895, Jolles [7] tested the activity of a potassium soap against typhoid and coli bacilli in bouillon culture. He found that a 6% soap solution killed the typhoid bacteria within 15 minutes at 4 to 8 °C and the killing time increased with decreasing soap concentration. The killing time was longer at 18 °C than at lower temperatures. The coli bacteria were more resistant than the typhoid bacilli and were killed in one hour by a 5 to 7% soap solution at 4 to 8 °C. The killing time decreased with increasing soap concentration, but was slightly longer at 18 and 30 °C than at the lower temperatures. Using a recent isolate from a fatal case of cholera, Nijland in 1893 [8] tested the killing of cholera bacilli in water by a number of different commercial soaps, either white hard sodium soap or green soft potassium soap. He found that the potassium soap was more active than the sodium soap, totally killing the bacteria in the water in 10 minutes at a concentration of 0.24%. In 1896, Beyer [9] tested a number of commercial soaps with various free alkali contents against cholera bacilli, typhoid bacilli, *Bacterium coli*, *S. pyogenes aureus* and diphtheria bacilli, all recent isolates from patients. The soaps were tested in a concentration of 3% at various temperatures. They were found to differ in their germicidal activities but, in contrast to Behring's results, showed no correlation with the free alkali contents of the soaps.

Reithoffer [2] pointed out the conflicting results of previous studies on the germicidal activity of soaps, which he considered to be partly caused by faulty test methods, such as adding the soap solutions to bacteria in bouillon culture. He therefore used a standardized protocol to study the activity of three different commercial soaps, in which a 24-hour bacterial culture was suspended in sterile distilled water and then filtered to remove clumps of bacteria. An equal volume of a soap solution was then added to the bacterial suspension and mixed at room temperature. Samples were removed from the mixture at intervals, inoculated into sterile bouillon and then regularly passed into a second bouillon. The bouillon cultures were incubated for eight days at 37 °C. A 10% soap solution killed cholera bacilli within half a minute under these experimental conditions. Sodium or potassium hard soaps[1] in 2% solutions killed the bacteria within one minute, whereas the killing time for a 2% solution of potassium soft soap[2] was two to five minutes. The potassium hard soap was even highly germicidal in 1% solution, killing the bacteria in less than one minute. The killing time was five minutes for a 0.5% soap solution and then increased rapidly at lower concentrations so that a 0.1% solution of potassium hard soap did not kill all the bacteria in two hours. A different strain of cholera bacilli isolated in Africa (Massaua) was also killed by low concentrations of soap, but in a slightly longer time. Thus, Reithoffer concluded that soaps are highly germicidal against cholera bacilli. The same method was used to test the germicidal activity of the soaps against typhoid bacilli and *B. coli*. The activity against these bacteria was found to be less than against cholera bacilli and a 10% concentration

[1] Hard soaps include a variety of soaps of different ingredients, usually sodium soaps, and are hard and compact.
[2] Soft soaps are of a green or brownish yellow color and of semifluid consistence. They are made from potash or the lye from wood ashes. They are strongly alkaline and often contain glycerin.

was needed to kill the bacteria within one minute. As for cholera bacilli, the potassium hard soap was the most active and killed within three minutes in a 5% solution. On the other hand, the three soaps were almost without germicidal activity against *S. pyogenes aureus*, even in a 20% solution for one hour.

Serafini [3] studied the germicidal activity of a number of commercial soaps against the Massaua strain of cholera. In his experiments, 1 ml of a 24- to 48-hour bouillon culture was added to 10 ml of sterile soap solution in either distilled or tap water and the mixture was kept at 10 to 15°C. Samples were removed at intervals and inoculated into culture tubes with bouillon culture medium. The different soaps, both hard sodium soaps and soft potassium soaps, were found to kill the bacterium, the killing time decreasing from a few hours to a few minutes when the soap concentration increased from 0.25% to 5%. There was, however, a distinct difference in the germicidal activities of the soaps. Serafini speculated about the possible cause for this difference and, therefore, which constituent of the commercial soaps was responsible for their germicidal activity. Based on his results he concluded, in agreement with Beyer, that the active ingredient was not free alkali but rather the neutral sodium and potassium salts of fatty acids – that is, the soap proper – as also suggested by Koch [1]. However, based on his own results and the results of others, Serafini could not detect a difference in the germicidal activities of sodium and potassium soaps. On the other hand, Konrádi [4] in 1902 concluded, after studying a commercial perfumed soap, that its germicidal activity against anthrax spores was due to an odorous essential oil, terpineol, not to salts of fatty acids.

In 1908, Reichenbach [10] addressed the question of the conflicting results of studies on germicidal activity of soaps and claimed that they were not due to different research methods but rather to a difference in the composition of the commercial soaps used by the various investigators. He emphasized the importance of determining separately the germicidal activities of the main components of commercial soaps, that is 1) salts of fatty acids, 2) free alkali and 3) other additives, and he was the first to carry out a systematic work with chemically pure compounds. For his studies, Reichenbach synthesized the potassium salts of purified fatty acids, which were commonly found in soaps, and tested each salt against *B. coli*. A 24-hour culture of the bacterium was suspended in water. After filtering, three drops of the suspension were added to 10 ml of a soap solution, which was thoroughly mixed, and samples were removed at intervals and inoculated into culture medium. Of the salts of saturated fatty acids, the potassium palmitate showed the highest activity, a 0.72% concentration killing the bacteria in less than five minutes. Stearate and particularly myristate were somewhat less active, and the activities of salts of shorter saturated fatty acids decreased with the chain length. In contrast, potassium salts of long-chain unsaturated fatty acids showed very little germicidal activity against *B. coli*, even at concentrations as high as 13 to 14%. From these results, Reichenbach concluded that the difference in activity of individual potassium salts against *B. coli* partly explained the difference in activity of various commercial soaps, which are made from a variety of fatty acids. Reichenbach also separated free alkali from soaps, which was mostly in the form of carbonates, and tested the activity against *B. coli*. In agreement with the results of Serafini [3], the free alkali could not explain the germicidal activity of soaps. The same was true of additives such as terpineol, although they were found to have a slight activity against typhoid bacilli.

In all studies of the germicidal action of soaps, care was taken to prevent carry-over of residual germicidal activity from the bacterial soap mixtures to the assay medium, such as bouillon, by dilution of the soap mixture. Geppert in 1889 [11] first drew attention to the false results that might be caused by such carry-over and advised the use of a neutralizing agent to stop the action of a disinfectant.

2.2.1   Antibacterial Action of Fatty Acids and Their Derivatives

Following the work of Reichenbach [10] on the bactericidal action of salts of purified fatty acids, studies were increasingly focussed on the fatty acids, which are commonly found in soaps, and their derivatives. In the literature, the salts of fatty acids were commonly referred to as 'soaps'. In 1911, Lamar [12] reported the results of a study of the action of sodium oleate on a virulent strain of pneumococcus. When sodium oleate was added to a broth culture of the pneumococcus in 0.5 to 1% concentration the diplococci were killed within 15 to 30 minutes. Higher dilutions of the sodium oleate were less active and 0.1% solutions did not kill the pneumococci in one hour.

In an extension of the previous study, Lamar [13] tested the haemolytic activity of a number of free fatty acids and their potassium and sodium salts. He used a mixture of unsaturated fatty acids, mostly oleic and linoleic acids, extracted from the livers of rabbits poisoned with phosphorus, and also several single isolated acids from other sources, including oleic, linoleic and linolenic acids. All of the unsaturated fatty acids were strongly haemolytic in 1 : 20 000 to 1 : 40 000 dilution and caused a complete lysis of red cells from dogs in one hour. The potassium and sodium salts were haemolytic in a higher dilution than the corresponding acids, with the highest acivity shown by potassium linolenate in a 1 : 150 000 dilution. The potassium salts were tested against pneumococci in the same way as previously reported for sodium oleate [12]. There was a close relationship between their germicidal activity against virulent pneumococci and their lytic activity for red blood cells. Potassium linoleate and linolenate were the most germicidal, being four and six times as destructive as sodium oleate.

Nichols [14] tested the bactericidal activity of chemically pure sodium salts of oleic and stearic acids and of brown soap, which is commonly used for dish washing. He found that a 0.5% solution of sodium oleate killed *Streptococcus pyogenes*, *Bacillus influenzae* and pneumococci in less than half a minute at room temperature, whereas *Staphylococcus aureus* and typhoid bacilli were not killed in 10 minutes. In contrast, sodium stearate had no or very little germicidal effect on any of the bacteria tested. Brown soap had a similar germicidal effect as the oleate, although it took two minutes to reduce the number of streptococci from innumerable to 15. The bactericidal activity of the soaps was dependent on the pH and was lost if it fell from about pH 8.5 to 7 by addition of acid, probably due to precipitation of the soap from solution.

Walker, in a series of papers [15–18], studied the germicidal properties of chemically pure soaps. He emphasized the practical importance, from a hygienic point of view, of knowing which soaps are germicidal against the different pathogens. In addition he pointed out other aspects which make studies of fatty acids and their salts worthwhile, namely their possible role in the defence of the body against bacteria and their potential as therapeutic compounds, for example in treatment of wounds.

Potassium and sodium salts were made from a series of chemically pure fatty acids and diluted in distilled water. Bacteria in broth culture were added to the soap solution at 20 °C and samples were taken after 2.5 and 15 minutes and tested for viability by inoculation into broth. The soaps were tested against typhoid bacilli, haemolytic streptococci, pneumococci and staphylococci. They were found to vary markedly between themselves and with regard to their germicidal activity against the different bacteria. Of the soaps of saturated fatty acids, laureate was the most active, a 0.002% solution killing the pneumococci in 15 minutes. The shorter- and longer-chain saturated soaps were less active. Among the 18-carbon unsaturated soaps the activity increased with the number of double bonds, from oleates to linolenates. There was no or very little difference in the activities of sodium and potassium salts, indicating that the germicidal activity resides in the fatty acid moiety. The pneumococci were the most susceptible to the germicidal activity of all the soaps and the streptococci were somewhat less sensitive. The typhoid bacilli were not killed by the unsaturated soaps and only by low dilutions of the saturated soaps, of which 0.7% palmate was the most active. The staphylococci were resistant to all of the soaps. These results [15] were therefore in agreement with those of Nichols [14]. Of the soaps studied, sodium and potassium laurates seemed to have the most general germicidal activities, since they killed pneumococci and streptococci in high dilutions and also had an appreciable activity against typhoid bacilli, which might be sufficient to prevent the spread of the bacteria and thus be of hygienic value. Addition of serum, and to a lesser extent of broth, was found to inhibit the germicidal activity of soaps, as demonstrated by the effect of serum on the activity of sodium laurate against pneumococci. The serum effect was less pronounced in the case of typhoid bacilli. In contrast to the results of Jolles [7], it was found that the germicidal activity of soaps increased with temperature.

In a paper published in 1925 [16], Walker reported a study of the germicidal properties of commercial soaps as well as of neutral soaps prepared in the laboratory from known fats and oils. The experiments were carried out in the same way as in the previous study and the same bacteria were used, except that diphtheria bacilli were tested instead of pneumococcci. Since all of the commercial soaps contained mixtures of different fatty acids they showed similar activities against each type of bacterium, although coconut-oil soap was the most active against the typhoid bacillus at 20 °C. On the other hand, there was a distinct difference in the activities of the pure soaps against the different germs. The diphtheria bacillus was the most susceptible, similar to pneumococci, followed by the streptococci. The typhoid bacillus was much more resistant and *S. aureus* was not killed by any of the soaps under the experimental conditions. All the soaps were more germicidal at 35 °C than at 20 °C. Sodium laurate and potassium palmitate were tested against coli bacillus and found to be slightly less active against this bacterium than against the typhoid bacillus. Both types of bacterium were resistant against sodium oleate and sodium linolate at 20 °C for 15 minutes. However, high concentrations of coconut-oil soap manufactured in the laboratory killed the coli bacilli within one minute and a commercial coconut-oil soap within 2.5 minutes. The germicidal activity of coconut-oil soaps against typhoid bacilli and to a lesser degree against coli bacilli seemed to be due to their high contents of saturated fatty acids, particularly lauric acid, and very low amounts of unsaturated acids.

In a later publication [17], Walker extended his studies of soaps to include meningococcus and gonococcus. He found both cocci to be susceptible to chemically pure soaps of both saturated and unsaturated fatty acids tested at 20 °C for 2.5 and 15 minutes. Since meningococcus was somewhat less susceptible than gonococcus to the germicidal action of all the chemically pure soaps, its susceptibility to four common commercial soaps was also tested. As expected, the meningococcus was killed by all of the soaps in 2.5 minutes at 20 °C at a soap concentration of approximately 0.04%. Chemically pure soaps were also tested against dysentery and paratyphoid bacilli. Only soaps of saturated fatty acids were found to kill these intestinal bacteria and their germicidal activities were almost identical to those previously found for the typhoid bacillus. A commercial soap prepared from coconut oil, and therefore mostly containing salts of saturated fatty acids, was found to kill these intestinal bacilli in 15 minutes at 20 °C in a concentration of approximately 0.4 to 1.6%.

From these studies, Walker [17, 18] concluded that the bacteria he had studied could be placed into two groups. One group, which included pneumococci, gonococci, meningococci, streptococci and diphtheria bacilli, was highly susceptible to sodium and potassium salts of both saturated and unsaturated fatty acids and was readily killed by any ordinary soap. In the other group were the enteric bacteria, that is typhoid, paratyphoid, dysentery and coli bacilli, which showed a moderate susceptibility to salts of saturated fatty acids. They were only killed by soaps which contained these salts, such as coconut-oil soap, but were completely resistant to soaps of unsaturated acids at ordinary temperatures.

Following the study by Nichols [16], who found that the sodium salt of oleic acid was most germicidal at alkaline pH, Eggerth in 1926 [19] did a thorough study of the effect of pH on the germicidal activity of chemically pure potassium soaps on a series of fatty acids. Serial dilutions of the soaps were made in distilled water and the pHs of the mixtures of bacteria and soap were controlled by buffers ranging from pH 3.8 to 10.0. The soap solutions were tested against *Streptococcus pyogenes*, *S. aureus*, *V. cholerae*, *Bacillus diphtheriae* and *Bacillus typhosus*. The tests were performed by inoculating samples of bacterial broth cultures into the appropriate buffer, then adding the soap solutions. The mixtures were incubated at 37 °C and samples were removed at intervals ranging from 30 minutes to 18 hours and inoculated on to agar plates.

The somewhat complicated data emerging from this study showed the germicidal activity of soaps to vary not only with the pH, but also with their chain length and saturation, and with the species of bacterium studied. Generally it was found that the short- and medium-chain saturated soaps were most active in acid solution and the long-chain soaps at alkaline pH, although the point of change varied with the bacterium. From the generally higher activity of the short- and medium-chain soaps at acid pH it was concluded that the corresponding fatty acids were more germicidal than their potassium salts. It was considered likely that the low activity of long-chain soaps at acid pH was due to the diminishing solubility of fatty acids with increasing chain length and therefore that their germicidal activity failed because they were not soluble in the needed concentration. On the other hand, when a fatty acid was sufficiently soluble it was more active than the corresponding soap. It was suggested that the greater germicidal activity of fatty acids might be due to the fact that they were less dissociated than their salts and could therefore penetrate more easily into a bacterium.

Dresel in 1926 [20] reported a study on the germicidal action of a number of chemically pure fatty acids on both Gram-positive and Gram-negative bacteria. Emulsions of fatty acids were diluted into bouillon cultures and samples were removed at intervals and tested for bacterial viability. The medium-chain fatty acids showed the highest germicidal activity against Gram-positive bacteria, such as diphtheria, anthrax bacilli, streptococci, pneumococci and staphylococci, but had no activity against most Gram-negative bacteria, such as typhoid and coli bacilli. A microscopic examination of the killed bacteria showed that the fatty acids caused a complete lysis of the bacterial cells.

Hettche in 1934 [21] compared the bactericidal effects of a number of chemically pure sodium salts of fatty acids against both Gram-positive and Gram-negative bacteria. He found that 0.1% sodium salts of palmitic and oleic acids had no effect on coli bacilli in bouillon. *Staphylococcus* was tested in the same way with sodium salts of fatty acids. The salts of capric, stearic and palmitic acids had little or no effect on the bacterium, whereas 0.1% sodium oleate caused a 10-fold reduction in viable bacteria in one hour and linoleate and linolenate killed all the bacteria. When tested in lower concentrations, linolenate killed the highest number of bacteria. The bactericidal activity therefore increased with the number of double bonds in the 18-carbon chain, as did the haemolytic activity.

In 1934, Barnes and Clarke [22] determined the minimum amounts of sodium ricinoleate and oleate needed to kill pneumococci of types I, II and III in broth culture. Series of dilutions of the two soaps were inoculated with broth cultures and incubated at 37.5 °C. Microscopic examination and subcultures of samples taken after incubation for 24 hours showed that 0.0004% oleate and 0.004% ricinoleate were effective in killing all three types under these conditions. This study was done with the aim of determining the concentration of soap which would kill the bacteria without impairing the bacterial antigens needed for preparation of a vaccine.

Bayliss in 1936 [23] studied the germicidal action of a number of sodium soaps and other chemicals at pH 8.0 against four different bacteria, *S. aureus*, *Streptococcus lactis*, *Escherichia coli* and *Diplococcus pneumoniae*. Of the saturated soaps, sodium myristate was found to be most active against pneumococci, with sodium laurate second in activity. Soaps of unsaturated long-chain fatty acids were much more active than the corresponding saturated soaps against pneumococci. The activity was not found to increase with the number of double bonds but the positions of the double bonds markedly affected the germicidal activity. As in the case of pneumococci, myristate and laurate were active in killing *S. lactis*, and sodium oleate was equally active. In contrast to pneumococci, the presence of a second double bond, as in linoleate, increased the activity. Addition of a hydroxyl group, as in ricinoleate, enhanced the activity against the streptococcus but decreased the activity against pneumococcus. The effectiveness of most soaps against the staphylococci was very low, sodium undecylenate being the most active of the soaps tested. The results for *E. coli* were similar to those for staphylococci, with sodium undecylenate being the most active soap. Thus, the bacteria showed a greatly variable susceptibility to the different soaps. Bayliss [23] stated that 'studies on the properties of soaps are of importance because of their common use as detergents, in the preparation of vaccines, in the treatment of disease and because of their natural occurrence in the animal body where they may have some effect in determining the resistance of the animal organism to infection'.

In 1941, Cowles [24] studied the bactericidal action of the lower fatty acids, from acetic to caprylic acids, against *S. aureus* and *E. coli* at low pH to ascertain the relative importance of the hydrogen ion and the undissociated fatty acid in this action. The results showed that at pH above about 2.6 the hydrogen ion lost its bactericidal activity against *S. aureus*, and at pH above 1.75 the activity was lost against *E. coli*. For *E. coli*, the bactericidal action seemed to be due almost entirely to the undissociated acids. For *S. aureus*, the action of the lowest acids, for example acetic acid, seemed to be partly caused by hydrogen ions, whereas the action of the higher acids, caproic and caprylic acids, was largely due to the un-ionized molecule. The anions did not seem to play a significant role in the bactericidal action.

Weitzel in 1952 [25] tested the effect of a large number of fatty acids in low concentrations on long-term growth of tubercle bacilli in culture medium. Of the saturated fatty acids, the medium-chain undecylic and lauric acids were the most active. The unsaturated fatty acids studied were mostly of 18-carbon chain length. Oleic acid and other 18-carbon acids with one double bond of different configurations showed similar activities, which were not different from the saturated stearic acid. Linoleic acid and other 18-carbon fatty acids with two double bonds were distinctly more active than either oleic acid with one or linolenic acid with three double bonds. This did not agree with the results from other studies on tubercle bacilli, which found the activity of unsaturated fatty acids to increase with the number of double bonds. Free fatty acids from chaulmoogra oil showed by far the highest activity as compared with free fatty acids extracted from various other natural sources, such as linseed oil and cod-liver oil, although they had higher contents of fatty acids with a large number of double bonds. The high antitubercular activity of chaulmoogra oil was most likely due to its content of the 18- and 16-carbon cyclopentenyl fatty acids, chaulmoogric and hydnocarpic acids.

2.3 Inhibition of Lipids and Serum Albumin against the Antibacterial Action of Soaps

As shown by Noguchi [26], Landsteiner and Ehrlich [27], Lamar [12] and Walker [15], serum inhibits the germicidal activity of soaps. Eggerth in 1927 [28] studied the inhibiting effect of serum at various pHs. As in his previous study [19], the soaps were made from chemically pure fatty acids and tested against the same bacteria, except for *V. cholerae*, and the experiments were carried out in the same way, either without serum or with sheep serum in a series of dilutions. In general, addition of serum reduced the activity of soaps at all pH values tested, the activity becoming gradually lower with increasing serum concentration. This was illustrated in mixtures of streptococci with either myristate or oleate, where the serum inhibited the soap almost equally at both high and low pH. A combination of *S. aureus* with laurate showed a somewhat different pattern. Here the activity of the soap rose steeply when the pH of the control buffer solution was lowered from pH 9 to 5. The addition of 1 and 5% of serum had no effect on the germicidal titre at pH 8 and 9, but greatly reduced the titre at lower pH, thus flattening the curve. In an attempt to explain the mechanism of serum inhibition, Eggerth tested a number of protein substances not found in serum, such as crystalline egg white, gelatin and casein, in streptococcus–oleate and staphylococcus–laurate mixtures at various pHs and found them all to be inhibitory. However, the germicidal curves over the pH range were different from those with sheep serum, suggesting a different action. On the other hand, serum with or without calcium gave identical results, ruling out the role of calcium in the serum. Emulsions were made of cholesterol, lecithin and other lipoids,[3] which were either extracted from serum or prepared from other sources. Such emulsions, particularly those containing lecithin, were found to inhibit the action of soaps against bacteria and gave curves which were identical to those of serum. Lipoid emulsions were soluble in casein solutions and such mixtures were more inhibitory to soaps than casein

---

[3] Lipoids include fats, their derivatives (e.g. lecithin) and their components (e.g. fatty acids and cholesterol) [28].

alone. Eggerth concluded that the inhibitory effect of serum on soaps was a complex action of proteins and lipoids which was probably due to adsorption or interactions of soaps with these substances, thus reducing the germicidal activity of the soaps.

In 1947 Dubos [29] studied the effect of lipids and serum albumin on the growth of bacteria in liquid media. The growth of human tubercle bacillus was completely inhibited by 0.0001% oleic acid but not by methyl oleate. The effect of oleic acid was completely neutralized by the addition of crystalline serum albumin. The inhibiting effect of serum albumin seemed to be due to formation of a lipoprotein complex between the fatty acid and the albumin. This was supported by the fact that a cloudy fatty acid emulsion became clear upon addition of an adequate amount of serum albumin and at the same time became inactive against bacteria. Of many proteins tested, only native serum albumin had this property.

In a following paper, Davis and Dubos [30] studied the interaction between fatty acids and serum albumin. They found that the albumin was able, over a wide concentration range (0.1 to 1.6%), to bind 1 to 2 % of its weight of oleic acid. The binding was firm enough to neutralize the bacteriostatic effect of the fatty acid against tubercle bacillus in liquid medium. Serum albumin was also found to protect sheep red blood cells from haemolysis by oleic acid. Tween 80 (polyoxyethylene sorbitan monooleate) had the same protective effect. Serum albumin was found to have a greater capacity for tightly binding long-chain fatty acids than any other substance tested, such as human serum globulin which had no binding effect. It was speculated that a specific configuration of amino-acid residues on the surface of albumin molecules led to the interaction with fatty acids. A positively charged quaternary nitrogen atom on an amino acid, such as lysin, would be likely to bind to the negatively charged carboxyl ion on the fatty acid. Similarly, a properly located nonpolar residue on an amino acid, such as leucin, might attract the nonpolar chain of the fatty acid by van der Waals forces and thus contribute to a firm binding to the albumin molecule. The affinity of serum albumin for chemotherapeutic agents with anionic groups was considered to be of practical importance.

2.4 Diverse Actions of Fatty Acids and Their Salts on Bacteria

In a study by Avery in 1918 [31], addition of a solution of 0.1% sodium oleate was found to stimulate the growth of *B. influenzae* on blood agar plates, as well as in oleate–haemoglobin bouillon, whereas the oleate completely inhibited the development of pneumococci and streptococci in the same medium. This study demonstrated a strong difference in the effects of a lipid, namely sodium oleate, on different bacteria. The diverse actions of lipids on various bacteria were later shown in a number of publications and reviewed by Nieman in 1954 [32].

Kodicek and Worden in 1945 [33] studied the effects of long-chain fatty acids on the growth and acid production of *Lactobacillus helveticus*. Linolenic, linoleic and to a lesser degree oleic acids were inhibitory when added to the growth medium in a concentration of 160 μg per 10 ml medium, whereas stearic and palmitic acids stimulated the acid production of the bacterium. The inhibitory action of the unsaturated fatty acids on growth and acid production was reversed by addition of lecithin, cholesterol and a number of other compounds to the medium. Their action was therefore bacteriostatic rather than bactericidal.

Linoleic acid was found to exert an inhibitory effect on a number of other Gram-positive bacteria but not on Gram-negative bacteria tested, such as *E. coli*.

In a later work, Kodicek [34] studied the bacteriostatic effect of low concentrations of free fatty acids on Gram-positive bacteria, particularly *Lactobacillus casei*. Of the saturated acids, the lauric acid depressed the growth of this bacterium but the ethyl ester had no effect. The growth-inhibiting activity of the unsaturated fatty acids increased with the number of double bonds and was limited to the naturally occurring *cis*-forms of the free acid. As in the case of lauric acid, esterification abolished the activity of the acid but unlike with lauric acid, the effect of the unsaturated fatty acids was reversed by cholesterol. In addition to cholesterol, other surface-active agents, especially lecithin, abolished the bacteriostatic effect of unsaturated acids.

Humfeld in 1947 [35] observed antibiotic activity in culture medium containing wheat bran and hypothesized that it was due to hydrolysis of fatty constituents of the medium caused by cultured microorganisms. Potassium salts of fatty acids extracted from wheat bran inhibited the growth of *S. aureus*, *Streptococcus faecalis* and *Micrococcus conglomeratus*, similarly to the inhibition caused by potassium linoleate. Potassium salts extracted from the wheat bran stimulated the growth of *E. coli* in the medium, in contrast to their inhibitory effect on the other bacteria tested.

Foster and Wynne in 1948 [36] studied the effect of 10 saturated and unsaturated fatty acids on spore germination of the anaerobic Gram-positive bacterium *Clostridium botulinum* and found that oleic, linoleic and linolenic acids were strongly inhibitory in germination medium. Curiously, the degree of inhibition depended on the lot of medium used, and starch neutralized the inhibitory action of the fatty acids in medium. The inhibition was specific for the germination process and fatty acids had no effect on the growth of vegetative cells after germination. The germination of other clostridium species tested was less inhibited by the fatty acids than the germination of *C. botulinum*, and the germination of endospores of aerobic bacilli was not inhibited by the fatty acids. Notably, oleic acid at the highest concentration studied did not kill the spores of *C. botulinum* in distilled water, and apparently showed inhibitory action only in concert with other inhibitors in the medium. Interestingly, oleic acid had previously been found to be an essential growth factor for *Clostridium tetani* as well as for the diphtheria bacillus [37].

Hardwick and coworkers in 1951 [38] studied factors in peptone medium which inhibited sporulation of various aerobic bacilli, such as *Bacillus mycoides*. By fractionation of the medium the antisporulation factors were identified as saturated fatty acids, most of the activity being present in the nonvolatile acid fraction. By comparing the antisporulation activities of pure fatty acids, capric (C10), lauric (C12), tridecylic (C13) and myristic (C14) acids were found to be the most active, whereas the low-carbon-chain volatile C1 to C7 fatty acids were much less active. The activity of saturated fatty acids with 16- and 18-carbon chains was also lower, whereas unsaturated long-chain fatty acids had considerable activity, particularly linoleic and ricinoleic acids. Oleic acid, on the other hand, had little antisporulation activity. All of the saturated fatty acids studied inhibited vegetative growth of the bacilli to various extents, but at concentrations well above those exhibiting antisporulation activity. The unsaturated acids, linoleic and ricinoleic acids, were equally active in inhibiting vegetative growth and sporulation of these aerobic bacilli.

Nieman [32] reviewed the literature on the influence of small quantities of fatty acids on microorganisms, particularly bacteria. He discussed the remarkable fact that fatty acids, particularly unsaturated at very low concentrations, are able to either stimulate or inhibit the growth of microorganisms. This 'double effect' or 'dualism' was most clearly demonstrated by the fact that the effect of a fatty acid as a growth factor could be changed into an inhibitory action by increasing its concentration above a critical level. The inhibitory action of low concentrations of fatty acids in the 5 to 50 ppm (parts per million) range, or even lower, had been known for a long time, particularly for Gram-positive bacteria. This inhibitory action, demonstrated by depression of either respiration or growth rate, was in many cases reversible and can therefore be characterized as bacteriostatic. It was influenced by both fatty acid structure and concentration, as well as by the bacterial species. As reported by many investigators, the inhibitory properties of fatty acids are generally most pronounced for longer-chain unsaturated acids. This was particularly true for bacteria. However, saturated fatty acids have also been found to act as growth inhibitors, especially acids with chain lengths around 12 carbon atoms.

Dubos had found in 1948 [29] that fatty acids, both saturated and unsaturated, had a bacteriostatic as well as bactericidal effect on tubercle bacilli in protein-free medium. Stearic acid was the least active and the 18- and 22-carbon unsaturated fatty acids were the most active, even in concentrations as low as 0.00 001 to 0.0001%. The bacteriostatic/bactericidal activity of the fatty acids was very much decreased or completely abolished by addition of serum albumin. However, the effect of the fatty acids depended greatly on the experimental conditions, such as the composition of the medium and the size of the bacterial inoculum. Small concentrations of some fatty acids were even found to stimulate bacterial growth under certain conditions, for example when their bacteriostatic effect was inhibited by serum albumin.

These studies emphasized the importance of distinguishing between a reversible bacteriostatic (growth-inhibiting) effect and an irreversible bactericidal (killing of bacteria) effect.

2.5   The Nature of the Bactericidal Action of Fatty Acids

Several studies have addressed the question of how fatty acids and soaps kill bacteria. In a series of publications [39–41], Eggerth studied the germicidal/bactericidal activity of various soap derivatives with the purpose of throwing light on the relationship between their chemical structure and their germicidal action. In the first study, a hydrogen atom was replaced with bromine on the carbon atom next to the carboxyl group of the fatty acid chain, thus making an alpha-brom fatty acid, which was then changed into the corresponding soap by addition of potassium hydroxide. The brominated soaps of saturated fatty acids were much more soluble in water than the unsubstituted soaps and were generally more germicidal and haemolytic. This was particularly true for soaps of long-chain saturated fatty acids. Thus, potassium palmitate had no activity against *S. aureus*, whereas the bacterium was killed by bromopalmitate in high dilutions. All of the Gram-positive bacteria tested were most effectively killed by soaps of 16 or 18 carbon atoms and the Gram-negative bacteria by soaps of 12 or 14 carbon atoms. These findings were explained by the hypothesis that the germicidal activity of the soaps for all species of bacteria increases rapidly with the length of the fatty acid chain. However, the ability to penetrate the limiting membranes of the bacteria decreases with equal rapidity, the Gram-negative bacteria being less permeable than the Gram-positive. The point of maximal germicidal activity would then be determined by the permeability of bacterial membranes to soap molecules. With low permeability, only soaps with smaller molecules can enter and the maximal activities will be low, whereas with high permeability the longer-chained soaps can act, consequently with higher maximal activities [39]. Like the unsubstituted soaps, the brominated soaps were more germicidal at acid than at alkaline pH, strengthening the previous conclusion that fatty acids are more germicidal than the corresponding soaps [28, 39].

In a study of alpha-hydroxy soaps [40], Eggerth saw a similar change in germicidal activity, compared to unsubstituted soaps, as that observed for the alpha-brom soaps; that is, a shift of the maximal germicidal activities towards longer-chain soaps. However, the germicidal activity of alpha-hydroxy soaps varied greatly with the different bacteria tested, which prompted the author to suggest that they could be used to separate mixtures of bacteria by selective germicidal action [40]. In a paper in 1931 [41], Eggerth pursued the idea of selective action of soaps in a study of the germicidal action of alpha-mercapto and alpha-disulfo soaps. Similar to alpha-brom and alpha-hydroxy substituted soaps, the germicidal activities of these soaps increased rapidly with the length of the fatty acid chain until a maximum was reached, and then fell off. This was taken to support the previous hypothesis that the germicidal action increases with the molecular weight of the soap up to a point where the molecule is too large to penetrate the outer membrane of the bacterial cell. Like the previously studied alpha-substituted soaps, the alpha-mercapto and alpha-disulfo soaps showed a highly selective action, with no two bacterial species having exactly the same action profile.

Hotchkiss [42], in a review of the bactericidal action of surface-active agents, proposed the following scenario: the first stage of the interaction is a combination of the ionic group of a hydrophobic–hydrohilic surface-active molecule, such as a fatty acid, with oppositely charged ionic groups on the bacterial surface. As pointed out by A.W. Ralston in a discussion following the review, the surface-active ions might take the form of ionic micelles, which are the highest-charged particles in an aqueous collodial system. The ionic combination might be prevented or reversed by competition from suitably charged ionic sites on proteins or lipids, or other molecules, in the environment. If the hydrophobic parts of the surface-active agent had a specific affinity for nonpolar structures on the bacterial surface they would become anchored at these sites and thereby determine whether the interaction with the bacterium would be permanent. In any event, a successful ionic combination must precede the second step of hydrophobic interaction. A permanent binding of the surface-active agent would result in irreversible damage to the bacterial surface that could initiate lysis of the bacteria. Rupture of the cellular membrane would cause leakage of soluble nitrogen and phosphorus compounds out of the cell, leading to irreparable autolysis. This process appeared to be analogous to the haemolysis of red blood cells by surface-active agents. Previously, Dresel had shown by a microscopic examination of bacteria killed by fatty acids that the fatty acids caused a complete lysis of the bacterial cells [20].

An earlier microscopical examination by Lamar [12] revealed that 0.5 to 1% sodium oleate caused a disintegration of pneumococcal cells, which turned into a mucin-like material. Pneumococci exposed to much lower concentrations of the oleate – 1 to 10 000 or 1 to 20 000 – did not show detectable changes in their morphology or in their ability to grow on culture medium. However, they underwent autolysis much more rapidly than untreated bacteria. More notably, pneumococci treated with such diluted oleate solutions became much more susceptible to lysis by immune antipneumococcus serum than untreated pneumococci. Lamar speculated that sodium oleate caused a weakening of the lipoidal barrier on the outer surface of the coccus, making it more pervious to the immune serum and its destructive action.

Kodicek in 1949 [34] speculated about the nature of the inhibiting effect of fatty acids on bacterial growth and tried to explain it in physicochemical terms. He found former explanations, such as changed surface tension, too simplistic and reasoned that the effect was far more complex and must involve a change in the permeability of the bacterial wall. A simple calculation showed that linoleic acid in the lowest effective concentration had enough surface area to cover the whole surface of the inactivated bacteria. This supported a direct interaction of the fatty acid with the outer wall of a bacterium in which the polarity of the carboxyl group must be of importance. Certain proteins such as albumins can form complexes with fatty acids and it was well known that sterols, lecithins and other surface-active agents also formed complexes with unsaturated fatty acids. Such molecules might act as competitors to the lipoproteins of the bacterial cell wall and thus counteract the bacteriostatic or bactericidal effect of the fatty acids, as observed in many studies. Accordingly, the action of fatty acids would depend on the composition of the cell wall, as shown by the difference in susceptibility of various bacterial species to the effect of fatty acids. In spite of a physicochemical mechanism being the most likely explanation of the action of fatty acids on bacteria, Kodicek did not exclude a possible metabolic effect.

2.6 A Possible Role of Soaps and Fatty Acids in Host Defence against Bacteria

Using specific staining technics, Klotz in 1905 [43, 44] showed that calcarious tissue degeneration is preceded or accompanied by deposits of fatty or soapy material, and that such material is present in chronically inflamed tissues as neutral fats, fatty acids or soaps. Thus, prior to calcification of tissues, the cells undergo so-called fatty degeneration; that is, accumulation of fat granules which occupy the greater part of the cell. This infiltration of fat seems to play a part in the pathological process through conversion of neutral fat into fatty acids or their sodium or potassium salts, which in turn are transformed into insoluble calcium soap. Noguchi [26] made alcohol and ether extracts from various animal organs which contained mixtures of fatty acids, particularly oleic acid, and their soaps. These extracts showed haemolytic activity mostly due to their contents of oleates. The haemolytic activity of the soaps was inhibited by addition of serum. The oleate soaps were found to be highly active against *B. anthracis* and *V. cholerae* but were less active against typhoid and dysentery bacilli. In all cases, the germicidal activity of the soaps was inhibited by addition of serum. Noguchi concluded that the presence of fatty acids and soaps in blood and organs supported the opinion that they play a certain role in the natural defence mechanism of the organism. Landsteiner and Ehrlich [27] also addressed the question of the bactericidal effect of lipoid extracts from organs, particularly from leukocytes and lymphoid tissues. They confirmed a bactericidal activity of bovine spleen extracts and ovary extracts from frogs against *B. anthracis*, *Vibrio Massauah* and *S. aureus* and showed that serum inhibited the activity. They repeated these experiments using oleic acid emulsified in 1% saline, with or without serum added, and demonstrated bactericidal activity of the oleic acid in line with the organ extracts. Like Noguchi, Landsteiner and Ehrlich suggested that lipids found in organ extracts are a part of the complement system which in junction with antiserum cause lysis of bacteria and red blood cells. They concluded that further research should elucidate the possible role of lipids in the defence of the organism against infection.

In the introduction to a paper by Lamar in 1911 [12], Flexner stated that recovery from bacterial infections had not been fully explained by the antibacterial activities of serum and phagocytes. Thus, he raised the question of how chemical substances, other than antibodies, might play a role in the recovery from a local bacterial infection. He quoted earlier studies by several investigators which showed that substances extracted from leucocytes are capable of killing many different types of bacteria. These intracellular substances were not only bactericidal but often also haemolytic and were characterized by their thermostability and solubility in alcohol. As mentioned above, Noguchi [26] determined in 1907 that the activity of these extracts depended on their content of certain higher unsaturated fatty acids or their alkaline soaps and Klotz [43] had shown in 1905 that soaps occur in inflammatory foci. Previously, lipase had been found in such foci, which could account for the presence of fatty acids. Disintegrating tissues were known to contain neutral fat and higher phosphorized fats, which under the influence of lipase in mononuclear leucocytes or lymphocytes were decomposed into free fatty acids. Thus Lamar [12] extracted large quantities of lipoids and fatty acids from human lung with resolving lobar pneumonia and suggested that the decrease in the number and virulence of pneumococci present in the exudate depended in part upon the presence of germicidal fatty acids and soaps. This assumption was further strengthened by the fact that pneumococci treated *in vitro* with soap become gradually less virulent. The quantity of bactericidal and haemolytic substances released by disintegrating cells in inflammatory foci could be considerable and was assumed to exert a destructive action upon the infecting bacteria and thus contribute to recovery from the infection. Flexner suggested that the bactericidal substances consist of soaps of fatty acids. The defensive processes acting upon bacteria in the site of infection might be complicated by the inhibiting action of serum proteins on the bactericidal action of unsaturated fatty acid soaps. It was suggested that this might be overcome by the proximity of the bacteria to the nascent fatty acids.

It had been known since the beginning of the twentieth century that the human skin is able to control the microflora on its surface and to get rid of harmful bacteria coming from the environment. This ability to destroy exogenous microbes was referred to as the self-disinfecting power of the skin [45]. In the 1930s and 1940s there was an increased interest in the self-disinfecting ability of human skin and its appendages: hair, nails and cerumen. Initially, the disinfection was thought to be caused directly by the acidity of the skin but studies by Burtenshaw [46] demonstrated antibacterial activity in lipids extracted from the skin. Most of the activity was in the fractions containing oleic acid and other long-chain fatty acids and their soaps. In nearly all the experiments, the disinfecting extracts and fractions thereof were far more bactericidal at an acid than at a more alkaline pH. It was concluded from these experiments that the disinfecting effect of acidity on the skin was, at least partly, due to release of antimicrobial fatty acids from their soaps at low pH.

In their study of self-disinfection (self-sterilization) of the skin of young adult volunteers, Ricketts and his coworkers [47] confirmed and extended the results of Burtenshaw. Their study showed that the removal of lipids from the surface of the skin reduced its self-sterilizing power against *S. pyogenes* and to a lesser degree against *S. aureus*, but had no effect on resident micrococci or the Gram-negative *B. coli* and *Pseudomonas pyocyanea*.

For *S. pyogenes*, the main disinfecting factor was the bactericidal effect of unsaturated fatty acids in the skin lipids, particularly oleic acid. On the other hand, drying of the skin was the major factor in the killing of micrococci and the Gram-negative bacilli. For *S. aureus* both factors seemed to have an effect.

The role of antimicrobial lipids in host defence was reviewed in 2007 [48] and is discussed in detail in Chapter 6.

2.7   Studies of Prophylactic and Therapeutic Applications of Soaps and Fatty Acids

Since a certain similarity seemed to exist between the destruction of pneumococci in pneumonic exudate and in artificial soap–serum mixtures, Lamar [12] speculated whether the changes in the pneumococci caused by nonbactericidal dilute sodium oleate might also occur in the animal body in the course of infection and play a role in natural defence against the infection and its final resolution. Based on these speculations, he found that a mixture of soaped pneumococci and immune serum applied by intraperitoneal inoculation was harmless to mice, in contrast to pneumococci treated with dilute oleate alone or in mixtures with normal serum. When immune serum was injected into mice within one hour after inoculation with soaped pneumococci the infection was prevented, but not if the serum was injected later. In order to ascertain whether or not a mixture of sodium oleate and immune serum could prevent infection in already infected animals, mice were inoculated with untreated virulent pneumococci and then injected with a mixture of oleate, immune serum and boric acid. The latter was needed to eliminate the inhibiting effect of serum proteins on the oleate. If the mixture was injected within one hour of inoculation it prevented the pneumococcal infection and all of the animals survived. Injection after two hours only saved half of the animals. Based on these results, Lamar speculated on whether soap–serum mixtures could be used for treatment of local pneumococcus infections in humans.

Considering the possibility that *Treponema pallidum*, the causative agent of syphilis, could inadvertently be transmitted to small cuts in the face, for example in a barbershop, and thus cause extragenital lesions, Reasoner in 1917 [49] tested the germicidal activity of soap against this bacterium. Testicular juice from infected rabbits rich in treponemata was mixed with equal parts of a solution of various soaps prepared in distilled water. Both Ivory soap and shaving creams from various manufacturers were tested. After 15 to 30 seconds the juice was examined by dark-field microscopy. Almost no motile spirochetes could be found, and those that were found were usually swollen and distorted, whereas in control samples without soap motility could be observed for at least four hours. There was no difference in the action of the different soaps. Thus, soap solutions or lather, such as that used in shaving, immediately killed the treponemata on contact and were suggested as a possible prophylactic measure against syphilis.

In a study of how germs which cause respiratory infections pass from one person to another, Nichols [14] examined the route of travel of haemolytic streptococci in an army camp. One possible point of transfer between individuals was through contaminated dish-washing water. He found that colon bacilli were on several occasions present in the water under conditions which showed no streptococci, indicating that in contrast to bacteria which cause intestinal disease, streptococci and pneumococci are not spread to eating utensils through this route. The bactericidal activity of the soaps was dependent on the pH and was lost if it fell to 7 or lower, probably because of precipitation of the soap from solution. He concluded that for a maximum antibacterial action, the pH of washing-up water should be kept at 8 or higher.

In 1930, Renaud [50] published a paper on the possible therapeutic application of soaps. In this study he favourably used a 2% solution of sodium oleate for rinsing and as a wet dressing for wounds and abscesses, such as genital sores and tuberculous and pyogenic abscesses.

Following his studies of the germicidal action of sodium salts of 18-carbon unsaturated fatty acids on Gram-positive bacteria, Hettche in 1934 [21] speculated about a possible therapeutic application for the highly bactericidal linolenate and suggested a possible use as throat wash in diphtheria or for cleaning infected wounds. In his opinion, a parental and particularly an intravenous application would not be possible because of the haemolytic activity of the lipids, whereas oral intake might be considered because of experimental evidence that oleic acid was well tolerated by mice.

Chaulmoogra oil, made from the seeds of a tree common in Southeast Asia, was used for centuries to treat leprosy. Due to the reputation of this oil in the treatment of leprosy, Walker and Sweeney [51] carried out a series of experiments to test its activity *in vitro* against a number of acid-fast bacteria, such as the bacilli of rat and human leprosy and the bacilli of human, avian and bovine tuberculosis.The activities of the sodium salts of the total fatty acids of chaulmoogra oil were first investigated against the bacillus of rat leprosy and found to be completely antiseptic (bacteriostatic) up to a dilution of 1 : 125 000 and fully bactericidal up to a dilution of 1 : 100 000. The antiseptic and bactericidal actions of the total sodium salts were then tested against the human leprosy bacillus and the three tubercle bacilli, with results comparable to those for the rat leprosy bacillus. Notably, the bactericidal activity of sodium chaulmoograte against acid-fast bacteria was about 100-fold greater than that of the nonspecific antiseptic compound phenol.

In order to identify the active ingredients of chaulmoorga oil, chemically distinct fatty acids were isolated and found to consist mostly of the cyclic 18- and 16-carbon fatty acids chaulmoogric and hydnocarpic acids. Their sodium salts were tested against the rat leprosy bacillus and found to account for the antiseptic activity of the total fatty acids, whereas palmitic acid, which constitutes only about 10% of the total fatty acids compared with 90% for the chaulmoogric acids series, had very little activity. While chaulmoogric acid was highly active against acid-fast bacteria it was inactive against eight different nonacid-fast bacteria tested, even in dilutions as low as 1 : 1000. Also, in contrast to the findings of Rogers [52], Walker and Sweeney found the sodium salt of the noncyclic 18-carbon linoleic acid as well as sodium salts of the total fatty acids of cod-liver oil, sodium morrhuate, to be much less active than sodium chaulmoogrates against leprosy and tubercle bacilli. The authors concluded that their results provided a scientific basis for the use of chaulmoogra oil and its products in the treatment of leprosy and furnished grounds for their application in the therapy of tuberculosis.

Walker and Sweeney speculated about the nature of the specific action of chaulmoogric acid against acid-fast bacteria and found it unlikely that it was due to a direct effect on their protective fatty capsule, manifest by irregular acid-fast staining, as suggested by Rogers [52]. The authors hypothesized that the bactericidal action was due to an aberration in lipid metabolism, in which chaulmoogric acid molecules accumulated in the fatty capsules and exerted a toxic effect on the bacterial cell as a consequence of their cyclic structure.

Based on an apparently successful treatment of leprosy by intravenous injection of chaulmoogra oil, Lindenberg and Pestana in 1921 [53] decided to test the activity of the oil, and fatty acids extracted from it, against leprosy bacilli, in order to establish whether or not the oil had a specific action against the causative agent. Since a culture technic for leprosy bacillus was not available to them at that time, they used cultures of human and avian tubercle bacilli and other acid-fast bacilli in their experiments. High dilutions of fatty acids from chaulmoogra oil were found to be active in preventing the growth of tubercle bacilli and other acid-fast bacteria, whereas they had much less or no effect on antrax bacilli and *S. aureus*. Cottonseed oil, linseed oil and cod-liver oil were also tested against tubercle bacilli. All three were found to be almost as active as chaulmoogra oil in preventing growth of the bacilli. This was considered of interest since cod-liver oil had for a long time been used as a remedy for tuberculosis. An effort was made to determine the chemical nature of the active ingredients of the oils and they were found to be mostly unsaturated fatty acids. Attempts to treat experimental animals and patients with tuberculosis with fatty acids from chaulmoogra oil, cod-liver oil and cottonseed oil by intravenous or subcutaneous injection were not successful.

As reviewed by Stanley and his coworkers [54], chaulmoogra oil was used for centuries in the treatment of leprosy. Rubbing of lesions with the oil or taking it by mouth seemed to give some relief to lepers, but the latter caused side effects such as nausea. Subcutaneous or intramuscular injection of the oil was somewhat more effective than oral administration, but had the disadvantage of causing severe pain. Treatment with ethyl esters or sodium salts of the fatty acids of chaulmoogra oil was more successful, so that a small percentage of advanced cases and a much larger percentage of incipient leprosy cases became negative for the bacteria and could be discharged as cured. Stanley and coworkers studied the chemistry of the active principles of chaulmoogra oil, the 18- and 16-carbon cyclopentenyl fatty acids, chaulmoogric and hydnocarpic acids, and confirmed that neither the double bond of the ring structure nor the carboxyl group were essential for the bactericidal activity against *Mycobacterium leprae*. The analogous straight-chained fatty acids had no activity against the bacterium, so it was concluded that the ring structure of the molecules and the number of carbon atoms were largely responsible for the bactericidal action. Efforts by these authors to synthesize derivatives of chaulmoogric acids with higher *in vitro* activities against acid-fast bacteria, with the aim of improving the treatment of leprosy, were, however, unsuccessful. The use of sulfones to treat leprosy in the 1940s put an end to experiments with chaulmoogra-oil treatments.

In the late 1940s, on the other hand, Hänel and Piller [55] reported clinical improvement in patients with pulmonary tuberculosis treated by mouth with an emulsion of unsaturated fatty acids from cod-liver oil or with pharmaceutical preparations, which had been shown to be highly active in preventing growth of the tubercle bacillus *in vitro*. These clinical trials suggested that treatment of patients with concentrated emulsions of unsaturated fatty acids from fish oil might be beneficial. The authors reasoned that unsaturated fatty acids, in conjunction with phosphatides such as lecithin, might play a role in the natural defence of the body against the tubercle bacillus and that therapy with high doses of the fatty acids would therefore be helpful. In this context it is of interest that native populations in tropical areas have used fats against mycobacterial infections, especially leprosy and tuberculosis.

In Nigeria palmkernel oil has been used, while turtle oils have been used by Mexican Indians. Turtle oils are rich in 10-, 12- and 14-carbon saturated fatty acids, which have been shown to be strongly bacteriostatic for *Mycobacterium tuberculosis* and other mycobacteria *in vitro* [32].

2.8 Conclusions

One hundred years ago there were few resources to fight bacterial infections. Therefore, the discovery that common soaps kill many types of pathogenic bacteria raised the hopes of health professionals and researchers that soaps might be used as disinfectants or even for prevention and treatment of infectious diseases. For more than half a century the germicidal properties of potassium and sodium salts of fatty acids were therefore actively studied. Although the results of these studies may have helped to improve sanitary practices, they did not contribute much to treatment of infections or infectious diseases. The advent of antibiotics around the middle of the twentieth century led to a breakthrough in the therapy and cure of many previously fatal infectious diseases and consequently interest in the germicidal action of fatty acids and other lipids decreased. After 1960, there was an increase in studies of antimicrobial lipids, particularly milk lipids, largely due to the pioneering work of Kabara and his associates and of Welsh and May on antiviral lipids in human milk. These studies will be described in later chapters.

References

1. Koch, R. (1881) Ueber Desinfection. *Mittheil. des kaiserl. Gesundheitsamtes*, **1**, 234–282.
2. Reithoffer, R. (1896) Ueber die Seifen als Desinfektionsmittel. *Arch. f. Hyg.*, **27**, 350–364.
3. Serafini, A. (1898) Beitrag zum experimentellen Studium der Desinfectionsfähigkeit gewöhnlicher Waschseifen. *Arch. f. Hyg.*, **33**, 369–398.
4. Konrádi, D. (1902) Über die baktericide Wirkung der Seifen. *Arch. f. Hyg.*, **44**, 101–112.
5. Behring, E.A. (1890) Ueber Desinfection, Desinfectionsmittel und Desinfectionsmethoden. *Zeitschr. f. Hyg.*, **9**, 395–478.
6. Jolles, M. (1893) Ueber die Desinfectionsfähigkeit von Seifenlösungen gegen Cholerakeime. *Zeitschr. f. Hyg.*, **15**, 460–473.
7. Jolles, M. (1895) Weitere Untersuchungen über die Desinfectionsfähigkeit von Seifenlösungen. *Zeitschr. f. Hyg.*, **19**, 130–138.
8. Nijland, A.H. (1893) Ueber das Abtödten von Cholerabacillen in Wasser. *Arch. f. Hyg.*, **18**, 335–372.
9. Beyer, T. (1896) Ueber Wäschedesinfection mit dreiprocentiger Schmierseifenlösungen und mit Kalkwasser. *Zeitschr. f. Hyg.*, **22**, 228–262.
10. Reichenbach, H. (1908) Die desinfizierenden Bestandteile der Seifen. *Zeitschr. f. Hyg.*, **59**, 296–316.
11. Geppert, J. (1889) Zur Lehre von den Antisepticis. Eine Experimentaluntersuchung. *Berl. klin. Wochenschr.*, **26**, 789–794, 819–821.
12. Lamar, R.V. (1911) Chemo-immunological studies on localized infections. First paper. Action on the pneumococcus and its experimental infections of combined sodium oleate and antipneumococcus serum. *J. Exp. Med.*, **13**, 1–23.
13. Lamar, R.V. (1911) Chemo-immunological studies on localized infections. Second paper. Lysis of the pneumococcus and hemolysis by certain fatty acids and their alkali soaps. *J. Exp. Med.*, **13**, 380–386.
14. Nichols, H.J. (1919–1920) Bacteriologic data on the epidemiology of respiratory diseases in the army. *J. Lab. Clin. Med.*, **5**, 502–511.

15. Walker, J.E. (1924) The germicidal properties of chemically pure soaps. *J. Infect. Dis.*, **35**, 557–566.

16. Walker, J.E. (1925) The germicidal properties of soap. *J. Infect. Dis.*, **37**, 181–192.

17. Walker, J.E. (1926) The germicidal properties of soap. *J. Infect. Dis.*, **38**, 127–130.

18. Walker, J.A. (1931) The germicidal and therapeutic applications of soaps. *J. Am. Med. Assoc.*, **97**, 19–20.

19. Eggerth, A.H. (1926) The effect of the pH on the germicidal action of soaps. *J. Gen. Physiol.*, **10**, 147–160.

20. Dresel, E.G. (1926) Bakteriolyse durch Fettsäuren und deren Abkömmlinge. *Zentralbl. f. Bakt. I. Orig.*, **97**, 178–181.

21. Hettche, H.O. (1934) Die Wirkung von Fettsäuresalzen auf Erythrozyten und Bakterien. *Zeitschr. Immunitätsforsch.*, **83**, 506–511.

22. Barnes, L.A. and Clarke, C.M. (1934) The pneumococcicidal powers of sodium oleate and sodium ricinoleate. *J. Bacteriol.*, **27**, 107–108.

23. Bayliss, M. (1936) Effect of the chemical constitution of soaps upon their germicidal properties. *J. Bacteriol.*, **31**, 489–504.

24. Cowles, P.B. (1941) The germicidal action of the hydrogen ion and of the lower fatty acids. *Yale J. Biol. Med.*, **13**, 571–578.

25. Weitzel, G. (1952) Antituberkuläre Wirkung ungesättigter Fettstoffe in vitro. *Zeitschr. physiol. Chem.*, **290**, 252–266.

26. Noguchi, H. (1907) Über gewisse chemische Komplementsubstanzen. *Biochem. Zeitschr.*, **6**, 327–357.

27. Landsteiner, K. und Ehrlich, H. (1908) Ueber bakterizide Wirkungen von Lipoiden und ihre Beziehung zur Komplementwirkung. *Centralbl. f. Bakt.*, **45**, 247–257.

28. Eggerth, A.H. (1927) The effect of serum upon the germicidal action of soaps. *J. Exp. Med.*, **46**, 671–688.

29. Dubos, R.J. (1947) The effect of lipids and serum albumin on bacterial growth. *J. Exp. Med.*, **85**, 9–22.

30. Davis, B.D. and Dubos, R.J. (1947) The binding of fatty acids by serum albumin, a protective growth factor in bacteriological media. *J. Exp. Med.*, **86**, 215–228.

31. Avery, O.T. (1918) A selective medium for B. influenzae. Oleate-hemoglobin agar. *Am. Med. Assoc.*, **71**, 2050–2051.

32. Nieman, C. (1954) Influence of trace amounts of fatty acids on the growth of microorganisms. *Bacteriol. Rev.*, **18**, 147–163.

33. Kodicek, E. and Worden, A.N. (1945) The effect of unsaturated fatty acids on *Lactobacillus helveticus* and other Gram-positive micro-organisms. *Biochem. J.*, **39**, 78–85.

34. Kodicek, E. (1949) The effect of unsaturated fatty acids on Gram-positive bacteria. *Soc. Exp. Biol. Symp.*, **3**, 217–232.

35. Humfeld, H. (1947) Antibiotic activity of the fatty-acid-like constituents of wheat bran. *J. Bacteriol.*, **54**, 513–517.

36. Foster, J.W. and Wynne, E.S. (1948) Physiological studies on spore germination, with special reference to *Clostridium botulinum*. IV. Inhibition of germination by unsaturated C18 fatty acids. *J. Bacteriol.*, **55**, 495–501.

37. Feeney, R.E., Mueller, J.H. and Miller, P.A. (1943) Growth requirements of *Clostridium tetani*. II. Factors exhausted by growth of the organism. *J. Bacteriol.*, **46**, 559–562.

38. Hardwick, W.A., Guirard, B. and Foster, J.W. (1951) Antisporulation factors in complex organic media. II. Saturated fatty acids as antisporulation factors. *J. Bacteriol.*, **61**, 145–151.

39. Eggerth, A.H. (1929) The germicidal and hemolytic action of α-brom soaps. *J. Exp. Med.*, **49**, 53–62.

40. Eggerth, A.H. (1929) The germicidal action of hydroxy soaps. *J. Exp. Med.*, **50**, 299–313.

41. Eggerth, A.H. (1931) The germicidal action of α-mercapto and α-disulfo soaps. *J. Exp. Med.*, **53**, 27–36.

42. Hotchkiss, R.D. (1946) The nature of the bactericidal action of surface active agents. *Ann. NY Acad. Sci.*, **46**, 479–493

43. Klotz, O. (1905) Studies upon calcareous degeneration. I. The process of pathological calcification. *J. Exp. Med.*, **7**, 633–675.
44. Klotz, O. (1906) Studies upon calcareous degeneration. II. The staining of fatty acids and soaps in the tissues by Fischler's method, and a modification of the same. *J. Exp. Med.*, **8**, 322–336.
45. Arnold, L., Gustafson, C.J., Hull, T.G. *et al.* (1930) The self-disinfecting power of the skin as a defense against microbic invasion. *Am. J. Epidemiol.*, **11**, 345–361.
46. Burtenshaw, J.M.L. (1942) The mechanism of self-disinfection of the human skin and its appendages. *J. Hyg.*, **42**, 184–210.
47. Ricketts, C.R., Squire, J.R. and Topley, E. (1951) Human skin lipids with particular reference to the self-sterilising power of the skin. *Clin. Sci.*, **10**, 89–111.
48. Thormar, H. and Hilmarsson, H. (2007) The role of microbicidal lipids in host defense against pathogens and their potential as therapeutic agents. *Chem. Phys. Lip.*, **150**, 1–11.
49. Reasoner, M.A. (1917) The effect of soap on *Treponema pallidum*. *J. Am. Med. Assoc.*, **68**, 973–974.
50. Renaud, M. (1930) Sur le traitement des plaies et surfaces ulcéres par les savon. *Bull. et mém. Soc. méd. d. hop. de Paris*, **46**, 1584–1589.
51. Walker, E.L. and Sweeney, M.A. (1920) The chemotherapeutics of the chaulmoogric acid series and other fatty acids in leprosy and tuberculosis. I. Bactericidal action, active principle, specificity. *J. Infect. Dis.*, **26**, 238–264.
52. Rogers, L. (1919) A note on sodium morrhuate in tuberculosis. *Brit. Med. J.*, **1** (3032), 147–148.
53. Lindenberg, A. und Pestana, B.R. (1921) Chemotherapeutische Versuche mit Fetten an Kulturen säurefester Bacillen. *Zeitschr. f. Immunitätsf.*, **32**, 66–86.
54. Stanley, W.M., Coleman, G.H., Greer, C.M., Sacks, J. and Adams, R. (1932) Bacteriological action of certain synthetic organic acids toward *Mycobacterium leprae* and other acid-fast bacteria. *J. Pharm. Exp. Therap.*, **45**, 121–161.
55. Hänel, F. und Piller, S. (1950) Ungesättigte Fettsäuren in der Therapie der Lungentuberkulose. *Beitr. Klin. Tuberk.*, **103**, 239–245.

3

Antibacterial, Antiviral and Antifungal Activities of Lipids

Gudmundur Bergsson[1], Hilmar Hilmarsson[2] and Halldor Thormar[2]

[1] *Institute of Molecular Medicine, Trinity College Dublin, Dublin, Ireland*
[2] *Faculty of Life and Environmental Sciences, University of Iceland, Reykjavik, Iceland*

Lipids and Essential Oils as Antimicrobial Agents   Halldor Thormar

3.1   Introduction

Antimicrobial lipids are widespread in nature, where they are found in species that range from plants to animals and act as both potent antiinflammatory and antimicrobial effector molecules [1–8]. Simple natural lipids, such as fatty acids and monoglycerides, exhibit antimicrobial properties and are capable of killing Gram-negative and Gram-positive bacteria, enveloped viruses, yeast, fungi and parasites that infect both the skin and the mucosa of animals. Fatty acids and/or 1-monoglycerides are therefore considered important as target molecules for drug applications, such as antimicrobial formulations [9–11], alternatives or adjuncts to antibiotics against antibiotic-resistant bacteria [8, 12–14], emulsions [11] and potent effector molecules of innate immunity furthering bacterial clearance after skin infections [15].

The antimicrobial *in vitro* effect of lipids has been studied for over a century [1, 16] and although most of the early studies focussed on the bactericidal effects of soaps, long- and medium-chain fatty acids and monoglycerides were also shown to kill microbes. Fatty acids and 1-monoglycerides have been found to kill pathogens that are known to infect mucosa and skin, where they are considered to be potent inhibitors of human pathogens. The main subject of this chapter is to review the *in vitro* antimicrobial activities of simple lipids – that is, straight-chain fatty acids, fatty alcohols and glycerides – with special focus on the comparison of the antimicrobial effects of fatty acids and their corresponding 1-monoglycerides on various groups of microbes. An overview of the most common lipids tested for antimicrobial activities is given in Table 3.1. Furthermore, possible mechanisms behind the antimicrobial effect of these lipids will be described and *in vivo* studies of their possible role in host defence against bacterial and fungal infections will be discussed.

3.2   Antibacterial Activities of Fatty Acids and Monoglycerides

3.2.1   Activities against Gram-Positive Bacteria *In Vitro*

Kabara and coworkers [17, 18] determined the minimal inhibitory concentrations (MIC) of medium-chain fatty acids against a number of Gram-positive and Gram-negative bacteria at 35 °C for 18 hours. Of all the saturated fatty acids that ranged from 6 to 18 carbons in chain length, lauric acid was found to exhibit the greatest antibacterial effect against the Gram-positive bacteria that were tested. The most susceptible were pneumococci and group A streptococci, while *Staphylococcus aureus*, *Staphylococcus epidermidis* and group D streptococci were the least susceptible. Similar results were reported by Heczko *et al.* [19], who determined the minimal inhibitory concentrations of capric (10 : 0), lauric (12 : 0), linoleic (18 : 2) and linolenic (18 : 3) acids for 242 strains of *S. aureus* and 117 strains of streptococci of groups A, B, C and G. *S. aureus* proved to be generally less susceptible to all four fatty acids than the streptococcal strains and lauric acid was the most inhibitory fatty acid against the bacteria.

To determine which lipids kill Gram-positive bacteria the fastest, Bergsson *et al.* investigated short incubation times of 1 to 10 minutes [20]. The results are summarized in Table 3.2, which represents the concentrations of fatty acids and monoglycerides that cause more than 100-fold (greater than 99%) reduction of viable bacteria after incubation for 10 minutes or less. The Gram-positive streptococci, group A (GAS) and group B (GBS), and

Table 3.1 *Antimicrobial fatty acids, monoglycerides and fatty alcohols.*

Fatty acids	Monoglycerides	Fatty alcohols	Carbon atoms : double bonds
Formic	—	—	1 : 0
Acetic	—	—	2 : 0
Propionic	—	—	3 : 0
Butyric	—	—	4 : 0
Valeric	—	—	5 : 0
Caproic	Monocaproin	—	6 : 0
Caprylic	Monocaprylin	—	8 : 0
Pelargonic/Nonanoic	—	—	9 : 0
Capric	Monocaprin	Decanol/n-decyl	10 : 0
Undecanoic	—	—	11 : 0
Undecylenic/Undecylic	—	—	11 : 1
Lauric	Monolaurin	Dodecanol/lauryl	12 : 0
Myristic	Monomyristin	Tetradecanol	14 : 0
Palmitic	Monopalmitin	—	16 : 0
Palmitoleic	Monopalmitolein	—	16 : 1
Stearic	Monostearin	—	18 : 0
Oleic	Monoolein	—	18 : 1
Linoleic	—	—	18 : 2
Linolenic	—	—	18 : 3
Arachidonic	—	—	20 : 4

S. aureus were incubated with selected fatty acids at 37 °C for 10 minutes. GAS and GBS were killed by 5 mM lauric acid and palmitoleic acid (16 : 1). When compared at lower concentrations, palmitoleic acid and monocaprin (10 : 0) showed considerable activities at 0.63 mM. Other fatty acids and monoglycerides had small or negligible effects. The lipids showed less activity and somewhat different antibacterial profile against *S. aureus*, which was susceptible to 10 mM capric acid. Similar effect has been demonstrated against *Streptococcus agalactiae*, *Streptococcus dysgalactiae* and *Streptococcus uberis* after as little as 1 minute incubation in milk supplemented with high concentrations (100 mM) of caprylic acid (8 : 0) [13].

It was established that antibacterial activities of long-chain fatty acids are increased by addition of a double bond [17, 18]. Addition of a second double bond to oleic acid (18 : 1) further increased the antibacterial activity but a third double bond decreased the activity. Linolenic acid has been found to exhibit strong antibacterial activity against the Gram-positive bacteria *Bacillus cereus* and *S. aureus*, an activity which can be enhanced by emulsifying agents such as monolaurin(12 : 0) and monomyristin (14 : 0) [21]. Knapp and Melly [22] investigated the antibacterial activity of polyunsaturated fatty acids and found that Gram-positive bacteria, such as *Lactobacillus acidophilus* and *S. epidermidis*, were inactivated by 0.01 mM of the essential fatty acid (EFA) arachidonic acid (20 : 4) after incubation at 37 °C for one hour. More recently, several other Gram-positive bacteria have been found susceptible to low concentrations of the medium-chain lauric acid, including *Propionibacterium acnes* (MIC 20 μM) [8] and *Bacillus anthracis Sterne* (MIC 228 μM) [14].

Table 3.2 Concentrations (mM) causing more than 100-fold (>99%) reduction of colony or inclusion-forming units at physiological pH in 10 minutes or less.

	GAS	GBS	S. aureus	E. coli	Salmonella spp.	H. pylori	N. gonorrhoeae	C. trachomatis	C. jejuni	C. albicans
Fatty acids										
Caprylic acid (8:0)	NA	NA	NA	NA	NA	10	NA	NA	NA	NA
Capric acid (10:0)	NA	NA	10	NA	NA	2.5	2.5[a]	10	10[a]	10
Lauric acid (12:0)	5	5	10	NA	NA	0.63	0.63	2.5[a]	NA	5
Myristic acid (14:0)	NA	NA	NA	NA	NA	2.5	NA	NA	NA	NA
Palmitoleic acid (16:1)	0.63	0.63	NA	NA	NA	0.31[a]	0.63	NA	ND	NA
Oleic acid (18:1)	5[a]	5[a]	NA	NA	NA	10[a]	NA	NA	NA	NA
1-monglycerides										
Monocaprylin (8:0)	NA	NA	NA	NA	NA	NA	NA	NA	NA	NA
Monocaprin (10:0)	0.63	0.63	1.25	NA	NA	0.31	0.63	2.5[a]	10[a]	10[a]
Monolaurin (12:0)	5[a]	NA	NA	NA	NA	0.15[a]	0.31	NA	NA	NA
Monomyristin (14:0)	NA	NA	NA	NA	NA	10[a]	NA	NA	NA	NA
Monopalmitolein (16:1)	5[a]	NA	NA	NA	NA	5[a]	NA	NA	NA	NA
Monoolein (18:1)	NA	NA	NA	NA	NA	NA	NA	NA	NA	NA
Reference	[20]	[20]	[20]	[26]	[26]	[26]	[29]	[30]	[11]	[102]

[a]Lowest concentration tested.
Highest concentration tested was 10 mM.
Number of carbon atoms : Number of double bonds.
Salmonella spp. and E. coli were tested for 30 minutes.
NA, not active (less than 100-fold reduction); ND, not done.

The antibacterial activities of mono-, di- and triglycerides of fatty acids against a variety of bacteria have been thoroughly studied by Conley and Kabara [18] and Kabara and coworkers [17]. The spectrum of activity of monoglycerides was found to be narrower as compared to that of free fatty acids. The authors determined the MIC for glycerides against a number of Gram-positive bacteria: *S. aureus*, *S. epidermidis* and *Streptococcus pyogenes*. Of the monoglycerides tested, monolaurin was the most active; it was slightly more active than monocaprin against most of the bacteria and more active than the free acid. *S. pyogenes* was the most susceptible bacterium, particularly to monocaprin, monolaurin and monomyristin. Monolaurin and to a lesser degree monocaprin were the only monoglycerides that showed activity against *S. aureus* and were more active than capric acid. In addition, monolaurin was shown to kill planktonic and sessile *Listeria monocytogenes* cells with higher efficiency than exponential cells [23]. A comparison of the bactericidal effect of 1-monoglycerides against Gram-positive cocci after shorter incubation times [20] is given in Table 3.2. The results revealed an equal killing effect of palmitoleic acid and monocaprin against group A and group B streptococci. By using longer incubation times, lower concentrations of monocaprin were needed to kill group B streptococcus and a complete killing of the bacteria was observed after seven-hour incubation at 0.63 mM. By further testing the bactericidal effect against *S. aureus* [20], monocaprin proved to be the most effective of all lipids tested, causing more than 99% killing at 1.25 mM concentration. This is a similar activity profile to that observed when 1-monoglycerides were added to human milk or milk formulas [24].

3.2.2 Activities against Gram-Negative Bacteria *In Vitro*

Marounek *et al.* studied the antimicrobial effect of fatty acids ranging from 2 to 18 carbons in chain length against *Escherichia coli* [25]. It was demonstrated that when the pH was dropped from 6.5 to 5.2, the medium-chain fatty acids caprylic acid and capric acid exhibited considerable inhibition and a reduction in the number of viable *E. coli* cells. In contrast, fatty acids ranging in chain length from 2 to 6 or from 12 to 18 carbons had negligible effect. This effect was confirmed by Nair *et al.*, who found that 25 to 100 mM caprylic acid added to milk reduced *E. coli* populations [13]. As presented in Table 3.2, no fatty acids were found to exhibit bactericidal activity against *E. coli* or *Salmonella* spp. when tested at low concentrations and short incubation times for up to 30 minutes at physiological pH [26]. In contrast, the number of colony-forming units of the Gram-negative pathogen *Campylobacter jejuni*, which is another foodborne pathogen, was effectively reduced by 10 mM capric acid [11]. By measuring minimum inhibitory concentrations, Skrivanova *et al.* [27] demonstrated inhibitory effect by 21 mM caprylic acid against both *E. coli* and *Salmonella* sp. and by 29 mM capric acid against *E. coli*.

Petschow *et al.* [28] found that lauric acid was the only medium-chain fatty acid that exhibited antibacterial activity against *Helicobacter pylori* after incubation at 1 mM for one hour. The susceptibility of *H. pylori* to medium-chain saturated and long-chain unsaturated fatty acids was also studied by Bergsson *et al.* [26]. They found 5 mM capric, lauric, myristic and palmitoleic acids caused more than one million-fold reduction in the number of viable bacteria after 10-minute incubation at 37 °C. They also found 10 mM caprylic acid showed high activity against *H. pylori* that varied among bacterial strains. Lauric acid at a concentration of 1.25 mM and palmitoleic acid at 0.625 mM had a significant antimicrobial effect against *H. pylori* after 10-minute incubation. Another Gram-negative bacterium,

Neisseria gonorrhoeae, was found to be killed by capric, lauric and palmitoleic acids [29]. Further analysis by exposure for one minute to a 2.5 mM concentration identified lauric and palmitoleic acids as the most effective fatty acids against *N. gonorrhoeae* [29]. Similar studies found 10 mM capric and lauric acids to cause more than 10 000-fold reduction in the number of *Chlamydia trachomatis* inclusion forming units in 10 minutes [30]. When tested at 5 mM concentration for five minutes, lauric acid was the most active lipid, causing more than 100 000-fold inactivation of *C. trachomatis*. The effect was followed by capric acid and monocaprin, but caprylic, myristic, palmitoleic and oleic acids had negligible effects against *C. trachomatis* [30]. The long-chain polyunsaturated arachidonic acid (20 : 4) also killed the Gram-negative bacteria *Neisseria* and *Haemophilus* spp. In contrast, *E. coli* was not susceptible to arachidonic acid, even after six-hour incubation with 0.3 mM arachidonate [22].

As to studies on the effect of glycerides of fatty acids on Gram-negative bacteria, Conley and Kabara [18] and Kabara *et al.* [17] found that di- and triglycerides did not exhibit antibacterial effects and none of the compounds tested inactivated *E. coli* or a number of other Gram-negative bacteria. Further studies by Isaacs *et al.* [24] showed that when medium-chain saturated monoglycerides were added to human milk and milk formulas the mixtures inactivated *Haemophilus influenzae*. The same was not true for polyunsaturated long-chain monoglycerides, which were less active. Petschow *et al.* [28] also studied the antibacterial effect of medium-chain monoglycerides against *H. pylori* and found that incubation of the bacteria with 1 mM monoglycerides of 10-, 12- and 14-carbon fatty acids for one hour at 37 °C caused a 10 000-fold reduction in the number of viable bacteria. Lower activities were observed after incubation with monoglycerides with shorter- or longer-chain fatty acids, that is 9-, 15- and 16-carbon chains [28]. The antibacterial effect of 1-monoglycerides against Gram-negative bacteria was also tested at incubation times of 1 to 10 minutes by Bergsson *et al.* [26, 29, 30], as shown in Table 3.1. Monocaprin, monolaurin and monopalmitolein (16 : 1) at 5 mM concentration were all highly effective in killing *H. pylori* in 10 minutes. At lower concentrations (0.625 mM) and after one minute incubation time, monocaprin and monolaurin were the only lipids showing significant activities against *H. pylori* [26]. The Gram-negative bacterium *N. gonorrhoeae* was found to be killed by 0.31 mM monocaprin and 0.63 mM monolaurin in 10 minutes [29]. All other monoglycerides showed much smaller effect. The Gram-negative pathogens *H. pylori* and *N. gonorrhoeae* were found to be exceptionally susceptible to lipids, both fatty acids and 1-monoglycerides, when compared to other bacteria studied, either Gram-negative or Gram-positive. The effects of 1-monoglycerides were also studied against *C. trachomatis* [30]. The studies found monocaprin to be the most effective glyceride, causing more than 100 000-fold reduction in viable bacteria after five minutes at 5 mM concentration. A 50% effective concentration of 0.12 mM was found by incubation of *C. trachomatis* with monocaprin for two hours. *C. trachomatis* was not susceptible to 10 mM monocaprylin, monolaurin, monopalmitolein or monoolein (18 : 1). A study of the susceptibility of *Campylobacter jejuni* to a series of monoglycerides found monocaprin to be particularly active in killing this bacterium [11].

In contrast to the above Gram-negative bacteria, *E. coli* and *Salmonella* spp. were found to be resistant to all monoglycerides tested at neutral pH [26]. However, in an acidified buffer at pH < 5, monocaprin was highly effective in killing both *E. coli* and *Salmonella*, reducing the number of colony-forming units by more than one million-fold in 10 minutes

[11]. Furthermore, Nair *et al.* [13, 31] showed that when monocaprylin was added in high concentrations (50 mM) to milk or at 5 mM to apple juice, it effectively killed *E. coli*. On the other hand, Preuss *et al.* [14], in a study of the bactericidal effect of essential oils and monolaurin on bacteria, confirmed that *E. coli* and *Klebsiella pneumoniae* were very resistant to this monoglyceride, whereas *H. pylori* was susceptible.

In general, the bactericidal activities of fatty acids and monoglycerides produced by different laboratories are in agreement. Any discrepancies may be due to differences in experimental conditions, particularly differences in lipid concentrations and incubation times. Gram-positive and many Gram-negative bacteria have been found to be highly susceptible, with the exception of *Enterobacteriacae*, which in most studies have proved to be resistant at neutral pH.

3.2.3 Antibacterial Mechanism

Kabara [1, 16] made generalizations concerning the relationship between lipid structure and antibacterial activity. He concluded that 1) lauric acid is the most active saturated fatty acid, palmitoleic acid the most active monounsaturated acid, and linoleic acid the most active polyunsaturated fatty acid against Gram-positive bacteria, 2) monoglycerides of medium-chain saturated fatty acids are more active than the free acids, particularly the monoglyceride of lauric acid, 3) fatty acids have very low activity against Gram-negative bacteria except when having a very short chain (6 carbons or less), and 4) yeasts are affected by fatty acids with short chain lengths (10 to 12 carbons). The most important generalization is probably that monoesters of fatty acids, such as monolaurin, are very active, while di- and triglycerides are not. From these studies it was concluded that the antibacterial effect of short-chain (6 to 10 carbons) saturated fatty acids is related to the degree of their dissociation, and they were found to be more active at pH 6.5 than at pH 7.5. These early studies established that unsaturated fatty acids are active against Gram-positive bacteria and indicated that they become more active with an increasing number of double bonds [1, 16]. Further studies by Kabara and coworkers demonstrated that the activities of long-chain fatty acids were increased by addition of a double bond. Addition of a second double bond to oleic acid (18 : 1) further increased the antibacterial activity, but a third double bond decreased the activity [17, 18].

The ability of lipids to disrupt cellular membranes has been demonstrated for bacterial cells, both Gram-negative and Gram-positive. As mentioned above, monocaprin and monolaurin had little or no activity against Gram-negative bacteria such as *E. coli* at 30 °C [32] and physiological pH [26]. However, when *E. coli* was incubated with lauric acid, monocaprin and monolaurin at 50 °C for five minutes, or with monocaprin at lower pH [11], a remarkable inactivation was observed, much more than that produced by heating alone. Cells growing in the logarithmic phase were more sensitive than stationary-phase cells. Taken together, this suggests a reduction in the permeability barrier in the outer membrane caused by heating, allowing penetration of the lipid through the outer membrane and the cell wall. Increased transfer of monolaurin into bacterial cells was seen when incubating *E. coli* in a mixture of monocaprin or monolaurin with citric acid. These studies suggest that the acid removes a barrier in the outer membrane of the cell wall, giving the lipid access to the inner membrane, or that lower pH will increase the concentration of the undissociated form of the fatty acids. In summary, saturation of natural antimicrobial lipids affects the lipid ability to cause disintegration of microbial membranes. This effect can be enhanced by altering environmental factors.

Hypotheses concerning the antimicrobial mechanism of fatty acids and their 1-monoglycerides have been tested by several authors. Complete cell disorganization and disruption were seen by Knapp and Melly in *N. gonorrhoea* cells after incubation with 0.05 mM arachidonic acid for one hour [22].These authors obtained similar results for *H. influenzae*, while no morphological changes were noted in *E. coli*, which is not susceptible to arachidonic acid. Peripheral cytoplasmic condensations were visualized in *S. aureus* and *Lactobacillus acidophilus*, while no changes were visible in the cell walls. As illustrated in Figure 3.1, the viability of the Gram-negative bacterium *C. trachomatis* was irreversibly lost after treatment with monocaprin [30]. The negatively stained elementary bodies of *C. trachomatis* appeared deformed and partly disintegrated after 10 minutes, suggesting that monocaprin kills the bacteria by disruption of the cellular membrane(s) (Figure 3.1) [30]. This effect is similar to that of glyceryl ethers, 1-*O*-hexyl-sn-glycerol [33]. The effect of monocaprin on the structure of Gram-positive cocci was studied by transmission electron microscopy (TEM) (Figure 3.2), and with a two-colour fluorescent bacterial viability kit to stain monocaprin-treated group B streptococci [20]. A treatment with 10 mM mono-caprin for 10 minutes caused damage to the cell membrane, which became permeable to propidium iodide. Examination of thin sections of bacterial cells in TEM confirmed this finding (Figure 3.2), since after treatment with monocaprin the plasma membrane was no longer visible, indicating a disintegration of the membrane. No changes were detectable on the surface or in the structure of the bacterial cell wall by scanning electron microscopy (SEM). This suggests that the lipids can penetrate the cell wall of Gram-positive bacteria and cause disintegration when it reaches the bacterial cell membrane.

The differences in the susceptibilities of Gram-negative bacteria to antimicrobial lipids are notable, particularly between *H. pylori* and *Enterobacteriacae* [26], and may be due to differences in the outer membrane of the bacteria. In this regard, the external leaflet of the outer membrane of *E. coli* and *Salmonella* is almost entirely composed of lipopolysac-charides (LPS) and proteins. These bacteria have a hydrophilic surface due to the O-side chains of the LPS, making it difficult for hydrophobic molecules, like lipids, to enter the bilayer. Moreover, molecules may have difficulty in penetrating these membranes because of the low fluidity of the hydrocarbon chains in the LPS leaflet and the strong interactions between the LPS molecules. On the other hand, Gram-negative bacteria with a relatively hydrophobic surface, such as *N. gonorrhoeae* [34], are easily killed by medium-chain fatty acids and monoglycerides [29]. The outer membrane of *N. gonorrhoeae* is not composed of LPS but of lipooligosaccharides (LOS) that lack the O-antigen side chain. Even if LOS is mainly expressed on mucosal bacteria, it can also be expressed on enteric bacteria such as *Campylobacter* spp. [34], which have been found to be susceptible to monocaprin [11].

Desbois and Smith in 2010 [35] reviewed the mechanisms of antibacterial actions of fatty acids. A variety of detrimental effects can be caused by partial solubilization of the cell membrane by fatty acids, leading to release of membrane proteins. This can result in interference with the electron transport chain and uncoupling of oxidative phosphorylation, or to inhibition of enzyme activity and nutrient uptake, causing inhibition of bacterial growth. At higher concentrations, the detergent effect of fatty acids can kill bacteria by cell lysis. Thus the action can be either bacteriostatic and reversible or bactericidal and irreversible.

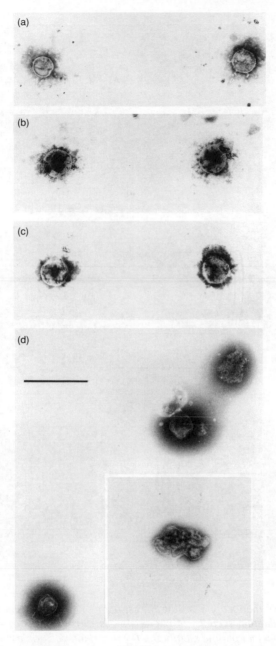

Figure 3.1 *Electron micrographs of negatively stained elementary bodies (EBs) of C. tra-chomatis. The EBs were untreated (a) and treated with 10 mM monocaprin for 1 minute (b), 5 minutes (c) and 10 minutes (d). After treatment for 10 minutes, the EBs appear deformed and shrunken or partially disrupted (d, inset). Bars, 1 μm. (Reprinted from [30] with permission from American Society for Microbiology.)*

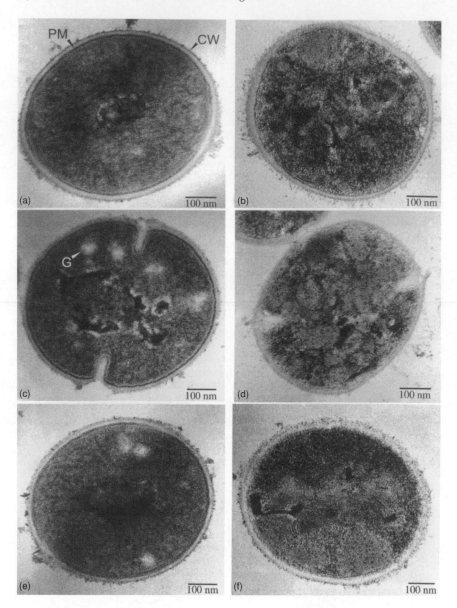

Figure 3.2 *Effects of monocaprin on the ultrastructure of group B streptococci as demonstrated by TEM of ultrathin sections of bacterial cells. (a), (c) and (e) show cells from the control samples with intact plasma membrane (PM) and intact cytoplasmic granules (G). (b), (d) and (f) show cells from samples treated with 10 mM monocaprin for 30 minutes, demonstrating disappearance of plasma membranes and cytoplasmic granules. Some changes can be seen in the cell wall (CW). (Reprinted from [20] with permission from John Wiley and Sons, Ltd.)*

Lipids such as fatty acids have to permeate the cell wall of Gram-positive bacteria and the outer membrane of Gram-negative bacteria to gain access to the bacterial cell membrane. As discussed above for Gram-negative bacteria, differential susceptibility of bacterial species to the action of lipids is most likely due to differences in these outer structures, which can act as a barrier and protect the inner cell membrane. Exposure of strains of the Gram-positive bacterium *S. aureus* to long-chain unsaturated fatty acids has been shown to upregulate the expression of genes associated with peptidoglycan synthesis, thus increasing the thickness of the bacterial cell wall. An adaptive response which reduced the surface hydrophobicity of the wall was also observed, making the bacteria more resistant to the action of the fatty acids [36, 37].

3.2.4   A Possible Role of Antimicrobial Lipids in Host Defence against Bacteria

It has been known for decades that sweat, sebum [38] and breast milk [3, 39] contain lipids which are highly microbicidal and it has been suggested that they play an important role in the innate defence mechanism against microbial infections on the surface of skin and in the stomach of suckling infants. These lipids are mostly fatty acids released from sebaceous triglycerides by lipases of the normal bacterial flora [40, 41], fatty acids synthesized by sebocytes [42–44] and keratinocytes [45], and fatty acids in the cream fraction [46] of breast milk, which becomes activated upon enzymatic digestion of triglycerides by lipases [47].

Aly *et al.* [48] demonstrated that *S. aureus*, *S. pyogenes* and *Candida albicans* diminished in number when deposited on normal human skin. In contrast, acetone-washed forearms gave greater persistence of these bacteria and when these extracts/washes were replaced on the skin, the recolonization was prevented. These authors further pointed out a correlation between *in vitro* studies on acetone lipid extracts from skin and the microbes localized to skin. Interestingly, *S. pyogenes*, which rarely colonizes skin, is extremely sensitive to skin lipids [49]. Free fatty acids of human stratum corneum, as well as both polar lipids and glycosphingolipids, exhibit antistaphylococcal activity [45] at concentrations 10 to 100 times lower than the concentrations calculated to be present in the stratum corneum intercellular domain of the skin [50]. This suggests that stratum corneum lipids are effective in physiologically relevant quantities. More recently, lipids isolated from sebum have been shown to reduce the growth of Gram-positive bacteria such as *S. aureus*, *Streptococccus salivarius* and *Fusobacterium nucleatum*, while most Gram-negative bacteria are resistant to sebum lipids [51]. It was shown that both saturated and unsaturated fatty acids are responsible for the antimicrobial activity and that lauric acid is the most potent of the saturated fatty acids. However, a palmitoleic acid isomer, an exclusive lipid of sebum [52], was found to account for most of the fatty acid antimicrobial activity.

The antimicrobial effect of milk lipids, fatty acids and monoglycerides, has been demonstrated for a variety of both Gram-positive and Gram-negative bacteria [2, 24, 47]. It was demonstrated that fivefold-diluted stomach contents of infants fed human milk or infant formula had varying degrees of activity against both *S. epidermidis* and *E. coli* after 30 minutes at 30 °C [47]. The same authors further illustrated that infant formulas treated with lipases for one hour at 37 °C were active against the Gram-positive bacteria *S. epidermidis* but not against Gram-negative bacteria, including *E. coli* and *Salmonella enteritidis*. The discrepancies in the results on *E. coli* may reflect incomplete lipolysis of formula triglycerides in lipase-treated mixtures. Additional studies demonstrated inactivation of the Gram-negative bacterium *H. influenzae* by 1-monoglyceride-supplemented human milk and infant formulas after 30 minutes at 37 °C. Medium-chain saturated monoglycerides showed activity but polyunsaturated and long-chain monoglycerides were less active.

Lipids located to a cream-like substance that covers the skin of foetuses and newborn babies have been shown to potentiate the antimicrobial action of other effector molecules of innate immunity, namely antimicrobial peptides, *in vitro* [53], demonstrating the importance of the microenvironment for the function of antimicrobial components. This has been further studied in other recent studies, where enhanced synergistic effect in the antimicrobial activity of monolaurin has been documented when combined with various environmental and biological factors, such as temperature changes [32, 54, 55], organic acids [56], lower pH [25, 57], the chelating agent EDTA [58], antimicrobial peptides [53, 59] and the lactoperoxidase system [55], which also enhanced the effect of other antimicrobial lipids. The presence of natural antimicrobial lipids on skin and mucosa suggests that lipids might be useful as the active ingredients in pharmaceutical formulations for prevention and treatment of skin and mucosal infections [60].

Among the clinical symptoms suggesting the importance of fatty acids as part of the innate defence mechanism is that essential fatty acid deficiency causes increased susceptibility to infections and poor wound healing [61]. Essential fatty acids are unsaturated fatty acids of 18-, 20- or 22-carbon chain lengths that are not synthesized by the human body. Essential fatty acid deficiency was first described in 1929 by Burr and Burr, who noted that poor growth of rats and skin lesions could be prevented by linoleic acid [61]. A clear microbiological difference was observed between mice that were fed a diet lacking essential fatty acids and mice fed an ordinary diet including essential fatty acids. The mice fed a diet lacking essential fatty acids contained 100-fold higher levels of bacteria on their skin than mice fed an ordinary diet. Furthermore, the mice fed diet lacking essential fatty acids were commonly colonized by *S. aureus*. Surprisingly, *in vitro* experiments demonstrated that lipids isolated from mice fed a diet lacking essential fatty acids were more lethal to bacteria commonly found on skin than lipids isolated from mice fed a normal diet. This can possibly be explained by the results of Georgel *et al.* [15], who studied innate immunodeficiency in mice caused by a mutation in a fatty acid desaturase (stearoyl-CoA desaturase), which is an enzyme that is essential for synthesis of monounsaturated fatty acids in sebaceous glands. These authors demonstrated that this mutation resulted in a reduction in sebum production of palmitoleic acid and oleic acid and that it is responsible for the lack of bacterial clearance after skin infections. It was further shown that the expression of stearoyl-CoA desaturase is induced by *S. aureus*, a bacterium susceptible to monounsaturated fatty acids, but not by *E. coli* [15], which is resistant to monounsaturated fatty acids [26].

3.3   Antiviral Activities of Fatty Alcohols, Fatty Acids and Monoglycerides

3.3.1   Early Studies of Antiviral Lipids

The first reports that fatty acids and their soaps might inactivate viruses appeared in the 1930s. Begg and Aitken [62] reported that potent Rous sarcoma filtrates had been inactivated by addition of 1% solution of sodium oleate. Also, intratumoural injections of sodium oleate sometimes led to definite regression and occasionally to complete disappearance of Rous sarcomas in fowles. Helmer and Clowes in 1937 [63] found that an unsaturated fatty acid fraction from pig pancreatic tissue had a strong inhibiting activity against the agent of chicken sarcoma. A commercial oleic acid was shown to have a comparable activity. A number of free fatty acids were tested against the agent by addition to an aqueous extract of chicken tumour. The extract was then tested for infectivity by intradermal injection into chickens. The straight-chain 18-carbon unsaturated fatty acids had a much greater activity against the tumour agent than the saturated stearic acid. However, the number of double bonds had no effect on the activity, oleic, linoleic and linolenic acids all being equally active. The length of the carbon chain had a strong influence on the activity. The lower acids, either saturated or unsaturated, had little or no activity compared to the long-chain acids. However, the activity of the long-chain acids was limited by the lack of solubility in water. Only fatty acids and their soaps were active, whereas the glycerides found for example in olive oil, cod liver oil and sardine oil had no action. Pirie had observed in 1933 [64] that pancreatic extracts destroyed tumour viruses, but the active compounds were unknown. In a further study carried out by Pirie [65], preliminary experiments indicated that the virucidal activity was associated with a fairly unstable fatty acid. Several lipid fractions were therefore separated from dried pig pancreatic extract and tested for virucidal activity against vaccinia and Fujinami tumour virus. The lecithin and fatty acid fractions were found to account for the whole virucidal power of the dried extract and the acids were apparently responsible for most of the activity. The cephalin fraction had no effect. A number of commercially obtained fatty acids were shown to have virucidal activities. Of the higher fatty acids, only the unsaturated ones, such as oleic and linoleic acids, were virucidal. Pirie speculated that 'the fatty acids are adsorbed on the surface of the viruses and disrupt it to such an extent that the virus is lysed.'

In 1940, Stock and Francis [66] studied the inactivation of influenza virus by various fatty acids and their soaps. A virus suspension, made from the lungs of infected mice, was mixed with a fatty acid or soap solution at pH 7.5 and allowed to stand at room temperature. After 90 minutes and again after 24 hours, samples were tested for infectivity by intranasal inoculation into mice. Preliminary experiments showed that oleic and linoleic acids inactivated the virus while myristic and undecylenic (11 : 1) acids did not. A large number of fatty acids of increasing chain lengths, both saturated and unsaturated, and other lipids were then tested in the same way. The 18-carbon unsaturated fatty acids, oleic, linoleic, linolenic and chaulmoogric acids, were the most virucidal, whereas their isomers such as elaidic acid were ineffective. Of the shorter-chain saturated acids, the 12-carbon lauric acid was the most active, but the unsaturated 11-carbon undecylenic acid was ineffective. Other lipids, such as lecithin, had no effect. The results were interpreted to show a relationship between the configuration of the fatty acid molecule and the capacity to inactivate virus. The most likely mechanism for inactivation of the virus by long-chain unsaturated fatty acids was thought to be a more or less specific adsorption of fatty acid to proteins on the viral surface. This adsorption, which would probably be dependent on the configuration of both the fatty acid and the protein molecules, might result in dispersion of the protein layer on the viral surface. Experiments in which various dilutions of virus were mixed with different concentrations of oleic acid suggested that there exists a minimum concentration of oleic acid for effectively inactivating the infectivity of a given concentration of virus [67]. Notably, inactivation of the infectivity of influenza virus by oleic acid did not destroy the antigenicity of the virus, since mice vaccinated with the noninfective mixtures of virus and fatty acid showed a strong immune reaction [66]. However, the reaction was not as strong as in mice receiving untreated active virus [67]. On the other hand, influenza virus which had been inactivated by oleic acid was apparently more active as an antigen than other noninfectious preparations such as virus inactivated by formalin. Inactivation of lymphocytic choriomeningitis (LCM) virus by fatty acids and soaps was also studied by Stock and Francis [68]. The experiments were carried out in a similar way to the influenza virus experiments, except that the infectivity of samples was tested by intracerebral inoculation of mice. As in the previous studies of influenza virus, the 18-carbon unsaturated fatty acids were the most effective in inactivating the LCM virus. Of the saturated fatty acids, only myristic acid was effective, whereas in contrast to influenza virus, lauric acid was ineffective. As in the studies of influenza virus, the relative amounts of virus and oleic acid were found to be important in obtaining inactivation of LCM virus. In contrast to oleate-inactivated influenza virus, LCM virus inactivated in the same way failed to induce any detectable immunity in mice. This was in line with the failure to use formalinized LCM virus as an immunizing agent.

3.3.2   Antiviral Activities of Milk Lipids

Around 1960, human milk was shown to reduce the infectivity of a number of viruses. The activity against enveloped viruses, such as herpes simplex virus type 1 (HSV-1), was different from that against poliovirus, which has no lipid envelope. The antipoliomyelytic activity was bound to the protein fraction of the milk, which had no effect against enveloped viruses, while the cream fraction showed the antiherpetic activity [69]. Sarkar *et al*. [70] showed that whole, skimmed or cream-fraction human milk incubated with mouse mammary tumour virus for 18 hours at 37°C caused decreased viral infectivity. The cream fraction was more effective in the inactivation of the virons than the whole or skimmed milk. In the following years these findings where confirmed by the same authors and others [1]. Thus, the lipid fraction of milk and colostrum was found to inactivate dengue virus, which is an enveloped virus [71]. Free fatty acids and monoglycerides in milk showed antiviral activities against Semliki Forest virus, whereas diglycerides had no activity [39]. A series of studies of the antiviral properties of human milk showed that fresh milk samples are inactive against enveloped viruses such as HSV-1, vesicular stomatitis virus (VSV), measles virus and human immunodeficiency virus type 1 (HIV-1). However, most of the milk samples became highly antiviral after storage at 4 °C for a few days, reducing the virus titre 1000-fold in 30 minutes at 37 °C. Similar antiviral activities were found in stomach samples from suckling infants who had been fed fresh human milk or infant formula feeds one hour earlier. The antiviral activity was found in the cream fraction and it was concluded that it was due to milk lipase, particularly lipoprotein lipase, releasing fatty acids from triglycerides upon storage of human milk samples. In the stomachs of infants the activity was due to gastric and lingual lipases releasing fatty acids from milk fats [2, 3, 24, 47, 72, 73]. These findings confirmed and further extended the importance of lipases for antiviral activities of milk, which had been pointed out earlier by Welsh and coworkers [39]. A comparative study of purified fatty acids normally found in human milk and other food products showed that short-chain (4 to 8 carbon) and long-chain (16 to 18 carbon) saturated fatty acids had no or minor antiviral effects against the enveloped viruses VSV, HSV-1 and visna virus

(VV). In contrast, medium-chain saturated (10 to 14 carbon) fatty acids inactivated all the viruses, although myristic acid caused only 1.7 $\log_{10}$ reduction against VV after 30 minutes at 37 °C [74]. Long-chain (16 to 20 carbon) unsaturated fatty acids were also active and their activities increased with increasing number of double bonds. Thus, arachidonic acid was the most active fatty acid in lower concentrations. Corresponding 1-monoglycerides of several fatty acids were also tested and all, except monomyristin and monoolein, were 5 to 10 times more active against VSV and HSV-1 than the free fatty acids. These findings suggest that lipolysis of triglycerides from human milk, either in the gastrointestinal tract or by storage, releases two types of antiviral lipids: monoglycerides and free fatty acids [74, 75].

3.3.3 Antiviral Activities of Lipids and Fatty Alcohols

Sands [76] studied the effect of chemically purified fatty acids on the enveloped bacterio-phage Φ6. The results showed that unsaturated long-chain fatty acids, that is palmitoleic and oleic acids, inactivated the virus, while the saturated myristic and palmitic (16 : 0) acids were inactive against Φ6. The lipid-containing bacteriophages Φ6, PM2 and PR4 were fur-ther studied by Sands and coworkers [77] for antiviral susceptibilities to free fatty acids, fatty acid derivates, mono- and diglycerides and long-chain alcohols. Low concentrations (50 µg/ml) of medium-chain saturated and long-chain unsaturated alcohols caused at least a 99% inactivation of Φ6 after 30 minutes at 25 °C. Long-chain unsaturated fatty acids and corresponding monoglycerides were also antiviral, while diglycerides caused minor or no inactivation. PM2 and PR4 bacteriophages, which lack external envelopes, were much less susceptible to inactivation by the lipids. HSV type 2 (HSV-2) was inactivated by many of the lipids that were active against Φ6, where monoolein and monopalmitolein were found to be the most active.

Snipes and coworkers [78] studied inactivation by alcohols ranging from 4 to 18 carbons on lipid-containing viruses. Saturated alcohols with chain lengths from 10 to 14 carbons showed the most antiviral activity, suggesting that a proper balance between the polar and hydrophobic components can lead to maximal perturbation of the viral membrane. Decanol (10 : 0), dodecanol (12 : 0) and tetradecanol (14 : 0) at 0.5 mM readily inactivated HSV after 30-minute exposure, as well as the bacteriophage Φ6, even at lower concentrations. The lipid-containing virus PM2 was susceptible to decanol and dodecanol, while tetradecanol had neglible antiviral effects. Phytol, a naturally occurring branced-chain alcohol found in chlorophyll, was active against Φ6 in 0.001 mM and HSV in 1 mM concentration, but not against PM2 after 30-minute exposure. The nonenveloped polyoma virus and bacteriophage Φ23-1-a were not inactivated by any of the alcohols tested. At 0.05 mM concentration none of the alcohols showed cytotoxic effects on human embryonic lung cells, although several of the alcohols were active against HSV at this concentration.

Saturated alcohols and n-alkyl derivates of butylated hydroxytoluene (BHT) tested against Φ6 and HSV-2 showed that alcohols with 10 to 14 carbons in chain length were the most effective. Also, alcohols and BHT were all more active against Φ6 than against HSV-2. In addition, *in vivo* studies with dodecanol against HSV skin lesions in hairless mice revealed effective clearence of virus lesions when compared to untreated control mice [78, 79]. Thus, these early studies established that inactivation of enveloped viruses by lipids varies greatly, depending on both the nature of the lipid and the type of virus.

Kohn and coworkers [80] studied the incorporation of essential unsaturated fatty acids, that is oleic, linoleic and arachidonic acids, into phospolipids of animal cells, which induced a change in the fluidity of the membranes and loss of rigidity. Micromolar concentration of these fatty acids (10 to 50 μg/ml) applied for 15 minutes to enveloped viruses such as arbo-, myxo-, paramyxo- and herpesviruses caused a rapid loss of infectivity, as measured by plaque assay. However, the abovementioned fatty acids in concentrations up to 25 μg/10^6 cells/ml did not affect the growth of baby hamster kidney cells. No antiviral activities were found for linoleic acid when incubated against the non-enveloped poliovirus, siman virus 40 or encephalomyocarditis virus, suggesting destructive lipophilic effects of the fatty acids against enveloped viruses associated with disintegration of the virus envelope [81]

Thormar *et al.* [75] studied the effects of free fatty acids, monoglycerides and ethers against VV, which is a lentivirus related to HIV. Ethers of 6-, 7- and 8-carbon saturated fatty acids all showed activities against the virus after 30 minutes at 37 °C. The most effective ether was 1-*O*-octyl-*sn*-glycerol, which is an 8-carbon ether of caprylic acid and was much more active than caprylic acid alone or its monoglyceride ester, monocaprylin. Also 1-*O*-octyl-*sn*-glycerol was found to inactivate HIV in human plasma [82].

Hilmarsson *et al.* [83] studied the virucidal effects of medium- and long-chain (8 to 18 carbon) fatty alcohols and corresponding lipids against HSV-1 and HSV-2 at various concentrations, times and pH levels. After 10-minute incubation at 37 °C in 10 mM concentration, 14 of the lipids tested caused a significant reduction in HSV infectivity. N-decyl (10 : 0) alcohol, capric, lauric, myristic, palmitoleic, and oleic acids, and monocaprylin, monocaprin and monolaurin monoglycerides were all fully active – that is, no virus was detectable after the treatment – causing 100 000-fold or greater reduction in HSV titre. A generally good agreement was found between the activity profiles for HSV-1 and HSV-2. At shorter incubation time, monocaprin, monolaurin and all of the fatty acids except caprylic and capric acids acted rapidly, causing a significant reduction in virus titre after only one minute at 10 mM concentration. Further comparison of 10 to 12 carbon alcohols and the corresponding lipids at lower concentrations and increased incubation times against HSV-1 showed a complex relationship between concentration and duration of action. N-decyl alcohol acted rapidly at 5 mM concentration but at lower concentrations the activity was greatly reduced, even with increased incubation time. Lauryl alcohol acted more slowly compared to n-decyl alcohol, but was active at much lower concentrations (0.625 mM), particularly with increasing time, which is in agreement with previous studies by Snipes and coworkers [78, 79]. The three most active alcohols and the corresponding fatty acids and monoglycerides were further tested at pH 7 and 4.2. Lowering the pH to 4.2 caused a more rapid inactivation of HSV-1 by n-decyl and lauryl alcohols and capric acid, with a significant reduction of virus titre in one minute. The virucidal activities of all the alcohols, fatty acids and monoglycerides tested in 10 minutes at concentrations as low as 0.15 mM for monolaurin were greatly increased at pH 4.2 as compared with pH 7. It was hypothesized that the difference between the polar groups of alcohols and fatty acids, that is hydroxyl group versus carboxyl group, and the corresponding difference in their hydrophile–lipophile balance (HLB) might explain their different activities against HSV. Capric and lauric acids are hydrophilic at pH 7 with HLB numbers around 20, whereas n-decyl and lauryl alcohols are lipophilic with HLB numbers of 4.15 and 3.2, respectively. This may possibly explain the differences in activity profiles of fatty acids and alcohols seen in this study. However, the distinct differences in activity between various fatty acids, for example capric and lauric acids, and between various alcohols, for example n-decyl and lauryl alcohols, can hardly be explained by the small difference in HLB. In general, no correlation could be found between virucidal activities and HLB numbers. This was most obvious when activities were compared at different pHs. Alcohols and monoglycerides were much more active at pH 4.2 than at pH 7, although their HLB numbers did not change between these pH levels. The increased virucidal activity against HSV at low pH may be due to ionic changes in the surface of the host cell-derived viral envelope, which contains virus-encoded glycoproteins, somehow giving the alcohols, fatty acids and monoglycerides better access to the lipid bilayer of the envelope.

Alcohols and lipids were, in studies similiar to those of HSV-1 and 2, also tested against VV to further compare the virucidal profiles against viruses and possibly elucidate how lipids inactivate enveloped viruses [84]. In this comparative study it was shown that fatty acids were most active against VV, particularly lauric, palimitoleic and oleic acids, when tested either in a low concentration or at short incubation time. N-decyl and lauryl alcohols and monocaprin and monolaurin also had significant activities, with about 100- to 1000-fold reduction after 10 minutes in 10 mM concentration. Studies on virucidal activities at low pH against VV revealed that lipids became generally more virucidal at pH 4.2 compared with pH 7, which is in agreement with studies on HSV.

Previous studies by Thormar *et al.* [74,75] showed that certain fatty acids and monoglycerides were active against VV, which is compatible with these results. However, the data are not fully comparable since in the earlier study the incubation time was longer, that is 30 minutes. Previous studies on other enveloped viruses, such as HSV, showed that lipids and the corresponding alcohols with 10- and 12-carbon chain-length were the most virucidal, especially monocaprin [83]. This is in contrast with the results on VV, where only fatty acids had high virucidal activites.

Hilmarsson *et al.*[85] also studied the virucidal activities of alcohols and corresponding lipids against respiratory syncytial virus (RSV) and human parainfluenza virus type 2 (HPIV2) at different concentrations, times and pH levels. Monocaprin, which has been found to exhibit broad-spectrum and fast-acting virucidal activities against a number of enveloped viruses, was also tested against influenza A virus. The comparison of virucidal activities of fatty alcohols and lipids against RSV showed that eight of the compounds tested at 10 mM concentration for 10 minutes reduced the infectivity titres up to 4.5 $\log_{10}$ or greater. A good agreement was found between the activity profiles for RSV ATCC (American-type culture collection) and three recent clinical isolates of RSV. Lauric, palmitoleic and oleic acids and monocaprin all acted rapidly in 10 mM concentration, causing a significant titre reduction in only one minute against all RSV strains. The most active monoglyceride tested was monocaprin, which was still highly active after dilution to 1.25 mM. Lauric acid was the most active fatty acid and N-decyl alcohol the most active alcohol in lower concentrations tested for 10 minutes. A generally good agreement was seen in the virucidal activities of the 10- and 12-carbon compounds against HPIV2 and RSV, and also for monocaprin tested against influenza A virus in decreasing mM concentrations, which revealed about 10 000-fold reduction in influenza A virus infectivity at 1.25 mM monocaprin in 10 minutes. By changing the pH from 7 to 4.2, the activities of most of the compounds were greatly increased at both lower concentrations and shorter incubation times, which is in agreement with the previous studies on HSV and VV. Thus, lauric acid in 0.62 mM concentration was fully active against both RSV and HPIV2 at pH 4.2 after 10-minute exposure, whereas it had no activity at pH 7. However, there was a distinct difference in the effect of low pH on the activity of monocaprin and monolaurin against these viruses, since in contrast to RSV there was no increase in the activity of the monoglycerides against HPIV2 at pH 4.2. These findings, and the well-established difference in the envelope proteins of RSV and HPIV2, further supports the hypothesis that the increased virucidal activities of lipids at low pH are due to ionic changes in the glycoproteins on the surface of the viral envelope somehow leading to a better access to the lipid bilayer of the envelope.

In an extension of this study, the most active lipids were mixed in 10 mM concentration with milk products and fruit juices and the mixtures tested for virucidal effects against RSV. N-decyl alchol, lauric acid, monocaprin and monolaurin were found to preserve their activities against RSV to varying degrees in such mixtures. The experiments revealed that all four compounds were fully active in pear juice and monolaurin and monocaprin were also active in various milk products. The monoglycerides were fully active in pear juice and apple juice in concentrations as low as 1.25 mM, or about 0.03%. It is notable that the virucidal activities of the lipid compounds were better maintained in the juices than in the milk products. Most likely the compounds, particularly fatty acids and alcohols, bind to proteins and/or lipids in the milk, which may partly block their activity [82].

In the last few decades there has been an increase in research and development of antimicrobial agents which could preferably inactivate a number of human pathogens such as bacteria and viruses. The main reasons for this interest are the growing resistance of microorganism to known drugs, the worldwide threat of HIV and a possible pandemic caused by new virulent strains, for example of influenza virus. In addition to specific drugs there is also a need for new microbicides with less-specific actions against pathogenic microorganisms, which are therefore less likely to lead to the emergence of resistant bacterial or viral strains. For example, the drugs presently used against RSV and other viruses causing respiratory infections are only recommended for high-risk children or severely ill infants, due to efficacy and safety concerns [86, 87]. Therefore, new prophylactic or therapeutic compounds for general use against respiratory viruses would be desirable. The significant virucidal activities of fatty alcohols and lipids on RSV and HPIV2 raise the question of the feasibility of using the most active compounds as ingredients in pharmaceutical formulations against respiratory infections, or as food additives to protect the oral, nasal and respiratory mucosa of infants against viral infection.

In relation to a possible use of fatty acids as active antiviral ingredients, Loftsson *et al.* [88] studied the enhancement of fatty acid extract (FAE) from cod-liver oil on transdermal delivery of acyclovir, which is an HSV-specific antiviral drug, and the activity of the exctract against HSV-1. A 1% FAE caused a 50 000-fold or greater reduction in HSV-1 after 10 minutes, while an ointment containing 30% FAE inactivated the virus about 1.5 million-fold. Also, transdermal studies with variable ointments containing acyclovir showed that the largest acyclovir flux through the skin of hairless mice was obtained from a cream vehicle containing 30% FAE and about 30% propylene glycol. These findings suggest that FAE from cod-liver oil not only enhances skin penetration but might possibly also enhance the antiviral activity of the formulation.

Regarding safety concerns, lipids have been shown to be toxic to cell monolayers in cell cultures [30], although antimicrobial formulations containing monocaprin have been found to be harmless to skin and mucosa. Thus, hydrogels have been formulated which contain monocaprin at the highly virucidal concentration of 20 mM (0.5%) [89]. They did not cause irritation in the vaginal mucosa of mice and rabbits [10,90], in the skin of hairless mice [90] or in the vaginal mucosa of healthy human volunteers [91], even though they were toxic in cell culture. It is also noteworthy that monoglycerides of fatty acids are classified as GRAS (generally recognized as safe) by the United States Food and Drug Administration and are approved as food additives by the EU (E471).

3.3.4 Antiviral Mechanism of Lipids

Several attempts have been made to elucidate the mechanism behind the antiviral effects of fatty acids and monoglycerides. Sedimentation analysis of bacteriophage Φ6 inactivated by lipids showed that treatment with 10 μg/ml oleic acid for 30 minutes resulted in the loss of the viral envelope, whereas 50 μg/ml caused a complete disruption of the virus particle [76]. Another study, investigating the viral particles of mouse mammary tumour virus after inactivation by the cream fraction of human milk, also showed degradation of the viral envelope [70]. In yet another report [74] the effect of fatty acids on VSV particles was studied using negative staining and electron microscopy (EM). It was demonstrated that 0.5 mg/ml of linoleic acid partly disrupted the viral envelope and that it was completely disintegrated by 1 mg/ml of this particular fatty acid (Figure 3.3). The effect of human milk and linoleic acid on Vero cells was analysed by studies of the cells with SEM [74]. The cell monolayers were completely disintegrated by linoleic acid, as well as by antiviral milk samples.

As mentioned earlier, micromolar concentrations of linoleic acid reduced the infectivity of enveloped viruses without showing toxicity to animal cells [80]. Further studies by Kohn and coworkers showed that the loss of infectivity was due to disruption of the lipoprotein envelope of these virions, as observed in an EM. Thus, treatment of Sendai and Sindbis viruses with 10 μg/ml of linoleic acid for five to eight minutes caused damage or destruction of the viral envelopes. Kohn and coworkers further concluded that since naked viruses lacking host cell-derived envelope are unaffected by unsaturated fatty acids, the envelope must be the target of the virucidal activities of fatty acids. They suggested that the effect of polyunsaturated fatty acids on enveloped viruses is due to a decrease in the rigidity of the biological membranes, which can destroy the viruses by dissolving their envelopes [80,81].

Hilmarsson *et al.* [83] found a distinct difference in the activities of medium-chain monoglycerides against HSV, monocaprylin and monomyristin being markedly less effective than monocaprin and monolaurin upon contact for one minute. To further address this matter, a series of monoglycerides with chain lengths ranging from 8 to 12 carbons were tested in 10 mM concentration in one minute against HSV-1. Both monocaprin (10 : 0) and nonanoic acid monoglyceride (9 : 0) were fully active and reduced the virus titre 100 000-fold, while monocaprylin (8 : 0), undecanoic acid monoglyceride (11 : 0) and monolaurin (12 : 0) showed about 10 000-fold reduction. Thus, monoglycerides with 9- or 10-carbon chain length were highly virucidal in 10 mM concentration against HSV-1 in a short time, and addition or reduction of one carbon in the chain length seemed to reduce their virucidal activities [unpublished data]. Thus, a direct comparison of the virucidal activities of a series of saturated monoglycerides with increasing chain lengths suggests a possible relationship between the chain length and the thickness of the viral envelope. A difference in glyceride solubility is a less likely explanation of the difference in virucidal activity, since both monocaprylin (8 : 0), with the shortest chain length and highest solubility, and monolaurin

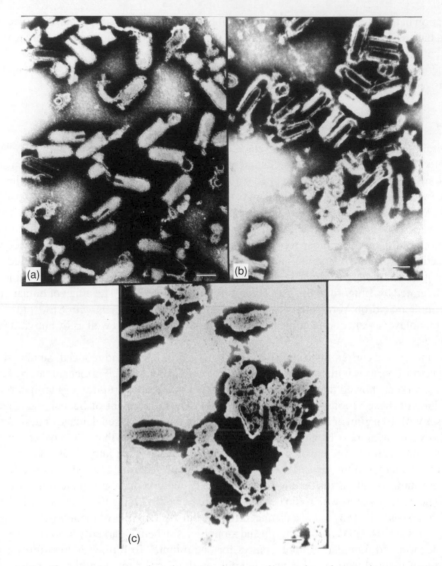

Figure 3.3 *Electron micrographs of negatively stained particles of vesicular stomatitis virus incubated at 37 °C for 30 minutes in (a) maintenance medium, showing normal intact particles covered with spikes, (b) 0.5 mg/ml linoleic acid, showing that the viral envelopes are no longer intact, allowing penetration of stain into most particles, and (c) 1 mg/ml linoleic acid, showing virus particles in various stages of disintegration. Bar, 0.1 μm. (Reprinted from [74] with permission from American Society for Microbiology.)*

(12 : 0), with the longest chain length and lowest solubility, were less active than monocaprin (10 : 0).

Exactly how lipids inactivate viruses is unknown, but it has been shown that fatty acids cause disintegration of the bilayer lipid envelope [70, 74, 76, 80, 81]. Medium- and long-chain alcohols may act in a similar way; that is, penetrating the envelope of the virus by hydrophobic effect, making it permeable to small molecules and thus inactivating the virus. The greater activities of 10- and 12-carbon saturated alcohols and lipids compared with other chain lengths, the higher activities of long-chain unsaturated versus saturated lipids, and the generally higher activities of monoglycerides than the corresponding fatty acids at low concentrations are all notable and have been found for both viruses and bacteria [20,74,75]. Possibly the different activities can be explained by a difference in the degree of penetration into lipid membranes due to the chain length of a lipid compared with the thickness of the membrane, as suggested by the effect of 8- to 12-carbon monoglycerides on HSV.

A different antiviral effect has been described for fatty acids, particularly lauric acid, against two enveloped viruses, Junin virus and VSV. In both viruses, micromolar concentrations of lauric acid significantly reduced the virus yield by altering the virus production. This was apparently due to stimulation of triacylglycerol production and increased incorporation into the host cell membrane, leading to decreased insertion of viral glycoproteins and inhibition of virus maturation and release [92,93].

3.4 Antifungal Activities of Fatty Acids and Monoglycerides

3.4.1 *In Vitro* Studies

Studies of the effect of acids, including acetic acid, on fungi go back to the work of Clark in 1899 [94] on the germination of a number of different moulds such as *Aspergillus* and *Penicillium.* Kiesel in 1913 [95] reported studies on the effect of fatty acids and their salts on the development of *Aspergillus niger* which showed that the activity of saturated fatty acids increased with the length of the carbon chain up to 11 carbon atoms. A few studies published in the early 1930s mostly confirmed his work.

Hoffman and others in 1939 [96] studied the fungistatic properties of fatty acids in buffered agar medium seeded with a mixed mould culture, mostly of *Aspergillus, Rhizopus* and *Penicillium* spp. obtained from mouldy food products. The agar plates were incubated for 48 hours at 37.5 °C and then observed for mould growth. All of the natural fatty acids containing from 1 to 14 carbon atoms were tested over a pH range of 2 to 8, as well as some unsaturated acids and isomeric forms. All of the saturated acids markedly inhibited mould growth and the activities varied with the pH and the chain length. All acids with chain lengths of 6 carbon atoms or lower were most active at pH <5, whereas the activities of acids with chain lengths of 7 to 12 carbon atoms were highest at pH 7 to 8. The isomeric forms were generally less active than the corresponding natural acids, but unsaturation increased the fungistatic activity of an acid and the effective pH range, as was found for the unsaturated 11-carbon hendecenoic acid (undecylenic acid).

Wyss and coworkers in 1945 [97] studied fungistatic and fungicidal actions of fatty acids and related compounds on *Trichophyton* and *Aspergillus.* The fungistatic effect was tested by growing the fungus in a medium containing the test compound. The compound concentration which just fully inhibited growth was measured. The fungicidal activity was measured by adding spore suspensions into solutions containing the prospective fungicidal compound. Samples were removed at intervals and tested for surviving spores. The results on the fungistatic actions were in agreement with the findings of previous workers in that the effect of fatty acids on both fungal species increased with the chain length, up to the point where solubility became the limiting factor. Unsaturated fatty acids were more active than the corresponding saturated acids. For *Aspergillus* the optimum chain length was 11 carbon atoms, whereas for the more sensitive *Trichophyton* spp. it was 13 to 14 carbon atoms. The fungistatic activity of the fatty acids increased with decreasing pH between 4.5 and 7.5, particularly for the short-chain acids. Undecylenic acid was found to be one of the most fungistatic fatty acids for all of the species tested and was most active at pH 5.5 against *Trichophyton* spp., which are a common cause of skin infections such as athlete's foot. The fungicidal activity of this unsaturated 11-carbon atom fatty acid was tested against *Trichophyton interdigitale* at pH 6.5 and 37 °C. The acid, in a concentration of 0.2%, killed 100 000 fungous spores in five minutes, but the killing time increased at lower concentrations.

The question of the difference between fungistatic and fungicidal action was addressed by Grunberg [98], who argued that these terms had been used too loosely in the past. According to his distinction, a fungistatic compound is one which inhibits or prevents the growth of fungi, while a fungicidal compound has a much more final effect in that it kills the fungi. Admitting that in some cases it may be difficult to differentiate between the two actions, he nevertheless developed methods designed to distinguish between inhibition and killing of fungi caused by chemotherapeutic agents *in vitro*. These methods were used to study the action of a series of acids from formic acid (1 : 0) to caprylic acid (8 : 0) against strains of *Trichophyton gypseum* and *Trichophyton purpureum* at various pHs. The results were mostly in agreement with previous studies by others, such as that both fungistatic and fungicidal actions of the fatty acids increased with the carbon chain length and decreased with an increase in pH. No appreciable difference was observed in the fungistatic and fungicidal effects of fatty acids against the two different strains of *Trichophyton*, although infections due to *T. purpureum* were more resistant to therapy than those caused by *T. gypseum*. Two important conclusions were drawn from this study, namely that the activity of the fatty acids was due to the undissociated molecules rather than to the ions and that emulsions of the higher fatty acids were more active than solutions at alkaline pH. This was explained by the fact that microorganisms tend to clump around globules of fatty acids in an emulsion, where their concentration is higher than in a solution.

Chattaway and his coworkers [99] studied the inhibiting effect on endogenous respiration of *Microsporum* spp. caused by saturated fatty acids of chain lengths ranging from 2 to 14 carbon atoms. The fatty acids strongly inhibited the respiration and the effect increased with the chain length and with decreasing pH, indicating that the undissociated form was most active. Release of soluble cell constituents from mycelium of *Microsporum canis*, caused by treatment with a fatty acid, was also measured. It was most rapid at low pH and correlated with the respiratory inhibition. Both effects were closely analogous to those found by Wyss *et al.* [97] for the fungistatic and fungicidal activity of fatty acids on *Trichophyton*.

Prince in 1959 [100] studied the effect of undecylenic acid and its calcium salt on the growth and respiration of various dermatophytes, such as *Trichophyton* and *Microsporum*, at various pHs. The acid and its salt were equally active at pH 4.5 to 6.0. The activity decreased at higher pH, more for the salt than for the acid. This pH effect was in agreement with data from other investigators, who had shown that the antifungal activity of fatty acids and their salts is greatest at low pH, where ionization is repressed. This supports the hypothesis that the undissociated fatty acid is the molecule most responsible for the antifungal effect.

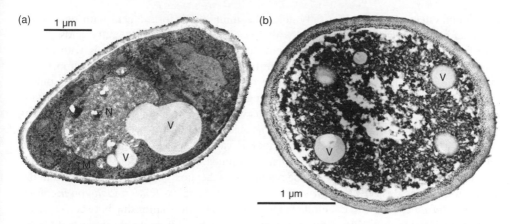

Figure 3.4 *Effects of capric acid on the ultrastructure of* C. albicans *as demonstrated by TEM of ultrathin sections of yeast cells. (a) Cell from the control sample with intact cytoplasm, nucleus (N), vacuoles (V) and mitochondria (M). (b) Cell from samples treated with 10 mM capric acid for 30 minutes, demonstrating disorganization of the cytoplasm. No visible changes are seen in the shape or size of the cell wall. (Reprinted from [102] with permission from American Society for Microbiology.)*

Candida albicans exhibits two cell morphologies: the round planktonic form, which functions in the spread of the yeast, and the elongated hyphae, which are the virulent form of *C. albicans* associated with active infection. Recent studies [101] have shown that undecylenic acid inhibits the appearance of germ tubes, which are the first hyphal formation emerging from a planktonic cell, at concentrations that do not inhibit cell proliferation. The acid therefore showed a specific inhibiting action on the morphogenesis of *C. albicans*, possibly through inhibition of enzymes involved in lipid metabolism. Bergsson *et al.* [102] studied the fungicidal effect of fatty acids and monoglycerides on the planktonic form of *C. albicans* and showed that 10 mM capric acid caused a more than 6 $\log_{10}$ reduction in viable cell number in 10 minutes. Lauric acid in lower concentrations also killed the yeast cells in large numbers after contact for 30 minutes. An EM examination of cells of *C. albicans* killed by capric acid showed disorganization of the cytoplasm, whereas no changes were seen in the shape of the cell wall (Figure 3.4). Undecylenic acid was not tested in this study.

In the late 1930s and the 1940s there was considerable interest in the antimycotic activity of fatty acids and related compounds, particularly in connection with treatment of dermatophytoses. These are fungal infections caused by dermatophytes, which is a common name for fungi that cause infections of the skin (ringworm), hair and nails in animals and humans, such as species of *Trichophyton*, *Microsporum* and *Epidermophyton*. Thus, at the First Annual Meeting of the Society for Investigative Dermatology held in New York in 1938, Peck and Rosenfeld [103] reported a comprehensive study on the effect of hydrogen ion concentration, fatty acids and vitamin C on the growth of dermatophytes in culture medium, particularly of *Trichophyton gypseum*, which is a pathogenic fungus causing dermatophytosis in the hair, skin and nails of humans. After determining that the fungus grew well *in vitro* at pH ranging from 4.0 to 10.0, Peck and Rosenfeld studied the fungicidal effect of low-carbon fatty acids, beginning with formic acid with one carbon atom and concluding with capric acid with 10 carbon atoms. All of the fatty acids studied had fungicidal effect at pH above 4.0, but differed in activities. The acids with an odd number of carbon atoms tended to be more fungicidal than acids with even numbers, the most powerful being valeric acid with five carbon atoms. The branched-chain isomers, which are not found in nature, were less fungicidal than the corresponding natural fatty acids and, with one exception, the sodium salts were less fungicidal than the corresponding acids. The unsaturated 11-carbon undecylenic acid, derived from castor-bean oil, was also studied and found to be almost as fungicidal as valeric acid. In contrast to sodium salts of other acids studied, sodium undecylenate was much more active than the acid, possibly due to its increased solubility. The undecylenate was the most fungicidal lipid tested in this study, not only against *Trichophyton gypseum*, but also against *Epidermophyton inguinale* and *Monilia albicans*. It was concluded that there is a definite relationship between structure and fungicidal action of fatty acids and that a double bond is probably important for the activity.

It had previously been observed that sweat had fungicidal properties, which seemed to be due to its content of fatty acids. Several of the fungicidal acids were shown to be present in sweat, such as acetic (2 : 0), butyric (4 : 0), caproic (6 : 0), caprylic (8 : 0) and probably capric (10 : 0) acid. In the discussion following his presentation at the meeting, Peck emphasized their possible role in the natural fungicidal action of sweat and the possibility of using some of these acids as therapeutic agents in a more physiological method of treating fungal infections. He proposed an interesting idea for treatment, namely ingestion of foods containing these fatty acids, or even some of the acids themselves, which from a theoretical point of view might increase the content of these substances in sweat and thus create a 'natural protective mantel'.

In a later publication [104], Peck and his coworkers studied the fungicidal effect of human sweat on three dermatophytes, *Trichophyton gypseum*, *Epidermophyton inguinale* and *Monilia albicans*. Small amounts of sweat were collected from the skin of people sitting for an hour under a carbon-filament lamp. Filtered sweat was added to Sabouraud's bouillon in quantities of 10 to 90%. Fifty per cent and higher concentrations of the heat sweat at pH <7 prevented growth of *T. gypseum* in the bouillon. Large quantities (275 to 340 ml) of perspiration were also collected from people sitting for hours inside a bag at a temperature of 41 °C. After being alkalinized by addition of sodium hydroxide, this sweat, which was much more dilute than normal sweat, had no activity against *T. gypseum*. Under normal conditions, evaporation of sweat from the skin tends to concentrate its ingredients. The collected sweat was therefore evaporated to dryness and a series of tubes of Sabouraud's bouillon containing from 1 to 10% of dry sweat was prepared. A 1% concentration of sweat had no effect on the growth of *T. gypseum*, whereas 4 to 7% completely inhibited the growth, depending on the sweat samples, which were either from single individuals or pooled. An analysis of two of the samples revealed similar amounts of acetic and propionic acids (3 : 0), whereas there was a considerable difference in the contents of caproic and caprylic acids.

The fungicidal action of individual constituents of sweat against *T. gypseum* was determined in bouillon cultures. Caproic acid was by far the most active in low concentrations, whereas formic, acetic, propionic, butyric and caprylic acids were all equally active, but less so than caproic acid. With the exception of sodium propionate, and to a certain extent sodium butyrate, the sodium salts were much less fungicidal than the corresponding acids. The activities of the sodium salts had to be considered because, depending on the pH on the surface of the skin, the sweat might be alkalinized and the acids thus converted into salts. This might make the sweat less fungicidal, although sodium propionate was equally active as the acid. Lactic acid (2-hydroxypropanoic acid), and particularly sodium lactate, were found to be much less fungicidal than the fatty acids and their salts. However, in view of the relatively large concentration of lactic acid in sweat, it might be of practical importance as a fungicide. *M. albicans* was much more resistant than *T. gypseum* to any of the fungicides present in sweat, whereas *E. inguinale* was generally more susceptible than *T. gypseum*.

3.4.2 *In Vivo* Studies

Because sodium propionate was found to be equally as fungicidal as its acid, it was tested therapeutically against various types of fungous infection in clinical trials [104]. The preparations tested were an alcoholic solution containing 10% propionate, a powder and an ointment containing 15% propionate. An alcoholic solution containing low amounts of propionic, butyric, caprylic, lactic and ascorbic acids was also used. Application of these preparations, given alone or in combinations, gave satisfactory results and led to either healing or improvement, depending on the condition treated. Tinea cruris (jock itch) responded well, as did other infections caused by *E. inguinale* and *T. gypseum*. The solution containing a mixture of compounds seemed to give the best results. Ringworm of the scalp, caused by *Microsporum audouini*, was not affected by any of the preparations; nor was epidermophytosis of the feet caused by *T. purpureum*.

The authors speculated about the role of sweat as a natural protective film on the skin against fungous infections. It had been suggested by other investigators that the hydrogen ion concentration of the sweat is in itself important in protecting the skin against infections and that the alkaline pH on some areas of the skin could explain their apparent susceptibility to pathogenic fungi. Arguing against this opinion, Peck and his coworkers [104] claimed that the pH effect is due to the fact that the acids in sweat are less fungicidal at alkaline pH and, more importantly, are changed into alkali salts, which are less fungicidal than the free acids. Quoting the work by other investigators who had studied the distribution of sweat on the surface of the body, Peck and his coworkers pointed out that areas such as the axillas, where sweating is profuse, are normally free of fungous infections, which attack the skin outside this area. In the groin, where fungous infections (tinea cruris/jock itch) are more common, there is relatively little sweat. The distribution of sweat on the feet was also considered in relation to fungous infection. There is little actual sweat between the toes and the moisture which is common in this area may cause maceration of the skin but has little fungicidal activity. As is commonly known, this area is an ideal location for fungal growth. From their own results and those of others, Peck and his coworkers suggested that there is a relation between the localization of fungous infection and the distribution of sweat on the surface of the body. Areas with the greatest concentration of sweat seem to be less prone to infection by fungi. According to this view, topical application of fungicidal ingredients of sweat would make sense as a treatment for some of these infections, as is also indicated by the results of the clinical experiments described in the paper. The authors proposed this type of treatment as a physiologic approach to therapy which might cause less irritation and complicating sequelae than treatment with other, less natural fungicidal agents. Thus, this form of therapy would simply augment the natural fungicidal action of the sweat. In their paper, the authors again proposed their idea of increasing the fungicidal elements in sweat, either by consumption of certain foods or by administration of fatty acids or their salts in concentrated form, either orally or parenterally.

In 1945, Shapiro and Rothman [105] published a paper on the treatment of dermatomy-coses (dermatophytoses) with undecylenate ointment containing 20% zinc undecylenate and 5% undecylenic acid. In a group of 113 patients with dermatomycosis of the feet, 86% achieved complete clinical cure within four weeks of treatment regardless of the species of causative fungi, which included *Trichophyton, Epidermophyton* and *Monilia*. This was slightly superior to older standard preparations used in treatment of this condition, but in contrast to them the undecylenate ointment was nonirritating. A rapid clinical cure with a complete absence of irritation of the skin was also observed in patients with tinea cruris and tinea axillaris. Treatment of ringworm of the body (tinea corporis) and the scalp (tinea capitis) was less successful. In general, treatment of intertriginous skin had a more favor-able effect compared with treatment of skin of nonintertriginous regions. The fact that the undecylenate ointment was strongly fungicidal and at the same time nonirritating to the skin was considered its main virtue.

Based on previous *in vitro* studies, which showed that sodium caprylate is particularly effective against *Monilia albicans* (*Candida albicans*), Keeney in 1946 [106] treated monil-iasis (candidiasis) of the skin and mucous membranes with preparations of this fatty acid sodium salt. In this clinical study, children with oral thrush were successfully treated with a 20% aqueous solution of caprylate. An adult female patient was treated with 10% capry-late ointment for extensive cutaneous moniliasis and with 10% caprylate vaginal jelly and talcum for vaginal moniliasis. Both treatments were effective but the vaginal preparations caused irritation, particularly the jelly.

Neuhauser and Gustus in 1954 [107] addressed the problem of the increased number of cases with intestinal infections due to *Candida albicans*. Large numbers of this organism appeared in the faeces of patients treated with antibiotics, because of a change in the normal intestinal flora caused by these antibacterial drugs. Although they most often disappeared upon discontinuation of antibiotic therapy, the fungous infections became chronic in some cases, with more or less severe symptoms and large numbers of *C. albicans* in the faeces, and sometimes with fatal outcome. Because oral administration of undecylenic acid had recently been successfully used to prevent monilial complications caused by the use of antibiotics, Neuhauser and Gustus decided to attempt to treat a seriously ill patient with a saturated complex of undecylenic acid and an anion-exchange resin. Such resins, in the form of a powder, had been successfully used for a number of years as carriers in the administration of gastrointestinal drugs, releasing the drug continuously at a controllable rate. Before giving the complex to the patient, each of the authors took large doses of the powdered product by mouth without experiencing any ill effects. The patient recovered fully from the intestinal moniliasis after being treated over a period of time. Generalized moniliasis of the skin was successfully treated at the same time with an ointment consisting of 10% calcium propionate and 10% sodium propionate. In order to determine whether other fatty acid–resin complexes were active against *C. albicans, in vitro* tests on Sabourauds's agar were carried out. Of all the fatty acid–resins tested, the caprylic acid–resin complex was found to be the most active in preventing the growth of *C. albicans* on agar medium. It was markedly more active than the undecylenic acid–resin, both at acid and at slightly alkaline pH. Treatment of several patients with caprylic acid–resin complex resulted in complete recovery from intestinal moniliasis without any undesired side effects. The recovery was more rapid than following treatment with undecylenic acid–resin.

Rothman and coworkers in 1946 and 1947 [108, 109] studied ringworm of the scalp hair, which was endemic among school children in the United States. This condition, which is caused by the fungus *Microsporon audouini*, clears up spontaneously during puberty and the scalps of adults are not susceptible to it. From clinical observation it had been known that under the influence of sex hormones during puberty the sebacious glands and their secretion changed profoundly, and there was a greatly increased amount of sebum on the scalp of adults, with a change of odour. Hypothesizing that the spontaneous cure of ringworm of the scalp during puberty was due to changes in the chemical composition of sebum, the fungicidal properties of adult hair fat were investigated and compared with those of hair fat from children. In the beginning, hair sweepings were collected from neighborhood barber shops, excluding children's hair. Ether extracts of the adult human hair were found to contain lipids which inhibited the growth of *M. audouini* on culture media, but were 10 times less active against *Trichophyton* and other pathogenic fungi which infect the hair of adults. The fungistatic action of lipids extracted from the hair of children under 11 years of age was more than five times less than that of adult lipids, as tested for prevention of *M. audouini* growth on Sabouraud's agar. A preliminary fractionation of the lipids showed that the free fatty acid components had the strongest inhibitory activity, and that a small fraction containing short-chain saturated fatty acids was the most active and prevented the growth of *M. audouini* on the agar medium in concentrations of 0.0002 to 0.0005%. It was shown that this threshold concentration of the fatty acids, which completely inhibits the growth of fungi, was likewise fungicidal.

With a possible therapeutic treatment in mind, Rothman and coworkers collected 45 kg of adult hair from more than 10 000 haircuts in large downtown barber shops. An ether extraction yielded 230 grams of free fatty acids, of which acids with chain lengths of 7, 9, 11 and 13 carbon atoms were 2.2 grams or about 1%. The 9-carbon pelargonic acid was prominent in this most active fraction. The more abundant, higher-molecular-weight fractions consisted of straight-chain saturated and unsaturated fatty acids with up to 22 carbon atoms, either odd- or even-numbered. The authors stated that human hair fat was, at that time, the only known natural source of straight-chained fatty acids with odd numbers of carbon atoms, with the exception of formic and propionic acids. However, odd-numbered acids with branched chains had previously been found in wool fat. As shown by Peck *et al.* [103], acids containing odd numbers of carbon atoms are more fungistatic than even-numbered acids, and this was strongly evident in the high activity against *M. audouini* of acids with 7 to 13 carbon atoms.

Therapeutic trials on children with ringworm of the scalp were carried out in which total hair fat, the most active fatty acid fractions, and pelargonic acid and its equally fungicidal sodium salt were all applied to the scalp by various methods. No therapeutic effect was obtained. This was shown to be due to the fact that neither fatty acids nor their salts were able to penetrate the keratin and therefore they could not reach the fungal spores inside the hair. However, it was suggested that local treatment of exposed but noninfected children might prevent the spread of this fungal disease in crowded facilites, such as in schools and orphanages. The glabrous skin of adults is not immune to *M. audouini* infections and adults who take care of children with ringworm of the scalp are exposed to the infection on other parts of their body. Apparently, the concentration of the active acids on the glabrous skin is below the critical level needed to protect against *M. audouini*, possibly because the sebacious glands are smaller there than on the scalp.

The most notable result of this study was the finding that with onset of puberty the sebaceous glands of the scalp begin to secrete a sebum containing an increased amount of highly fungistatic and fungicidal fatty acids with a selective activity against *M. audouini*, which makes adults naturally immune to infections by this fungus. This adult type of fat diffuses into the follicular canals and prevents infection of new hairs, which grow out following the shedding of old infected hairs. Consequently, this natural cure in puberty is a very slow process. This process is an example of the role of lipids in the natural defence against infections. Normal saturated fatty acids are thus fungicidal agents used by nature. Although only about 2% of the total fatty acid contents of the hair had a significant role in the immunity of adults to *M. audouini*, the remaining 98% were considered to be important in the physiology of the sebacious gland secretion, particularly in view of its complex mechanism and great number of specific products.

Following the study of Rothman and his coworkers, Perlman in 1949 [110] attempted treatment of children with ringworm of the scalp by oral administration of undecylenic acid in doses of 1 gram given three times daily. The treatment was unsuccessful and was discontinued. On the other hand, patients with chronic psoriasis and neurodermatis showed improvements after treatment with undecylenic acid for varying periods of time, with disappearance of lesions and relief of itching. The dose of 5 grams given three times daily caused no harmful effects but only some unpleasant side effects.

Between 1977 and 1983, several studies were carried out to evaluate the efficacy of undecylenic acid preparations in topical treatment of tinea pedis (athlete's foot) and of dermatophytoses of the glabrous skin and groin [111–115]. Treatments with ointments and powders containing either undecylenic acid or zinc undecylenate were found to be safe and effective and comparable to a common fungicidal cream. Desinex cream and a number of other preparations containing undecylenic acid as the active ingredient are on the market as over-the-counter antifungals [116].

3.5   Conclusions

In vitro studies have shown that microbes vary greatly in their susceptibility to microbicidal lipids: some are killed in a short time at low lipid concentration, others in a longer time or at higher concentrations. Which lipids are the most active against each bacterium, virus or fungus is also variable. Unsaturated fatty acids and medium-chain saturated fatty acids and their monoglycerides are generally most active against pathogenic bacteria and viruses and kill them rapidly and in large amounts on contact, whereas low- to medium-carbon fatty acids and their salts are in many cases active against fungi. The microbicidal action of lipids is affected by pH, the activity generally being higher in acid environment than at neutral or alkaline pH.

Several studies suggest that lipids, especially unsaturated fatty acids, play a role in the natural defence against bacterial infections in skin and mucosal membranes and that short- and medium-chain fatty acids in sweat are protective against fungal infections. Although a few attempts have been made to treat bacterial and viral infections in skin and mucosa with pharmaceutical formulations containing microbicidal lipids it has not yet been shown that such formulations are helpful in treatment or prevention of bacterial and viral infections in animals or humans. On the other hand, experiments have shown that treatment with formulations containing medium-chain fatty acids, particularly undecylenic acid and salts thereof, has a therapeutic effect in several types of fungal infection. Pharmaceutical dosage forms, such as creams, ointments and powders, are available as over-the-counter antifungals.

References

1. Kabara, J.J. (1980) Lipids as host-resistance factors of human milk. *Nutr. Rev.*, **38**, 65–73.
2. Isaacs, C.E., Kashyap, S., Heird, W.C. and Thormar, H. (1990) Antiviral and antibacterial lipids in human milk and infant formula feeds. *Arch. Dis. Childhood*, **65**, 861–864.
3. Isaacs, C.E. and Thormar, H. (1991) The role of milk-derived antimicrobial lipids as antiviral and antibacterial agents, in *Immunology of Milk and the Neonate* (eds J. Mestecki and P.L. Ogra), Plenum Press, New York, NY, USA, pp. 159–165.
4. Schlievert, P.M., Deringer, J.R., Kim, M.H. *et al.* (1992) Effect of glycerol monolaurate on bacterial growth and toxin production. *Antimicrob. Agents Chemother.*, **36**, 626–631.
5. Tsutsumi, K., Obata, Y., Takayama, K. *et al.* (1998) Effect of cod-liver oil extract on the buccal permeation of ergotamine tartrate. *Drug Dev. Ind. Pharm.*, **24**, 757–762.
6. Ogbolu, D.O., Oni, A.A., Daini, O.A. and Oloko, A.P. (2007) *In vitro* antimicrobial properties of coconut oil on *Candida* species in Ibadan, Nigeria. *J. Med. Food*, **10**, 384–387.
7. Do, T.Q., Moshkani, S., Castillo, P. *et al.* (2008) Lipids including cholesteryl linoleate and cholesteryl arachidonate contribute to the inherent antibacterial activity of human nasal fluid. *J. Immunol.*, **181**, 4177–4187.
8. Nakatsuji, T., Kao, M.C., Fang, J.Y. *et al.* (2009) Antimicrobial property of lauric acid against *Propionibacterium acnes:* Its therapeutic potential for inflammatory acne vulgaris. *J. Invest. Dermatol.*, **129**, 2480–2488.
9. Kristmundsdottir, T., Arnadottir, S.G., Bergsson, G. and Thormar, H. (1999) Development and evaluation of microbiocidal hydrogels containing monoglyceride as the active ingredient. *J. Pharm. Sci.*, **88**, 1011–1015.
10. Thormar, H., Bergsson, G., Gunnarsson, E. *et al.* (1999) Hydrogels containing monocaprin have potent microbicidal activities against sexually transmitted viruses and bacteria in vitro. *Sex. Transm. Infect.*, **75**, 181–185.
11. Thormar, H., Hilmarsson, H. and Bergsson, G. (2006) Stable concentrated emulsions of the 1-monoglyceride of capric acid (monocaprin) with microbicidal activities against the foodborne bacteria *Campylobacter jejuni, Salmonella* spp., and *Escherichia coli, Appl. Environ. Microbiol.*, **72**, 522–526.
12. Kitahara, T., Koyama, N., Matsuda, J. *et al.* (2004) Antimicrobial activity of saturated fatty acids and fatty amines against methicillin-resistant Staphylococcus aureus. *Biol.Pharm. Bull.*, **27**, 1321–1326.
13. Nair, M.K., Joy, J., Vasudevan, P. *et al.* (2005) Antibacterial effect of caprylic acid and mono-caprylin on major bacterial mastitis pathogens. *J. Dairy Sci.*, **88**, 3488–3495.
14. Preuss, H.G., Echard, B., Enig, M. *et al.* (2005) Minimum inhibitory concentrations of herbal essential oils and monolaurin for Gram-positive and Gram-negative bacteria. *Mol. Cell. Biochem.*, **272**, 29–34.
15. Georgel, P., Crozat, K., Lauth, X. *et al.* (2005) A toll-like receptor 2-responsive lipid effector pathway protects mammals against skin infections with Gram-positive bacteria. *Infect. Immun.*, **73**, 4512–4521.
16. Kabara, J.J. (1978) Fatty acids and derivatives as antimicrobial agents: a review, in *Symposium on the Pharmacological Effect of Lipids* (ed. J.J. Kabara), The American Oil Chemists' Society, Champaign, IL, USA, pp. 1–14.

17. Kabara, J.J., Swieczkowski, D.M., Conley, A.J. and Truant, J.P. (1972) Fatty acids and derivatives as antimicrobial agents. *Antimicrob. Agents Chemother.*, **2**, 23–28.
18. Conley, A.J. and Kabara, J.J. (1973) Antimicrobial action of esters of polyhydric alcohols. *Antimicrob. Agents Chemother.*, **4**, 501–506.
19. Heczko, P.B., Lütticken, R., Hryniewicz, W. *et al.* (1979) Susceptibility of *Staphylococcus aureus* and group A, B, C, and G streptococci to free fatty acids. *J. Clin. Microbiol.*, **9**, 333–335.
20. Bergsson, G., Arnfinnsson, J., Steingrímsson, Ó. and Thormar, H. (2001) Killing of Gram-positive cocci by fatty acids and monoglycerides. *APMIS*, **109**, 670–678.
21. Lee, J.Y., Kim, Y.S. and Shin, D.H. (2002) Antimicrobial synergistic effect of linolenic acid and monoglyceride against *Bacillus cereus* and *Staphylococcus aureus*. *J. Agric. Food Chem.*, **50**, 2193–2199.
22. Knapp, H.R. and Melly, M.A. (1986) Bactericidal effects of polyunsaturated fatty acids. *J. Infect. Dis.*, **154**, 84–94.
23. Chavant, P., Gaillard-Martinie, B. and Hébraud, M. (2004) Antimicrobial effects of sanitizers against planktonic and sessile *Listeria monocytogenes* cells according to the growth phase. *FEMS Microbiol. Lett.*, **236**, 241–248.
24. Isaacs, C.E., Litov, R.E. and Thormar, H. (1995) Antimicrobial activity of lipids added to human milk, infant formula, and bovine milk. *J. Nutr. Biochem.*, **6**, 362–366.
25. Marounek, M., Skrivanová, E. and Rada, V. (2003) Susceptibility of *Escherichia coli* to C2–C18 fatty acids. *Folia Microbiol. (Praha)*, **48**, 731–735.
26. Bergsson, G., Steingrímsson, Ó. and Thormar, H. (2002) Bactericidal effects of fatty acids and monoglycerides on *Helicobacter pylori*. *Int. J. Antimicrob. Agents*, **20**, 258–262.
27. Skrivanova, E., Marounek, M., Benda, V. and Brezina, P. (2006) Susceptibility of *Escherichia coli*, *Salmonella* sp. and *Clostridium perfringens* to organic acids and monolaurin. *Veterinarni Medicina*, **51**, 81–88.
28. Petschow, B.W., Batema, R.P. and Ford, L.L. (1996) Susceptibility of *Helicobacter pylori* to bactericidal properties of medium-chain monoglycerides and free fatty acids. *Antimicrob. Agents Chemother.*, **40**, 302–306.
29. Bergsson, G., Steingrímsson, Ó. and Thormar, H. (1999) In vitro susceptibilities of *Neisseria gonorrhoeae* to fatty acids and monoglycerides. *Antimicrob. Agents Chemother.*, **43**, 2790–2792.
30. Bergsson, G., Arnfinnsson, J., Karlsson, S.M. *et al.* (1998) *In vitro* inactivation of *Chlamydia trachomatis* by fatty acids and monoglycerides. *Antimicrob. Agents Chemother.*, **42**, 2290–2294.
31. Nair, M.K., Abouelezz, H., Hoagland, T. and Venkitanarayanan, K. (2005) Antibacterial effect of monocaprylin on *Escherichia coli* O157 : H7 in apple juice. *J. Food Prot*, **68**, 1895–1899.
32. Shibasaki, I. and Kato, N. (1978) Combined effects on antibacterial activity of fatty acids and their esters against Gram-negative bacteria, in *Symposium on the Pharmacological Effect of Lipids* (ed. J.J. Kabara), The American Oil Chemists' Society, Champaign, IL, USA, pp. 15–24.
33. Lampe, M.F., Ballweber, L.M., Isaacs, C.E. *et al.* (1998) Killing of *Chlamydia trachomatis* by novel antimicrobial lipids adapted from compounds in human breast milk. *Antimicrob. Agents Chemother.*, **42**, 1239–1244.
34. Preston, A., Mandrell, R.E., Gibson, B.W. and Apicella, M.A. (1996) The lipooligosaccharides of pathogenic Gram-negative bacteria. *Crit. Rev. Microbiol*, **22**, 139–180.
35. Desbois, A.P. and Smith, V.J. (2010) Antibacterial free fatty acids: activities, mechanisms of action and biotechnological potential. *Appl. Microbiol. Biotechnol.*, **85**, 1629–1642.
36. Clarke, S.R., Mohamed, R., Bian, L. *et al.* (2007) The *Staphylococcus aureus* surface protein IsdA mediates resistance to innate defenses of human skin. *Cell Host Microbe*, **1**, 199–212.
37. Kenny, J.G., Ward, D., Josefsson, E. *et al.* (2009) The *Staphylococcus aureus* response to unsaturated long chain free fatty acids: survival mechanisms and virulence implications. *PloS One*, **4**, e4344.
38. Burtenshaw, J.M.L. (1942) The mechanism of self-sterilization of human skin and its appendages. *J. Hyg.*, **42**, 84–209.
39. Welsh, J.K., Skurrie, I.J. and May, J.T. (1978) Use of Semliki forest virus to identify lipid-mediated antiviral activity and anti-alphavirus immunoglobulin A in human milk. *Infect. Immun.*, **19**, 395–401.

40. Kearney, J.N., Ingham, E., Cunliffe, W.J. and Holland, K.T. (1984) Correlations between human skin bacteria and skin lipids. *Br. J. Dermatol.*, **110**, 593–599.
41. Thielitz, A., Helmdach, M., Röpke, E.M. and Gollnick, H. (2001) Lipid analysis of follicular casts from cyanoacrylate strips as a new method for studying therapeutic effects of antiacne agents. *Br. J. Dermatol.*, **145**, 19–27.
42. Zouboulis, C.C., Xia, L.Q., Detmar, M. *et al.* (1991) Culture of human sebocytes and markers of sebocytic differentiation in vitro. *Skin Pharmacol.*, **4**, 74–83.
43. Zouboulis, C.C., Xia, L., Akamatsu, H. *et al.* (1998) The human sebocyte culture model provides new insights into development and management of seborrhoea and acne. *Dermatology*, **196**, 21–31.
44. Zouboulis, C.C., Seltmann, H., Neitzel, H. and Orfanos, C.E. (1999) Establishment and characterization of an immortalized human sebaceous gland cell line (SZ95). *J. Invest. Dermatol.*, **113**, 1011–1020.
45. Miller, S.J., Aly, R., Shinefeld, H.R. and Elias, P.M. (1988) In vitro and in vivo antistaphylococcal activity of human stratum corneum lipids. *Arch. Dermatol.*, **124**, 209–215.
46. Sabin, A.B. and Fieldsteel, A.H. (1962) Antipoliomyelitic activity of human and bovine colostrum and milk. *Pediatrics*, **29**, 105–115.
47. Isaacs, C.E., Litov, R.E., Marie, P. and Thormar, H. (2002) Addition of lipases to infant formulas produces antiviral and antibacterial activity. *J. Nutr. Biochem.*, **3**, 304–308.
48. Aly, R., Maibach, H.I., Shinefield, H.R. and Strauss, W.G. (1972) Survival of pathogenic microorganisms on human skin. *J. Invest. Dermatol.*, **58**, 205–210.
49. Aly, R., Maibach, H.I., Rahman, R. *et al.* (1975) Correlation of human in vivo and in vitro cutaneous antimicrobial factors. *J. Infect. Dis.*, **131**, 579–583.
50. Grayson, S. and Elias, P.M. (1982) Isolation and lipid biochemical characterization of stratum corneum membrane complexes: implications for the cutaneous permeability barrier. *J. Invest. Dermatol.*, **78**, 128–135.
51. Wille, J.J. and Kydonieus, A. (2003) Palmitoleic acid isomer (C16 : 1Δ6) in human skin sebum is effective against gram-positive bacteria. *Skin Pharmacol. Appl. Skin Physiol.*, **16**, 176–187.
52. Nicolaides, N. (1974) Skin lipids: their biochemical uniqueness. *Science*, **186**, 19–26.
53. Tollin, M., Bergsson, G., Kai-Larsen, Y. *et al.* (2005) Vernix caseosa as a multi-component defence system based on polypeptides, lipids and their interactions. *Cell. Mol. Life Sci.*, **62**, 2390–2399.
54. Takano, M., Simbol, A.B., Yasin, M. and Shibasaki, I. (1978) Bactericidal effect of freezing with chemical agents. *J. Food Sci.*, **44**, 112–115.
55. McLay, J.C., Kennedy, M.J., Orourke, A.L. *et al.* (2002) Inhibition of bacterial foodborne pathogens by the lactoperoxidase system in combination with monolaurin. *Int. J. Food Microbiol.*, **73**, 1–9.
56. Blaszyk, M. and Holley, R.A. (1998) Interaction of monolaurin, eugenol and sodium citrate on growth of common meat spoilage and pathogenic organisms. *Int. J. Food Microbiol.*, **39**, 175–183.
57. Mansour, M., Amri, D., Bouttefroy, A. *et al.* (1999) Inhibition of *Bacillus licheniformis* spore growth in milk by nisin, monolaurin, and pH combinations. *J. Appl. Microbiol.*, **86**, 311–324.
58. Branen, J.K. and Davidson, P.M. (2004) Enhancement of nisin, lysozyme, and monolaurin antimicrobial activities by ethylenediaminetetraacetic acid and lactoferrin. *Int. J. Food Microbiol.*, **90**, 63–74.
59. Chu-Kung, A.F., Bozzelli, K.N., Lockwood, N.A. *et al.* (2004) Promotion of peptide antimicrobial activity by fatty acid conjugation. *Bioconjug. Chem.*, **15**, 530–535.
60. Thormar, H. and Bergsson, G. (2001) Antimicrobial effects of lipids, in *Recent Developments in Antiviral Research*, Vol. **1** (ed. S.G. Pandalai), Transworld Research Network, Trivandrum, India, pp. 157–173.
61. Miller, S.J. (1989) Nutritional deficiency and the skin. *J. Am. Acad. Dermatol.*, **21**, 1–30.
62. Begg, A.M. and Aitken, H.A.A. (1932) The effect of tumour regression and tissue absorption on some properties of the serum. *Brit. J. Exp. Pathol.*, **13**, 479–488.

63. Helmer, O.M. and Clowes, G.H.A. (1937) Effect of fatty acid structure on inhibition of growth of chicken sarcoma. *Am. J. Cancer*, **30**, 553–554.
64. Pirie, A. (1933) CCLIX. The effect of enzymes on the pathogenicity of the Rous and Fujinami tumour viruses. *Biochem. J.*, **27**, 1894–1898.
65. Pirie, A. (1935) The effect of extracts of pancreas on different viruses. *Brit. J. Exp. Pathol.*, **16**, 497–502.
66. Stock, C.C. and Francis, Jr, T. (1940) The inactivation of the virus of epidemic influenza by soaps. *J. Exp. Med.*, **71**, 661–681.
67. Stock, C.C. and Francis, Jr, T. (1943) Additional studies of the inactivation of the virus of epidemic influenza by soaps. *J. Immunol.*, **47**, 303–308.
68. Stock, C.C. and Francis, Jr, T. (1943) The inactivation of the virus of lymphocytic choriomeningitis by soaps. *J. Exp. Med.*, **77**, 323–336.
69. Sabin, A.B. and Fieldsteel, A.H. (1962) Antipoliomyelitic activity of human and bovine colostrum and milk. *Pediatr.*, **29**, 105–115.
70. Sarkar, N.H., Charney, J., Dion, A.S. and Moore, D.H. (1973) Effect of human milk on the mouse mammary tumor virus. *Cancer Res.*, **33**, 626–629.
71. Falkler, Jr, W.A., Diwan, A.R. and Halstead, S.B. (1975) A lipid inhibitor of dengue virus in human colostrum and milk. *Arch. Virol.*, **47**, 3–10.
72. Isaacs, C.E., Thormar, H. and Pessolano, T. (1986) Membrane disruptive effect of human milk: inactivation of enveloped viruses. *J. Infect. Dis.*, **154**, 966–971.
73. Isaacs, C.E. and Thormar, H. (1990) Human milk lipids inactivate enveloped viruses, in *Breast-feeding, Nutrition, Infection and Infant Growth in Developed and Emerging Countries* (eds S.A. Atkinson, L.A. Hanson, and R.K. Chandra), ARTS Biomedical Publishers and Distributors, St. John's, Newfoundland, Canada, pp. 161–174.
74. Thormar, H., Isaacs, C.E., Brown, H.R. *et al.* (1987) Inactivation of enveloped viruses and killing of cells by fatty acids and monoglycerides. *Antimicrob. Agents Chemother.*, **31**, 27–31.
75. Thormar, H., Isaacs, C.E., Kim, K.S. and Brown, H.R. (1994) Inactivation of visna virus and other enveloped viruses by free fatty acids and monoglycerides. *Ann. NY Acad. Sci.*, **724**, 465–471.
76. Sands, J.A. (1977) Inactivation and inhibition of replication of the enveloped bacteriophage Φ6 by fatty acids. *Antimicrob. Agents Chemother.*, **12**, 523–528.
77. Sands, J.A., Auperin, D.D., Landin, P.D. *et al.* (1978) Antiviral effects of fatty acids and derivatives: lipid containing bacteriophages as a model system, in *Symposium on the Pharmacological Effects of Lipids* (ed. J.J. Kabara), The American Oil Chemists' Society, Champaign, IL, USA, pp. 75–95.
78. Snipes, W., Person, S., Keller, G. *et al.* (1977) Inactivation of lipid-containing viruses by long-chain alcohols. *Antimicrob. Agents Chemother.*, **11**, 98–104.
79. Snipes, W. and Keith, A. (1978) Hydrophobic alcohols and di-tert-butyl phenols as antiviral agents, in *Symposium on the Pharmacological Effects of Lipids* (ed. J.J. Kabara), The American Oil Chemists' Society, Champaign, IL, USA, pp. 63–73.
80. Kohn, A., Gitelman, J. and Inbar, M. (1980) Interactions of polyunsturated fatty acids with animal cells and enveloped viruses. *Antimicrob. Agents Chemother.*, **18**, 962–968.
81. Kohn, A., Gitelman, J. and Inbar, M. (1980) Unsaturated free fatty acids inactivate animal enveloped viruses. *Arch. Virol.*, **66**, 301–307.
82. Isaacs, C.E., Kim, K.S. and Thormar, H. (1994) Inactivation of enveloped viruses in human bodily fluids by purified lipids. *Ann. NY Acad. Sci.*, **724**, 457–454.
83. Hilmarsson, H., Kristmundsdóttir, T. and Thormar, H. (2005) Virucidal activities of medium- and long-chain fatty alcohols, fatty acids and monoglycerides against herpes simplex virus types 1 and 2: comparison at different pH levels. *APMIS*, **113**, 58–65.
84. Hilmarsson, H., Lárusson, L.V. and Thormar, H. (2006) Virucidal effects of lipids on visna virus, a lentivirus related to HIV. *Arch. Virol.*, **151**, 1217–1224.
85. Hilmarsson, H., Traustason, B.S., Kristmundsdóttir, T. and Thormar, H. (2007) Virucidal activities of medium- and long-chain fatty alcohols and lipids against respiratory syncytial virus and parainfluenza virus type 2: comparison at different pH levels. *Arch. Virol.*, **152**, 2225–2236.

86. Simoes, E.A. (2002) Immunoprophylaxis of respiratory syncytial virus: global experience. *Resp Res.*, **3**, 26–33.

87. Wyde, P.R., Chetty, S.N., Timmerman, P. *et al.* (2003) Short duration aerosols of JNJ 2408068 (R170591) administrated prophylactically or therapeutically protect cotton rats from experimental respiratory syncytial virus infection. *Antiviral Res.*, **60**, 221–231.

88. Loftsson, T., Thormar, H., Ólafsson, J.H. *et al.* (1998) Fatty acid extracts from cod-liver oil: activity against herpes simplex virus and enhancment of transdermal delivery of acyclovir *in vitro*. *Pharm. Pharmacol. Commun.*, **4**, 287–291.

89. Kristmundsdóttir, T., Árnadóttir, S.G., Bergsson, G. and Thormar, H. (1999) Development and evaluation of microbicidal hydrogels containing monoglyceride as the active ingredient. *J. Pharmaceut. Sci.*, **88**, 1011–1015.

90. Neyts, J., Kristmundsdóttir, T., De Clercq, E. and Thormar, H. (2000) Hydrogels containing monocaprin prevent intravaginal and intracutaneous infections with HSV-2 in mice: impact on the search for vaginal microbicides. *J. Med. Virol.*, **61**, 107–110.

91. Arnfinnsdóttir, A.V., Geirsson, R.T., Hilmarsdóttir, I. *et al.* (2004) Effects of monocaprin hydrogel on the vaginal mucosa and bacterial colonisation, 34th Congress of Nordic Federation of Societies of Obstetrics and Gynecology, *Helsinki, Finland*, p. 91.

92. Hornung, B., Amtmann, E. and Sauer, G. (1994) Lauric acid inhibits the maturation of vesicular stomatitis virus. *J. Gen. Virol.*, **75**, 353–361.

93. Bartolotta, S., García, C.C., Candura, N.A. and Damonte, E.B. (2001) Effects of fatty acids on arenavirus replicaton: inhibition of virus production by lauric acid. *Arch. Virol.*, **146**, 777–790.

94. Clark, J.R. (1899) On the toxic effect of deleterious agents on germination and development of certain filamentous fungi. *Bot. Gaz.*, **28**, 289–327.

95. Kiesel, A. (1913) Recherches sur l'action de divers acides et sels acides sur le développement de l'aspergillus niger. *Ann. de l'Inst. Pasteur*, **27**, 391–420

96. Hoffman, C., Schweitzer, T.R. and Dalby, G. (1939) Fungistatic properties of the fatty acids and possible biochemical significance. *Food Res.*, **6**, 539–545.

97. Wyss, O., Ludwig, B.C. and Joiner, R.R. (1945) The fungistatic and fungicidal action of fatty acids and related compounds. *Arch. Biochem.*, **7**, 415–425.

98. Grunberg, H. (1947) The fungistatic and fungicidal effects of the fatty acids on species of *Trichophyton*. *Yale J. Biol. Med.*, **19**, 855–876

99. Chattaway, F.W. and Thompson, C.C. (1956) The action of inhibitors on dermatophytes. *Biochem. J.*, **63**, 648–656.

100. Prince, H.N. (1959) Effect of pH on the antifungal activity of undecylenic acid and its calcium salt. *J. Bacteriol.*, **78**, 788–791.

101. McLain, N., Ascanio, R., Baker, C. *et al.* (2000) Undecylenic acid inhibits morphogenesis of *Candida albicans*. *Antimicrob. Agents Chemother.*, **44**, 2873–2875.

102. Bergsson, G., Arnfinnsson, J., Steingrímsson, Ó., Thormar, H. (2001) *In vitro* killing of *Candida albicans* by fatty acids and monoglycerides. *Antimicrob. Agents Chemother.*, **45**, 3209–3212.

103. Peck, S.M. and Rosenfeld, H. (1938) The effects of hydrogen ion concentration, fatty acids and vitamin C on the growth of fungi. *J. Invest. Dermatol.*, **1**, 237–265.

104. Peck, S.M., Rosenfeld, H., Leifer, W. and Bierman, W. (1939) Role of sweat as a fungicide. With special reference to the use of constituents of sweat in the therapy of fungous infections. *Arch. Dermatol. Syphilol.*, **39**, 126–148.

105. Shapiro, A.L. and Rothman, S. (1945) Undecylenic acid in the treatment of dermatocytosis. *Arch. Dermatol. Syphilol.*, **52**, 166–171.

106. Keeney, E.L. (1946) Sodium caprylate. A new and effective treatment for moniliasis of the skin and mucous membranes. *Bull. Johns Hopkins Univ. Hosp.*, **78**, 333–339.

107. Neuhauser, I. and Gustus, E.L. (1954) Successful treatment of intestinal moniliasis with fatty acid–resin complex. *Arch. Intern. Med.*, **93**, 53–60.

108. Rothman, S., Smiljanic, A.M. and Weitkamp, A.W. (1946) Mechanism of spontaneous cure in puberty of ringworm of the scalp. *Science*, **104**, 201–203.

109. Rothman, S., Smiljanic, A., Shapiro, A.L. and Weitkamp, A.W. (1947) The spontaneous cure of tinea capitis in puberty. *J. Invest. Dermatol.*, **8**, 81–98.

110. Perlman, H.H. (1949) Undecylenic acid given orally in psoriasis and neurodermatitis. *J. Am. Med. Assoc.*, **139**, 444–447.
111. Smith, E.B., Powell, R.P., Graham, J.L., and Ulrich, J.A. (1977) Topical undecylenic acid in tinea pedis: a new look. *Int. J. Dermatol.*, **16**, 52–56.
112. Tschen, E.H., Becker, L.E., Ulrich, J.A. *et al.* (1979) Comparison of over-the-counter agents for tinea pedis. *Cutis*, **23**, 696–698.
113. Chretien, J.H., Esswein, J.G., Sharpe, L.M. *et al.* (1980) Efficacy of undecylenic acid-zink undecylenate powder in culture positive tinea pedis. *Int. J. Dermatol.*, **19**, 51–54.
114. Fuerst, J.F., Cox, G.F., Weaver, S.M. and Duncan, W.C. (1980) Comparison between undecylenic acid and tolnaftate in the treatment of tinea pedis. *Cutis*, **25**, 544–546, 549.
115. Battistini, F., Cordero, C., Urcuyo, F.G. *et al.* (1983) The treatment of dermatophytoses of the glabrous skin: a comparison of undecylenic acid and its salt versus tolnaftate. *Int. J. Dermatol.*, **22**, 388–389.
116. Anon. (2002) Undecylenic acid. *Alt. Med. Rev.*, **7**, 68–70.

4

Antimicrobial Lipids in Milk

Charles E. Isaacs

Department of Developmental Biochemistry, New York State Institute for Basic Research in Developmental Disabilities, NY, USA

Lipids and Essential Oils as Antimicrobial Agents   Halldor Thormar
© 2011 John Wiley & Sons, Ltd

4.1   Introduction

Milk is not only a vehicle which delivers nutrients to the newborn, but also a natural functional food. It transports molecules and cells that are part of the secretory immune system and plays an important role in innate defence by protecting mucosal surfaces from infection [1]. Milk lipids are a class of molecule that have broad-spectrum antimicrobial activity in addition to their nutritional value. The antimicrobial activity of lipids, in particular fatty acids, has been studied since the late nineteenth century [2,3]. Lipid-dependent antimicrobial activity delivered in milk to the neonate results from the release of antimicrobial fatty acids and monoglycerides from milk triglycerides. These triglycerides are present in milk fat globules and represent 98% of the milk fat [4–6]. Monoglycerides, diglycerides and ether lipids are present in trace amounts. The triglyceride core of the milk fat globule is surrounded by a membrane of polar lipids, which prevents hydrolysis of the triglycerides by lipases in milk [7], and therefore human milk lipid-dependent antimicrobial activity is released by lipases in the infant's gastrointestinal tract [8]. The same is true for bovine milk. The digestion products of bovine milk triglycerides and membrane lipids inactivate *Escherichia coli* 0157 : H7, *Salmonella enteritidis*, *Campylobacter jejuni*, *Listeria monocytogenes* and *Clostridium perfringens* [9–12].

Milk lipids are not unique in possessing antimicrobial activity. Human epidermis-derived skin lipids, especially free fatty acids, have been found to inactivate bacterial pathogens, for example *Staphylococcus aureus* [13–17]. Studies at other mucosal surfaces showed that lung surfactant from humans, dogs, rats and guinea pigs contained free fatty acids that were bactericidal for pneumococci [18–20]. Inhaled pneumococci were killed extracellularly in rats [19] and the antibacterial activity resided in long-chain polyunsaturated free fatty acids in lung surfactant. Fractionation of porcine intestinal lipids showed that antibacterial activity against *Clostridium welchii* was primarily due to the activity of long-chain unsaturated fatty acids [21].

Free fatty acids isolated from nine species of brown algae [22] inactivated Gram-positive and Gram-negative bacteria. In another study examining Caribbean marine algae [23], over 70% of the lipid extracts from approximately 100 algae tested had antibacterial activity. Fatty acids and their antimicrobial derivatives not only have a protective role in milk and at mucosal surfaces of vertebrates but also appear to be utilized by simpler eukaryotes for antibacterial activity.

Lipids are one of a number of 'nonspecific' protective factors present in milk which function at mucosal surfaces [24, 25]. These 'nonspecific' protective factors are part of an innate defence system that phylogenetically precedes the adaptive immune system – the antibodies – and which is found solely in vertebrates [26]. In addition to lipids, nonspecific protective factors in milk include lactoferrin, lactoperoxidase, lysozyme, receptor oligosaccharides and antimicrobial peptides [3, 24, 25, 27–32]. Nonspecific protective factors such as lipids can prevent or inhibit the establishment, multiplication and spread of pathogenic microorganisms in the host [33]. In fact, microbial inactivation measured using individual purified antimicrobial factors may underestimate their *in vivo* effectiveness because antimicrobial factors in milk interact, for example sIgA, peroxidase and lactoferrin produce synergistic antimicrobial activity [34–36]. Further, studies that examine the antimicrobial potential of milk may not provide a complete measurement of its full protective activity, because the issue of co-infection is not addressed [37]. One pathogen may disrupt mucosal barriers, suppress the immune system or provide molecular cofactors, thereby facilitating the spread of a second infectious agent that would not establish, or as frequently establish, an infection by itself. Conversely, protective factors in milk, such as lipids, may indirectly reduce transmission of one pathogen by preventing another pathogen from establishing a successful infection. In summary, our present picture of the mechanisms by which human milk protects the suckling infant likely underestimates its antimicrobial potential.

4.2 Occurrence

4.2.1 Biosynthesis

The lipids in human milk (3 to 5%) occur as globules emulsified in the aqueous phase (87%) [5]. The globules contain nonpolar core lipids such as triglycerides, cholesterol esters and retinol esters. They are coated with bipolar compounds which form a loose layer referred to as the milk lipid globule membrane (MLGM). The MLGM prevents the globules from coalescing and acts as an emulsion stabilizer. The globules provide a large surface area for lipolytic enzymes in the gastrointestinal tract, facilitating lipolysis of milk triglycerides.

The lipid-dependent antimicrobial activity in milk is primarily present in two groups of fatty acids and their derivatives, which are released from milk triglycerides following lipolysis in the digestive tract: the long-chain unsaturated fatty acids and the medium-chain saturated fatty acids. The origin of fatty acids in milk is threefold: the diet, mobilization of endogenous stored fatty acids and synthesis in the mammary gland [5]. The relative proportion of milk fatty acids from each of these sources is dependent upon the mother's diet [38, 39]. Maternal diets which are high-energy and low-fat stimulate the production of milks with relatively high amounts of medium-chain fatty acids [39–41] as a result of increased medium-chain fatty acid synthesis in the mammary gland. Therefore, even when the fatty acid composition of human milk triglyceride varies as a result of diet [38], there will always be a sufficient concentration of antimicrobial fatty acids to inactivate susceptible pathogens. To measure the level of milk antimicrobial lipids, the lipid fraction can be extracted using chloroform-methanol (2 : 1), the extract dried and taken up in ethanol [42], and the extracted lipids can be tested against susceptible microorganisms.

4.2.2 Quantitative Assays

Lipids present in human milk can be quantified using methods developed and modified by Jensen *et al.* [43]. Lipids from other sources can be extracted as described above and quantified using thin-layer chromatography and flame ionization detection [44]. The method used to determine antimicrobial activity varies with the organism of interest.

4.2.2.1   Measurement of Antibacterial Activity

The minimum microbicidal concentration (MCC) assay is the current standard to determine the minimum concentration of lipid (or any antimicrobial agent) required to kill 99.99% of test organisms in 30 minutes [45, 46]. After various times of direct exposure to lipid in the *in vitro* assay, the organisms are tested for viability by inoculating them into the appropriate growth medium.

4.2.2.2   Measurement of Antiviral Activity

Antimicrobial lipids in milk inactivate enveloped viruses [8, 47–50]. Virucidal activity can be determined by incubating infectious virus particles in the presence of lipid and then determining the reduction in virus titre by plating the virus on susceptible cell cultures [47]. The decrease in virus titre is determined either by determining the cytopathic endpoint on 96-well tissue-culture plates [42] or by using a plaque assay [51].

The two main *in vitro* assay methods used to quantify infectious virus rely upon either cytopathic effect (CPE) [52] or formation of plaques [53]. The cytopathic endpoint titration with determination of the 50% tissue culture infective dose ($TCID_{50}$) has several advantages in that the scoring of wells as positive or negative can be more reliable and less time-consuming than counting plaques. Plaque assays have the advantage of producing a countable event (plaques) that increases with virus dose. The major disadvantage of plaque assays is that they typically have a very limited linear range. Test results producing plaque counts on either the high or low end of the linear range due to unexpected viral recovery or inactivation can lead to a successive-approximation approach which can be costly and time-consuming. Initial screening assays using endpoint titration can then be followed by plaque assay to obtain a more precise count.

Panels of clinical isolates and laboratory strains [54] can be titrated as previously described. The virus can be titrated by inoculation of 10-fold dilutions into monolayers of cells in 96-well microtitre tissue-culture plates. A virus dilution (0.1 ml) in maintenance medium (RPMI 1640, 1% foetal calf serum) can be inoculated into each well, with four wells per dilution. The plates are kept for two to seven days and examined daily for CPE. Virus titres are calculated by the method of Reed and Muench [52].

In the CPE assay, approximately 10^5 $TCID_{50}$s are mixed with maintenance medium. Pure antiviral compounds, or controls, are added and the mixture is incubated at 37 °C for 1 to 30 minutes. Virus mixed with the medium without added compounds serves as a control. After incubation, the infectivity of each mixture is titrated by the serial dilution endpoint method. Dilutions (10-fold) are made in maintenance medium. The 10^{-2} to 10^{-5} dilutions are inoculated into monolayers of cells and the virus titres are determined as described above. The difference between the titre (log_{10}) of the control virus and the titre of sample–virus mixture – the reduction of virus titre – is used as a measure of antiviral activity.

In the plaque assay, cells are seeded in 24-well plates using 10% foetal calf serum in RPMI 1640 medium and incubated at 37 °C in 5% CO_2. The medium is aspirated from the monolayer cultures and virus is inoculated on to cells. After one hour, the cells are washed three times with medium, and agar-medium overlay is dispersed on to each well. Plates are incubated at 37 °C for 48 hours and the cultures are then examined macroscopically and the plaques counted. Only wells with 20 to 100 plaques are used to determine the viral titre [53].

4.3   Molecular Properties

4.3.1   Lipids in Milk and Plant Oils

The major commercial source of antimicrobial fatty acids that are also found in milk is plant oils. These are primarily coconut, soy, palm, and high oleic safflower oils. Coconut

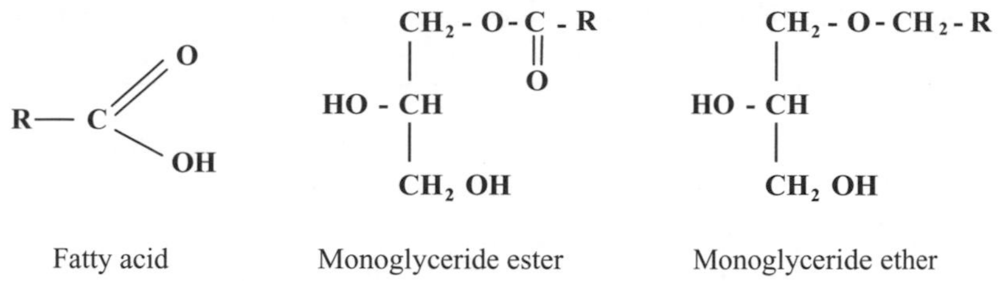

Figure 4.1   *The structures of a fatty acid, a monoglyceride ester and a monoglyceride ether.*

oil is a major source of lauric acid (12 carbons), which is a medium-chain fatty acid with strong antimicrobial activity [2,55]. The other plant oils are excellent sources of long-chain unsaturated fatty acid with antimicrobial activity. Antimicrobial fatty acids can be purchased commercially from a number of different suppliers. Antimicrobial monoglycerides in milk have ester linkages between the fatty acid and the glycerol backbone. This linkage can be replaced with an ether linkage using well-established methods of synthesis and provides a more stable compound with equal or greater antimicrobial activity [46,49,56].

4.3.2   Dependence of Antimicrobial Activity on Chain Length, Saturation and pH

The antimicrobial activity of short-chain fatty acids is due to the undissociated molecule and not the anionic form of the molecule [2,57] and therefore their antimicrobial activity is pH-dependent. Work by Kabara [2] shows that as the degree of dissociation of the acid increases, the activity decreases. The minimum inhibitory concentration of short-chain acids (6, 8 and 10 carbons) increases with increasing pH, while the minimum inhibitory concentration of medium-chain acids (12 and 14 carbons) decreases with increasing pH. The activity of long-chain unsaturated fatty acids is unaffected by changing pH values. Monoglyceride esters and ethers of short-chain fatty acids (see Section 4.4), as opposed to the free acids, are unaffected by pH alterations due to the attachment of the fatty acid carboxyl group to the glycerol backbone. The structures of a fatty acid, a monoglyceride ester and a monoglyceride ether can be seen in Figure 4.1.

4.4   Antimicrobial Activity

4.4.1   Mechanism of Action

The antimicrobial activity produced by the hydrolysis of milk triglycerides can be duplicated using purified fatty acids and monoglycerides [42]. Short-chain (butyric, caproic and caprylic) and long-chain saturated (palmitic and stearic) fatty acids have no or minimal activity against three enveloped viruses even at exceedingly high concentrations. On the other hand, medium-chain saturated and long-chain unsaturated fatty acids are strongly antiviral against enveloped viruses, regardless of whether the virus is a DNA, RNA or retrovirus (Table 4.1). In contrast, incubation of the non-enveloped poliovirus with capric, lauric, myristic, palmitoleic, oleic, linoleic and arachidonic acids does not significantly reduce viral infectivity [42].

Table 4.1   *Lipid-dependent inactivation of enveloped viruses by fatty acids in human and bovine milks. (Adapted from [5, 7, 42].)*

Fatty acid[a]	gram/100 gram fat		Relative effectiveness against enveloped viruses[b]
	Human	Bovine	
4 : 0	–	3.3	0
6 : 0	Trace	1.6	0
8 : 0	Trace	1.3	++
10 : 0	1.3	3.0	++++
12 : 0	3.1	3.1	++++
14 : 0	5.1	14.2	++++
15 : 0	Trace to 0.4	1.3	ND
16 : 0	20.2	42.7	+
16 : 1	5.7	3.7	++++
18 : 0	6.0	5.7	0
18 : 1	46.4	16.7	++++
18 : 2	13.0	1.6	++++
18 : 3	1.4	1.8	++++

[a]Designated by carbon chain length and number of double bonds.
[b]Antiviral activity shown reflects the inherent effectiveness of each fatty acid at equivalent concentrations, not that found in the milks.

These data suggest that the envelope of a virus such as herpes simplex virus (HSV) or human immunodeficiency virus (HIV) [58] is the target for lipid-dependent viral inactivation. Electron-microscopic studies done in our laboratory confirmed this and showed that the envelope of vesicular stomatitis virus (VSV) is completely destroyed by linoleic acid [42].

Other purified lipids, which are also the product of milk triglyceride hydrolysis by lipases, such as 1-monoglycerides of milk fatty acids, also effectively inactivate enveloped viruses [42]. In most instances the monoglyceride derivative of the fatty acid is antiviral at a concentration that is 5 to 10 times lower (mM) than the fatty acid itself. Results with purified monoglycerides indicate that monoglycerides produced by hydrolysis of milk triglycerides may have an antimicrobial role in the infant's gastrointestinal tract.

Further evidence that antimicrobial lipids destabilize membranes is provided by electron-microscopic studies of leukaemic cells treated with ester and ether lipids [59, 60]. These experiments showed that the plasma membrane of the leukaemic cells was the primary target of the lipids and that morphological damage induced by the lipids consisted of the formation of blebs, formation of holes and increased porosity of the membrane. *In vivo*, cells in the infant's gastrointestinal tract are protected from lipid damage by the mucosal barrier.

Depending upon the concentration, lipid-dependent inactivation of enveloped viruses can occur within a few seconds or over the course of a few hours. The antiviral effect of active fatty acids and monoglycerides is additive [12, 58]. When medium-chain saturated and long-chain unsaturated fatty acids were tested together, each at a concentration previously shown to be below that needed for antiviral activity, the mixture was antiviral because the total concentration of antiviral lipids was at or above a level sufficient to destabilize the lipid envelope of the virus. Regardless of whether fatty acids mixtures consisted of two medium-chain saturated fatty acids, a medium-chain saturated and a long-chain unsaturated fatty acid, or two long-chain unsaturated fatty acids, they were antiviral. Mixtures were made with as many as seven different fatty acids, each at a suboptimal concentration, but the lipid mixture itself was antiviral.

Milk lipids inactivate all enveloped viruses that have been examined, including HSV [47, 61], influenza virus [62, 63], respiratory syncytial virus [64, 65], measles virus, VSV and visna virus [47], mouse mammary tumour virus [66], dengue virus types 1 to 4 [67], cytomegalovirus,[68], Semliki Forest virus [25,68], Japanese B encephalitis virus [69], HIV [58] and simian immunodeficiency virus [70]. Milk lipids also inactivate both Gram-positive (e.g. *Staphylococcus epidermidis*) and Gram-negative (e.g. *Escherichia coli*) bacteria [8], as well as the protozoal pathogens *Giardia lamblia* [71–73] and *Trichomonas vaginalis* (Dorothy Patton, personal communication).

As mentioned above, the antimicrobial activity of free fatty acids may be altered by pH depending upon the chain length of the molecule, but monoglycerides maintain their antimicrobial activity across a wide pH range. The temperature at which lipids are incubated does, however, affect the rate at which they inactivate pathogens. The eight-carbon monoglyceride ether 1-0-octyl-sn-glycerol can decrease the titre of VSV by more than 10,000-fold at 37 °C in one hour but requires 18 hours for the same reduction at 4 °C [49].

The antibacterial and antifungal activity of lipids occurs by the same membrane-destabilizing mechanism as shown for enveloped viruses [2, 74, 75]. Mild heating, which makes bacterial membranes more fluid, increases the lipid-dependent killing of the Gram-negative bacterial pathogens *E. coli* and *Pseudomonas aeruginosa* by 1000 to 100 000 fold [74], providing further evidence that the bacterial membrane is the site of antimicrobial lipid activity.

4.4.2 Synergy between Antimicrobial Lipids and Antimicrobial Peptides

The immune system in milk uses a combination of direct-acting antimicrobial factors, anti-inflammatory factors and immunomodulators [1]. Direct-acting antimicrobial compounds in milk, such as lipids, attack pathogens at multiple points in their life cycle, while multiple protective factors attack each point in the pathogen's replication cycle, and thus the immune system in milk provides redundancy at multiple levels. Pathogens can be inactivated directly by antimicrobial lipids, antimicrobial peptides, antibodies and lysozyme [24, 37, 76].

4.4.2.1 Antimicrobial Peptides and Proteins

Antimicrobial peptides, the first novel class of antimicrobial agents since the introduction of nalidixic acid in 1970 [77], are, in general, short polypeptides and are produced by prokaryotes, plants and a wide variety of animals [31, 77–79]. In mammals they are expressed in a range of cells and in milk. Antimicrobial peptides present in milk are often derived from milk proteins. Lactoferrin is present in the milk of most species and human milk is particularly rich in it; lactoferrin represents about 20% of total human milk proteins [80]. While lactoferrin itself has antimicrobial activity, it has been shown that antimicrobial peptides derived from lactoferrin, in particular lactoferricin, which are produced by pepsin degradation and low pH in the stomach, can eradicate ingested pathogens and bind exotoxins in the stomach [32, 81–83]. In general, lactoferricin and other antimicrobial peptides cause lysis of microorganisms [81]. The exact mechanism of bacterial membrane permeabilization is not known.

4.4.2.2   *Potential Synergism between Antimicrobial Milk Lipids and Peptides*

The possibility of additive or synergistic interactions between antimicrobial lipids and peptides has been examined utilizing an *in vitro* system [76,84]. An antimicrobial lipid ether 1-0-octyl-sn-glycerol (OG) was combined with one of two synthetic antimicrobial peptides, D2A21 [84] and LSA5 [76]. Studies in our laboratory showed that combining OG with D2A21 reduced the titre of the enveloped viruses (HSV-1 and -2) by at least 1000-fold more than the sum of the inactivations produced by OG and D2A21 alone. Equally as important, since protection delayed may be protection denied, OG plus D2A21 reduced viral titres by more than 1000-fold in some instances in less than 10 minutes, whereas OG and D2A21 used individually produced almost no virus inactivation during this time period. When OG was combined with the LSA5 antimicrobial peptide, the combination was found to synergistically inactivate six clinical isolates of HSV-2 by 30- to 100-fold more than the sum of OG and LSA5 used individually within 30 minutes. These studies showed that combining antiviral lipids and peptides has a synergistic effect not only on the concentrations of active lipid and peptide required for viral inactivation but also on the time required to reduce the viral concentration. The results in this simplified system also demonstrated that combining multiple active components, lipid plus peptide, targeting two points on the HSV envelope by separate mechanisms, can potentially extend the time of protection over that provided by having only a single active component. Thus, persistent antimicrobial activity is maintained as the concentration of each active component diminishes over time – for example, in an infant's digestive tract – below its effective microbicidal level when used alone. These *in vitro* studies suggest strongly that the antimicrobial activity in human milk results from protective factors, such as lipids, working not only individually but also synergistically. This synergy can both decrease the concentrations of individual antimicrobial compounds required for protection of the neonate and, equally importantly, greatly reduce the time required for pathogen inactivation and extend the time that protection is provided between milk feedings.

4.4.3   Influencing Factors

The antimicrobial activity of fatty acids can be reduced in the presence of proteins, especially albumin, through specific and nonspecific binding [74, 85]. Microbicidal inhibition by unsaturated fatty acids can also be reduced by other surface-active agents such as cholesterol [2, 86]. Staphylococci, which are inactivated *in vivo* by long-chain unsaturated fatty acids, can develop resistance to fatty acid inactivation by producing an esterifying enzyme, which binds the fatty acids to cholesterol [87]. The use of monoglycerides would prevent this particular mechanism of inactivation.

The presence of serum or whole blood can reduce the effectiveness of lipid-dependent antimicrobial activity [45,46,84]. Albumin, the major serum protein, has well-characterized fatty acid-binding sites. Bacterial killing is reduced partially or completely depending upon the ratio of albumin to lipid. Experiments performed with essentially fatty acid-free human serum albumin suggest that there are six fatty acid-binding sites per albumin molecule.

When Lampe *et al.* [46] incubated the lipid monoglyceride ether 2-0-octyl-sn-glycerol with *Chlamydia trachomatis* in the presence of 10% whole blood the MCC was twofold higher than that without blood over a two-hour period. In this instance blood had a small but most likely insignificant effect on lipid antichlamydia activity. When 1-0-octyl-sn-glycerol was incubated with 1 or 10% serum in the presence of HSV-1 or HSV-2 [84] it was found that the time required for viral inactivation was increased as the serum concentration increased. Since viruses such as HSV and HIV establish permanent infections, decreasing pathogen concentrations as quickly as possible is essential and the presence of lipid-binding factors *in vivo* must be taken into account when quantifying lipid-dependent antimicrobial activity present at mucosal surfaces.

4.4.4   Spectrum

Antimicrobial lipids can inactivate both Gram-positive and Gram-negative bacteria. Lauric acid (12 : 0), as well as the monounsaturated palmitoleic acid (16 : 1) and the polyunsaturated linoleic acid (18 : 2), has been shown to be the most active against Gram-positive bacteria and those Gram-negative bacteria which can be inactivated by antimicrobial lipids [2]. A number of Gram-negative bacteria, including *E. coli*, are extremely resistant to inactivation by fatty acids, especially by long-chain unsaturated fatty acids [74], while others including *Neisseria gonorrhoeae* and *Haemophilus* spp. are as susceptible to lipid inactivation as Gram-positive bacteria [88, 89]. *C. trachomatis*, a common sexually transmitted pathogen, is also Gram-negative and is inactivated by eight-carbon fatty acid ethers [46]. Antimicrobial fatty acids present in milk also inactivate protozoa, fungi, yeast and all enveloped viruses tested thus far. The intestinal protozoan *G. lambia* [71, 73] is susceptible to inactivation by both medium-chain saturated and long-chain unsaturated fatty acids as well as by fatty acid derivatives including monoglycerides and lysophosphatidylcholine [73]. The fungi *Aspergillus niger* and *Trichoderma viride* are inactivated primarily by short-chain (three- to five-carbon) fatty acids and their derivatives [75], whereas the yeasts *Saccharomyces cerevisiae* and *Kluyveromyces marxiamus* are inhibited by medium-chain (eight- and 10-carbon) fatty acids [90]. Enveloped viruses are more readily inactivated by lipids than other microbes. They are susceptible to medium-chain saturated and long-chain unsaturated fatty acids and their monoglyceride derivatives.

4.5   Applications

4.5.1   Additive Advantage to Foods and Biological Products

Antimicrobial fatty acids and monoglycerides have been used as food preservatives [74] and in cosmetics [91]. Many of these lipids are considered GRAS (generally recognized as safe) by the United States Food and Drug Administration [92] and can therefore be used in food products to reduce foodborne infections. One potential new area where these lipids could be used is in infant formulas [93]. At present, the formulas contain triglycerides which are hydrolysed in the infant's gastrointestinal tract to produce antimicrobial fatty acids and monoglycerides in a similar fashion to milk triglycerides [94, 95]. Studies in our laboratory have shown that infant formulas containing varying proportions of triglycerides with medium-chain saturated and long-chain unsaturated fatty acids develop antiviral and antibacterial activity following incubation with lipases [55]. This is in agreement with our previous studies [42] showing that both purified medium-chain saturated (e.g. capric acid, 10 : 0) and long-chain unsaturated (e.g. oleic acid, 18 : 1) fatty acids have antiviral activity. When complete hydrolysis occurs, all formulas tested produce a total concentration of antimicrobial fatty acids ranging from 90 to 100 mmol/litre. Our studies with formula lipids confirmed our earlier hypothesis [48] that moderate changes in medium-chain saturated and long-chain unsaturated fatty acids in formula or in milk, as the result of maternal diet modifications, would not decrease the antimicrobial activity of lipid fractions because it is the total concentration of antimicrobial lipids that is critical and not one particular fatty acid.

Studies that examined the stomach contents of formula-fed infants suggest that following lipolysis in the infant's gastrointestinal tract all formulas develop comparable antiviral and antibacterial activity [8]. The lipid fraction taken from the stomach contents of infants fed infant formula or human milk inactivated both Gram-positive and Gram-negative bacteria. However, the persistence of antiviral activity following formula dilution varies with the formula and lipase combination utilized [55]. The maintenance of lipid-dependent antiviral activity with increasing formula dilution is likely reflective of the extent of triglyceride hydrolysis. This is an important observation because while there is complete hydrolysis of formula and milk triglycerides in the stomach and proximal small intestine, there is partial hydrolysis of triglycerides from salivary lipase, suggesting that there may be some lipid-dependent antimicrobial activity from formula and milk in the mouth [4]. Formula, however, is not protected from contamination between the time it is put into a bottle and when it is ingested by the infant, allowing time for potentially significant bacterial growth. The addition of these GRAS compounds to infant formulas could extend the time during which they can be safely transported. Caprylate (caprylic acid) effectively removes enveloped viruses during the purification of albumin for therapeutic use [96], as well as antivenoms [97] and human and horse immunoglobulins [98, 99]. Antiviral fatty acids and monoglycerides have recently been shown to inactivate respiratory syncytial virus when added to milk products and fruit juices [65].

4.5.2   Physiological Advantage

Antimicrobial lipids at mucosal surfaces would provide increased protection from infection by molecules that are naturally occurring *in vivo* and could be used for humans and animals [100, 101]. The transfer of lipid-based protection is already known to be successful because of that in milk [94, 95] and food [21]. The increased protection that would be provided by antimicrobial lipids at the host's mucosal surfaces would be especially important for stopping the spread of viruses such as HIV and HSV for which it has not been possible to develop a vaccine. Clinical trials have shown that a recombinant subunit vaccine containing two major HSV surface proteins induces neutralizing antibody production but does not prevent HSV infection [102–104]. HSV and HIV establish latent infections in the immuno-competent host, in contrast to a virus such as poliovirus which does not persist in the host and is eventually removed from host tissue by an immune system primed by vaccination. Vaccination does not prevent initial virus replication in host tissue [105]. Lipids provided to mucosal surfaces could inactivate HIV and HSV prior to their infection of cells at mucosal surfaces and prevent establishment of latent infection.

4.5.3 Sexually Transmitted Pathogens and Topical Activity

Most sexually transmitted pathogens are transmitted in sexual secretions at the time of contact. Many of them, including *C. trachomatis*, *N. gonorrhoeae*, HIV and human papillomavirus, are intracellular pathogens. Most can only replicate within eukaryotic cells and once they enter their target host cells they are resistant to killing by topical microbicides. However, for the short period of time before they are internalized, they may be susceptible to topical microbicide killing. The MCC assay, described above, is designed to test immediate, topical killing of bacteria by milk lipids. Results show that lipids are directly active on exposed extracellular organisms [45, 46]. Thus, milk lipids may be effective against sexually transmitted pathogens in general during the brief time after they are introduced into the genital tract and before they become intracellular. Once these pathogens have been internalized and initiated infection, they are more effectively treated by standard antibiotics and antiviral drugs. Topical activity of milk lipids can potentially reduce the vertical mother-to-child transmission of sexually transmitted pathogens. It is estimated that 40 to 80% of the children who acquire HIV-1 from their mothers become infected during delivery [106, 107] by exposure of the skin or mucous membranes to virus [108]. HSV is also transmitted from mother to infant, and infection in the perinatal period is responsible for approximately 80% of all instances of neonatal HSV infection [109]. Lipid protection during birth would not only be against vertically transmitted viruses but also against bacteria present in the birth canal. *C. trachomatis* has one of the highest incidences of infection of any sexually transmitted pathogen, and infants have a 33 to 50% risk of contracting the infection from an infected mother during vaginal delivery [110]. The antibacterial activity of milk lipids and related lipids could also be used to prevent bacterial vaginosis, which leads to preterm delivery of low-birth-weight infants [107, 111]. As a result of the negative impact of vaginal infections on the mother and infant, a simple, inexpensive, nontoxic intervention that can be used routinely is required [112]. Milk lipids meet all of these criteria.

4.6 Safety, Tolerance and Efficacy

Fatty acids and monoglycerides are already present in food and are considered GRAS compounds [92]. Monoglyceride esters, however, can be hydrolysed by lipases of either mammalian or microbial origin producing the free fatty acid, which is less active than the monoglyceride and can be inactivated by bacterial esterifying enzymes [87]. A major advantage of using an ether linkage is that this bond is stable toward enzymatic and chemical hydrolysis [113]. Therefore, intravaginally, for example, the monoglyceride ether structure will be maintained in the presence of lipases, whether of human or microbial origin, and antimicrobial activity will not be diminished. Ether lipids are fairly innocuous compounds which have been shown to be nontoxic even at high doses and can be used by mammals to synthesize membrane alkyl glycerolipids and plasmalogens [114, 115]. Ether lipids are also being developed as anticancer drugs for use against leukaemic cells [60]. Catabolism of ether lipids takes place primarily in the intestine and liver by oxidative cleavage of the ether bond [116]. Glycerol ether lipids are present in human and bovine milk and colostrum [117–119]. Previous studies have shown that ether lipids are not transported across the placenta from mother to infant [120]. Ether monoglycerides will potentially remain active longer *in vivo* than the comparable ester-linked compound but will not produce any increase in toxicity. Studies performed in our laboratory in conjunction with the National Institute of Allergy and Infectious Diseases have found that the lipid ether 1-0-octyl-sn-glycerol does not cause irritation in a rabbit model and is acceptable for vaginal administration.

4.7   Conclusions

Milk lipids are not only nutrients but also antimicrobial agents that are part of an innate defence system functioning at mucosal surfaces. Lipid-dependent antimicrobial activity in milk is due to medium-chain saturated and long-chain unsaturated fatty acids and their respective monoglycerides. The antimicrobial activity of fatty acids and monoglycerides is additive, so that dietary-induced alterations in the fatty acid distribution in milk do not reduce lipid-dependent antimicrobial activity. Fatty acids and monoglycerides can rapidly destabilize the membranes of pathogens. The antimicrobial activity of milk lipids can be duplicated using monoglyceride ethers, which very effectively inactivate enveloped viruses, bacteria and protozoa. Monoglyceride ethers are more stable than the ester linkages found in milk lipids since they are not degraded by bacterial or mammalian lipases. The sexually transmitted pathogens *C. trachomatis*, HIV, HSV, *N. gonorrhoeae* and *T. vaginalis* are effectively inactivated by the lipid ether 1-0-octyl-sn-glycerol. This compound could be used as part of a vaginal microbicide to reduce both the spread of sexually transmitted pathogens between adults and the vertical transmission of infectious agents from mother to infant during birth. Milk lipids and related GRAS antimicrobial lipids could also be added to infant formulas to reduce the risk of bacterial growth prior to consumption.

References

1. Goldman, A.S. and Goldblum, R.M. (1990) Human milk: immunologic-nutritional relationships. *Annals NY Acad. Sci.*, **587**, 236–245.
2. Kabara, J.J. (1978) Fatty acids and derivatives as antimicrobial agents: a review, in *Symposium on the Pharmacological Effect of Lipids* (ed. J.J. Kabara), The American Oil Chemists' Society, Champaign, IL, USA, pp. 1–14
3. Kabara, J.J. (1980) Lipids as host-resistance factors of human milk. *Nutr. Rev.*, **38**, 65–73.
4. Hamosh, M. (1991) Lipids metabolism, in *Neonatal Nutrition and Metabolism* (ed. W.W. Hay, Jr), Mosby Year Book, St Louis, MO, USA, pp. 122–142.
5. Jensen, R.G. (1996) The lipids in human milk. *Prog. Lipid Res.*, **35**, 53–92.
6. Hamosh, M. (1995) Lipids metabolism in pediatric nutrition. *Ped. Clinics N.A.*, **42**, 839–859.
7. Jensen, R.G., Bitman, J., Carlson, S.E. *et al.* (1995) Milk Lipids. A. Human milk lipids, in *Handbook of Milk Composition* (ed. R.G. Jensen), Academic Press, Inc., pp. 495–542.
8. Isaacs, C.E., Kashyap, S., Heird, W.C. and Thormar, H. (1990) Antiviral and antibacterial lipids in human milk and infant formula feeds. *Arch. Dis. Child.*, **65**, 861–864.
9. Sprong, R.C., Hulstein, M.F.E. and Der Meer, R.V. (2001) Bactericidal activities of milk lipids. *Animicrob. Agents Chemother.*, **4**, 1298–1301.
10. Petrone, G., Conte, M.P., Longhi, C. *et al.* (1998) Natural milk fatty acids affect survival and invasiveness of *Listeria monocytogenes*. *Lett. Appl. Microbiol.*, **27**, 362–368.
11. Wang, L. and Johnson, E.A. (1992) Inhibition of *Listeria monocytogenes* by fatty acids and monoglycerides. *Appl. Environ. Microbiol.*, **58**, 624–629.

12. Sun, C.Q., O'Connor, C.J. and Roberton, A.M. (2002) The antimicrobial properties of milk fat after partial hydrolysis by calf pregastric lipase. *Chem. Biol. Interact.*, **2**, 185–198.
13. Drake, D.R., Brogden, K.A., Dawson, D.V. and Werts, P.W. (2008) Antimicrobial lipids at the skin surface. *J. Lipid Res.*, **49**, 4–11.
14. Bibel, D.J., Miller, S.J., Brown, B.E. *et al.* (1989) Antimicrobial activity of stratum corneum lipids from normal and essential fatty acid-deficient mice. *J. Invest. Dermatol.*, **92**, 632–638.
15. Kearney, J.N., Ingham, E., Cunliffe, W.J. and Holland, K.T. (1984) Correlations between human skin bacteria and skin lipids. *Br. J. Dermatol.*, **110**, 593–599.
16. Miller, S.J., Aly, R., Shinefeld, H.R. and Elias, P.M. (1988) *In vitro* and *in vivo* antistaphylococcal activity of human stratum corneum lipids. *Arch. Dermatol.*, **124**, 209–215.
17. Aly, R., Maibach, H.I., Shinefield, H.R. and Strauss, W.G. (1972) Survival of pathogenic microorganisms on human skin. *J. Invest. Dermatol.*, **58**, 205–210.
18. Coonrod, J.D. and Yoneda, K. (1983) Detection and partial characterization of antibacterial factor(s) in alveolar lining material of rats. *J. Clin. Invest.*, **71**, 129–141.
19. Coonrod, J.D. (1986) The role of extracellular bactericidal factors in pulmonary host defense. *Sem. Respir. Infect.*, **1**, 118–129.
20. Coonrod, J.D. (1987) Role of surfactant free fatty acids in antimicrobial defenses. *Eur. J. Respir. Dis.*, **71**, 209–214.
21. Fuller, R. and Moore, J.H. (1967) The inhibition of the growth of *Clostridium welchii* by lipids isolated from the contents of the small intestine of the pig. *J. Gen. Microbiol.*, **46**, 23–41.
22. Rosell, K.-G. and Srivastava, L.M. (1987) Fatty acids as antimicrobial substances in brown algae, in *Hydrobiologia, Twelfth International Seaweed Symposium* (eds M.A. Ragan and C.J. Bird), Dr. W. Junk Publishers, Dordrecht, Netherlands, pp. 471–475.
23. Ballantine, D.L., Gerwick, H.W., Velez, S.M. *et al.* (1987) Antibiotic activity of lipid-soluble extracts from Caribbean marine algae, in *Hydrobiologia, Twelfth International Seaweed Symposium* (eds M.A. Ragan and C.J. Bird), Dr. W. Junk Publishers, Dordrecht, Netherlands, pp. 463–469.
24. Hosea Blewett, H.J., Cicalo, M.C., Holland, C.D. and Field, C.J. (2008) The immunological components of human milk. *Adv. Food Nutr. Res.*, **54**, 45–80.
25. Welsh, J.K. and May, J.T. (1979) Anti-infective properties of breast milk. *J. Pediatr.*, **94**, 1–9.
26. Fearon, D.T. (1997) Seeking wisdom in innate immunity. *Nature*, **388**, 323–324.
27. Goldman, A.S., Ham Pong, A.J. and Goldblum, R.M. (1985) Host defenses: development and maternal contributions, in *Advances in Pediatrics* (ed. L.A. Barnes), Year Book Publication, Chicago, IL, USA, pp. 71–100.
28. Ogra, P.L. and Losonsky, G.A. (1984) Defense factors in products of lactation, in *Nutritional and Immunological Interactions* (ed. P.L. Ogra), Grune and Stratton, Orlando, FL, USA, pp. 67–87.
29. Newburg, D.S. (1997) Do the binding properties of oligosaccharides in milk protect human infants from gastroinestinal bacteria? *J. Nutr.*, **125** (Suppl. 5), 980S–984S.
30. Newburg, D.S., Viscidi, R.P., Ruff, A. and Yolkcn, R.H. (1992) A human milk factor inhibits binding of human immunodeficiency virus to the CD4 receptor. *Pediatr. Res.*, **31**, 22–28.
31. Boman, H.G. (1998) Gene-encoded peptide antibiotics and the concept of innate immunity: an update review. *Scand. J. Immunol.*, **48**, 15–25.
32. Manzoni, P. (2009) Bovine lactoferrin supplementation for prevention of late-onset sepsis in very low-birth-weight neonates. *JAMA*, **302**, 1421–1428.
33. Mandel, I.D. and Ellison, S.A. (1985) The biological significance of the nonimmunoglobulin defense factors, in *The Lactoperoxidase System Chemistry and Biological Significance* (eds K.M. Pruitt and J.O. Tenovuo), Marcel Dekker, New York, NY, USA, pp. 1–14.
34. Maldovenau, Z., Tenovuo, J., Pruitt, K.M. *et al.* (1983) Antibacterial properties of milk: IgA-peroxidase-lactoferrin interactions. *Annals NY Acad. Sci.*, **409**, 848–850.
35. Watanabe, T., Nagura, H., Watanabe, K. and Brown, W.R. (1984) The binding of human milk lactoferrin to immunoglobulin A. *FEBS Lett.*, **168**, 203–207.
36. Reiter, B. (1981) The contribution of milk to resistance to intestinal infection in the newborn. in *Immunological Aspects of Infection in the Fetus and Newborn* (eds H.P. Lambert and C.B.S. Wood), The Beecham Colloquia, No. 2, Academic Press, London, UK.

37. Isaacs, C.E. (2005) Human milk inactivates pathogens individually, additively, and synergistically. *J. Nutr.*, **135**, 1286–1288.
38. Silber, G.H., Hachey, D.L., Schanler, R.J. and Garza, C. (1988) Manipulation of maternal diet to alter fatty acid composition of human milk intended for premature infants. *Am. J. Clin. Nutr.*, **47**, 810–814.
39. Francois, C.A., Connor, S.L., Wander, R.C. and Connor, W.E. (1998) Acute effects of dietary fatty acids on the fatty acids of human milk. *Am. J. Clin. Nutr.*, **67**, 301–308.
40. Spear, M.L., Bitman, J., Hamosh, M. *et al.* (1992) Human mammary gland function at the onset of lactation: medium-chain fatty acid synthesis. *Lipids*, **27**, 908–911.
41. Thompson, B.J. and Smith, S. (1985) Biosynthesis of fatty acids by lactating human breast epithelial cells: an evaluation of the contribution to the overall composition of human milk fat. *Ped. Res.*, **19**, 139–143.
42. Thormar, H., Isaacs, C.E., Brown, H.R. *et al.* (1987) Inactivation of enveloped viruses and killing of cells by fatty acids and monoglycerides. *Antimicrob. Agents Chemother.*, **31**, 27–31.
43. Jensen, R.G., Lammi-Keefe, C.J. and Koletzko, B. (1997) Representative sampling of human milk and extraction of fat for analysis of environmental lipophilic contaminants. *Toxicol. Environ. Chem.*, **62**, 229–247.
44. Peyrou, G., Rakotondrazafy, V., Mouloungui, Z. and Gaset, A. (1996) Separation and quantitation of mono-, di-, and triglycerides and free oleic acid using thin-layer chromatography with flame-ionization detection. *Lipids*, **31**, 27–32.
45. Moncla, B.J., Pryke, K. and Isaacs, C.E. (2008) Killing of Neisseria gonorrhoeae, streptococcus agalactiae (Group B Streptococcus). *Haemophilus* ducreyi, and vaginal lactobacillus by 3-0-octyl-sn-glycerol. *Antimicrob. Agents Chemother.*, **42**, 1577–1579.
46. Lampe, M.F., Ballweber, L.M., Isaacs, C.E. *et al.* (1998) Killing of *Chlamydia trachomatis* by novel antimicrobial lipids adapted from compounds in human breast milk. *Antimicrob. Agents Chemother.*, **42**, 1239–1244.
47. Isaacs, C.E., Thormar, H. and Pessolano, T. (1986) Membrane-disruptive effect of human milk: Inactivation of enveloped viruses. *J. Infect. Dis.*, **154**, 966–971.
48. Isaacs, C.E. and Thormar, H. (1991) The role of milk-derived antimicrobial lipids as antiviral and antibacterial agents, in *Immunology of Milk and the Neonate* (eds J. Mestecky and P.L. Ogra), Plenum Press, New York, NY, USA, pp. 159–165.
49. Isaacs, C.E., Kim, K.S. and Thormar, H. (1994) Inactivation of enveloped viruses in human bodily fluids by purified lipids. *Annals NY Acad. Sci.*, **724**, 457–464.
50. Thormar, H. and Hilmarsson, H. (2007) The role of microbicidal lipids in host defense against pathogens and their potential as therapeutic agents. *Chem. Phys. Lipids*, **150**, 1–11.
51. Prichard, M.N., Turk, S.R., Coleman, L.A. *et al.* (1990) A microtiter virus yield reduction assay for the evaluation of antiviral compounds against human cytomegalovirus and herpes simplex virus. *J. Virol. Meth.*, **28**, 106–106.
52. Reed, L.J. and Muench, H. (1938) A simple method of estimating fifty per cent endpoints. *Am. J. Hyg.*, **27**, 493–497.
53. Hierholzer, J. and Killington, R. (1996) in *Virology Methods Manual, Virus isolation and quantitation* (eds B.W.J. Mahy and H.O. Kangro), Academic Press, New York, NY, USA, pp. 25–46.
54. Isaacs, C.E., Wen, G.Y., Xu, W. *et al.* (2008) Epigallocatechin gallate inactivates clinical isolates of herpes simplex virus. *Antimicrob. Agents Chemother.*, **52**, 962–790.
55. Isaacs, C.E., Litov, R.E., Marie, P. and Thormar, H. (1992) Addition of lipases to infant formulas produces antiviral and antibacterial activity. *J. Nutr. Biochem.*, **3**, 304–308.
56. Baumann, W.J. (1972) The chemical syntheses of alkoxylipids, in *Ether Lipids Chemistry and Biology* (ed. F. Snyder), Academic Press, New York, NY, USA, pp. 51–79.
57. Lundblad, J.L. and Seng, R.L. (1991) Inactivation of lipid-enveloped viruses in proteins by caprylate. *Vox Sang.*, **60**, 75–81.
58. Isaacs, C.E. and Thormar, H. (1990) Human milk lipids inactivate enveloped viruses, in *Breastfeeding, Nutrition, Infection and Infant Growth in Developed and Emerging Countries* (eds S.A. Atkinson, L.A. Hanson and R.K. Chandra), ARTS Biomedical Publishers and Distributors, Newfoundland, Canada, pp. 161–174.

59. Noseda, A., White, J.G., Godwin, P.L. *et al.* (1989) Membrane damage in leukemic cells induced by ether and ester lipids: an electron microscopic study. *Exper. Mol. Pathol.*, **50**, 69–83.
60. Verdonck, L.F. and vanHeugten, H.G. (1997) Ether lipids are effective cytotoxic drugs against multidrug-resistant acute leukemia cells and can act by the induction of apoptosis. *Leukemia Res.*, **21**, 37–43.
61. Sands, J., Auperin, D. and Snipes, W. (1979) Extreme sensitivity of enveloped viruses, including herpes simplex, to long-chain unsaturated monoglycerides and alcohols. *Antimicrob. Agents Chemother.*, **15**, 67–73.
62. Kohn, A., Gitelman, J. and Inbar, M. (1980a) Interaction of polyunsaturated fatty acids with animal cells and enveloped viruses. *Antimicrob. Agents Chemother.*, **18**, 962–968.
63. Kohn, A., Gitelman, J. and Inbar, M. (1980b) Unsaturated free fatty acids inactivate animal enveloped viruses. *Arch. Virol.*, **66**, 301–307.
64. Laegreid, A., Kolstootnaess, A.-B., Orstavik, I. and Carlsen, K.H. (1986) Neutralizing activity in human milk fractions against respiratory syncytial virus. *Acta Paediatr. Scand.*, **7**, 696–701.
65. Hilmarsson, H., Traustason, B.S., Kristmundsdóttir, T. and Thormar, H. (2007) Virucidal activities of medium- and long-chain fatty alcohols and lipids against respiratory syncytial virus and parainfluenza virus type 2: comparison at different pH levels. *Arch. Virol.*, **152**, 2225–2236.
66. Sarkar, N.H., Charney, J., Dion, A.S. and Moore, D.H. (1973) Effect of human milk on the mouse mammary tumor virus. *Cancer Res.*, **33**, 626–629.
67. Falkler, Jr, W.A., Diwan, A.R. and Halstead, S.B. (1975) A lipid inhibitor of Dengue virus in human colostrums and milk; with a note on the absence of anti-Dengue secretory antibody. *Arch. Virol.*, **47**, 3–10.
68. Welsh, J.K., Arsenakis, M., Coelen, R.J. and May, J.T. (1979) Effect of antiviral lipids, heat, and freezing on the activity of viruses in human milk. *J. Infect. Dis.*, **140**, 322–328.
69. Fieldsteel, A.L. (1974) Nonspecific antiviral substances in human milk active against arbovirus and murine leukemia virus. *Cancer Res.*, **34**, 712–715.
70. Li, Q., Estes, J.D., Schlievert, P.M. *et al.* (2009) Glycerol monolaurate prevents mucosal SIV transmission. *Nature*, **458**, 1034–1038.
71. Hernell, O., Ward, H., Blackberg, L. and Pereira, M.E.A. (1986) Killing of *Giardia lamblia* by human milk lipases: an effect mediated by lipolysis of milk lipids. *J. Infect. Dis.*, **153**, 715–720.
72. Rohrer, L., Winterhalter, K.H., Eckert, J. and Köhler, P. (1986) Killing of *Giardia lamblia* by human milk is mediated by unsaturated fatty acids. *Antimicrob. Agents Chemother.*, **30**, 254–257.
73. Reiner, D.S., Wang, C.-S. and Gillin, F.D. (1986) Human milk kills *Giardia lamblia* by generating toxic lipolytic products. *J. Infect. Dis.*, **154**, 825–832.
74. Shibasaki, I. and Kato, N. (1978) Combined effects on antibacterial activity of fatty acids and their esters against Gram-negative bacteria, in *Symposium on the Pharmacological Effect of Lipids* (ed. J.J. Kabara), The American Oil Chemists' Society, Champaign, IL, USA, pp. 15–24.
75. Gershon, H. and Shanks, L. (1978) Antifungal activity of fatty acids and derivatives: structure activity relationship, in *Symposium on the Pharmacological Effect of Lipids* (ed. J.J. Kabara), The American Oil Chemists Society, Champaign, IL, USA, pp. 51–62.
76. Isaacs, C.E., Rohan, L., Xu, W. *et al.* (2006) Inactivation of herpes simplex virus clinical isolates by using a combination microbicide. *Antimicrob. Agents Chemother.*, **50**, 1063–1066.
77. Hancock, R.E. (1997) Antibacterial peptides and the outer membranes of Gram-negative bacilli. *J. Med. Microbiol.*, **46**, 1–3.
78. Lohner, K. (2009) New strategies for novel antibiotics: peptides targeting bacterial cell membranes. *Gen. Physiol. Biophys.*, **28**, 105–116.
79. Nissen-Meyer, J. and Nes, I.F. (1997) Ribosomally synthesized antimicrobial peptides: their function, structure, biogenesis, and mechanism of action. *Arch. Microbiol.*, **167**, 67–77.
80. Chierici, R. (2001) Antimicrobial actions of lactoferrin. *Adv. Nutr. Res.*, **10**, 247–269.
81. Clare, D.A. and Swaisgood, H.E. (2000) Bioactive milk peptides: a prospectus. *J. Dairy Sci*, **83**, 1187–1195.
82. Tomita, M., Bellamy, W., Takase, M. *et al.* (1991) Potent antibacterial peptides generated by pepsin digestion of bovine lactoferrin. *J. Dairy Sci.*, **74**, 4137–4142.

83. Kuwata, H., Yip, T.T., Tomita, M. and Hutchens, T.W. (1998) Direct evidence of the generation in human stomach of an antimicrobial peptide domain (lactoferricin) from ingested lactoferrin. *Biochim. Biophys. Acta*, **1429**, 129–141.

84. Isaacs, C.E., Jia, J.H. and Xu, W. (2004) A lipid-peptide microbicide inactivates herpes simplex virus. *Antimicrob. Agents Chemother.*, **48**, 3182–3184.

85. Kato, N. and Shibasaki, I. (1975) Comparison of antimicrobial activities of fatty acids and their esters. *J. Ferment. Technol.*, **53**, 793–801.

86. Ammon, H.V. (1985) Effects of fatty acids on intestinal transport, in *Symposium on the Pharmacological Effect of Lipids II* (ed. J.J. Kabara), The American Oil Chemists' Society, Champaign, IL, USA, pp. 173–181.

87. Kapral, F.A. and Mortensen, J.E. (1985) The inactivation of bactericidal fatty acids by an enzyme of *Staphylococcus aureus*, in *Symposium on the Pharmacological Effect of Lipids II* (ed. J.J. Kabara), The American Oil Chemists' Society, Champaign, IL, USA, pp. 103–109.

88. Knapp, H.R. and Melly, M.A. (1986) Bactericidal effects of polyunsaturated fatty acids. *J. Infect. Dis.*, **154**, 84–94.

89. Rabe, L.K., Coleman, M.S., Hillier, S.L. and Isaacs, C.E. (1997) The in vitro activity of a microbicide, octylglycerol, against Neisseria gonorrhoeae, *Trichomonas vaginalis*, and vaginal microflora. International Congress of Sexually Transmitted Diseases, Seville, Spain.

90. Viegas, C.A., Rosa, M.V., Sa-Correia, K. and Novais, J.M. (1989) Inhibition of yeast growth by octanoic and decanoic acids produced during ethanolic fermentation. *Appl. Environ. Microbiol.*, **55**, 21–28.

91. Kabara, J.J. (1991) Chemistry and biology of monoglycerides in cosmetic formulations. *Cosmet. Sci. Technol. Ser.*, **11**, 311–344.

92. Food and Drug Administration (1999) Code of Federal Regulations, Title 21, Vol. 3, Part 184, Sec. 184.1505, page 505, US Government Printing Office, Washington, DC.

93. Isaacs, C.E., Litov, R.E. and Thormar, H. (1995) Antimicrobial activity of lipids added to human milk, infant formula, and bovine milk. *J. Nutr. Biochem.*, **6**, 362–366.

94. Canas-Rodriguez, B. and Smith, H.W. (1966) The identification of the antimicrobial factors of the stomach contents of suckling rabbits. *Biochem. J.*, **100**, 79–82.

95. Smith, H.W. (1966) The antimicrobial activity of the stomach contents of suckling rabbits. *J. Pathol. Bacteriol.*, **91**, 1–9.

96. Johnston, E., Uren, E., Johnstone, D. and Wu, J. (2003) Low pH, caprylate incubation as a second viral inactivation step in the manufacture of albumin. Parametric and validation studies. *Biologicals*, **31**, 213–221.

97. Burnouf, T., Terpstra, F., Habib, G. and Seddik, S. (2007) Assessment of viral inactivation during pH 3. 3 pepsin digestion and caprylic acid treatment of antivenoms. *Biologicals*, **35**, 329–334.

98. Dichtelmüller, H., Rudnick, D. and Kloft, M. (2002) Inactivation of lipid enveloped viruses by octanoic acid treatment of immunoglobulin solution. *Biologicals*, **30**, 135–142.

99. Mpandi, M., Schmutz, P., Legrand, E. *et al.* (2007) Partitioning and inactivation of viruses by the caprylic acid precipitation followed by a terminal pasteurization in the manufacturing process of horse immunoglobulins. *Biologicals*, **35**, 335–341.

100. Clarke, S.R., Mohamed, R., Bian, L. *et al.* (2007) The *Staphylococcus aureus* surface protein IsdA mediates resistance to innate defenses of human skin. *Cell Host & Microbe*, **1**, 199–212.

101. Zinkermagel, A.S. and Nizet, V. (2007) *Staphylococcus aureus*: a blemish on skin immunity. *Cell Host & Microbe*, **1**, 161–162.

102. Corey, L., Langenberg, A.G.M., Ashley, R. *et al.* (1999) Recombinant glycoprotein vaccine for the prevention of genital HSV-2 infection. *JAMA*, **4**, 331–340.

103. Mascola, J.R. (1999) Herpes simplex virus vaccines: why don't antibodies protect? *JAMA*, **4**, 379–380.

104. Aurelian, L. (2004) Herpes simplex virus type 2 vaccines: new ground for optimism? *Clin. Diag. Lab. Immunol.*, **11**, 437–445.

105. Clements-Mann, M.L. (1998) Lessons for AIDS vaccine development from non-AIDS vaccines. *AIDS Res. Hum. Retrovir.*, **14** (Suppl. 3), S197-S203.

106. Mostad, S.B., Overbaugh, J., DeVange, D.M. *et al.* (1997) Hormonal contraception, vitamin A deficiency, and other risk factors for shedding of HIV-1 infected cells from the cervix and vagina. *Lancet*, **350**, 922–927.
107. Goldenberg, R.L., Andrews, W.W., Yuan, A.C. *et al.* (1997) Sexually transmitted diseases and adverse outcomes of pregnancy. *Infect. Perinatol.*, **24**, 23–41.
108. Baba, T.W., Koch, K.J., Mittler, E.S. *et al.* (1994) Mucosal infection of neonatal rhesus monkeys with cell-free SIV. *AIDS Res. Human Retrovir.*, **10**, 351–357.
109. Whitely, R.J. (1994) Herpes simplex virus infections of women and their offspring: Implications for a developed society. *Proc. Natl. Acad. Sci.*, **91**, 2441–2447.
110. Eng, T.R. and Butler, W.T. (1997) Medicine, in *The Hidden Epidemic: Confronting Sexually Transmitted Diseases*, National Academy Press, Washington, DC, USA, pp. 1–432.
111. Hillier, S.L., Nugent, R.P., Eschenback, D.A. *et al.* (1995) Association between bacterial vaginosis and preterm delivery of a low-birth-weight infant. *N. Engl. J. Med.*, **333**, 1373–1742.
112. Hofmeyr, G.J. and McIntyre, J. (1997) Preventing perinatal infections. *BMJ*, **315**, 199–200.
113. Paltauf, F. (1983) Ether lipids as substrates for lipolytic enzymes, in *Ether Lipids Biochemical and Biomedical Aspects* (eds H.K. Mangold and F. Paltauf), Academic Press, New York, NY, USA, pp. 211–229.
114. Mangold, H.K. (1983) Ether lipids in the diet of humans and animals, in *Ether Lipids Biochemical and Biomedical Aspects* (eds H.K. Mangold and F. Paltauf), Academic Press, New York, NY, USA, pp. 231–238.
115. Mangold, H.K. (1972) Biological effects and biomedical applications of alkoxylipid, in *Ether Lipids Chemistry and Biology* (ed. R. Snyder), Academic Press, New York, NY, USA, pp. 157–176.
116. Weber, N. (1985) Metabolism of orally administered ra-1-O[14C]dodecylglycerol and nutritional effects of dietary rac-1-O-dodecylglycerol in mice. *J. Lipid Res.*, **26**, 1412–1420.
117. Oh, S.Y. and Jadhav, L.S. (1994) Effects of dietary alkylglycerols in lactating rats on immune responses in pups. *Ped. Res.*, **36**, 300–305.
118. Hallgren, B., Niklasson, A., Stalberg, G. and Thorin, H. (1974) On the occurrence of 1-0-alkylglyceros and 1-0(2-Methoxyalkyl) glycerols in human colostrums, human milk, cow's milk, sheep's milk, human red bone marrow, red cells, blood plasma and a uterine carcinoma. *Acta Chem. Scand.*, **B28**, 1029–1034.
119. Hallgren, B. and Larsson, S. (1962) The glyceryl ethers in man and cow. *J. Lipid. Res.*, **3**, 39–38.
120. Das, A.K., Holmes, R.D., Wilson, G.N. and Hajra, A.K. (1992) Dietary ether lipid incorporation into tissue plasmalogens of humans and rodents. *Lipids*, **27**, 401–405.

5

Antimicrobial Lipids of the Skin and Tear Film

Carol L. Bratt[1], Phil Wertz[1,2], David Drake[1,3], Deborah V. Dawson[1,4] and Kim A. Brogden[1,5]

[1]Dows Institute for Dental Research
[2]The Department of Oral Pathology, Radiology & Medicine
[3]The Department of Endodontics
[4]Department of Preventive and Community Dentistry
[5]Department of Periodontics, College of Dentistry, The University of Iowa, Iowa City, IA, USA

Lipids and Essential Oils as Antimicrobial Agents Halldor Thormar
© 2011 John Wiley & Sons, Ltd

5.1   Introduction

Skin is a surprisingly durable and protective body-surface covering. It is a physical barrier that protects the underlying tissues against excessive temperature changes, mechanical injury, ultraviolet irradiation and environmental contamination [1]. Skin helps to maintain body temperature as the calibers of underlying capillaries are controlled to prevent heat loss and sweat is produced by sweat glands to dissipate heat via evaporation. It is also a permeability barrier that controls the transcutaneous movement of water and other electrolytes [2, 3]. The pH of the skin follows a sharp gradient across the stratum corneum, which is important in maintaining the permeability barrier, controlling enzymatic activities and cutaneous antimicrobial defence [4–6].

Skin also has distinct innate immune functions that limit and protect body tissues from colonized microorganisms [6]. Both keratinized and nonkeratinized cells respond to injury and exposure to microorganisms and microbial antigens by producing a variety of cytokines, chemokines, growth factors and antimicrobial peptides that serve to heal wounds, maintain integrity of the surface and initiate localized innate and adaptive immune responses.

Not surprisingly, epidermal and sebaceous lipids are involved in the physical-barrier, permeability-barrier and immunological-barrier functions of skin [3, 7]. Epidermal layers contain ceramides, free fatty acids and cholesterol [1, 7]. Sebaceous lipids contain a complex mixture of triglycerides, wax esters, squalene, cholesterol and cholesterol esters that contribute to 1) the transport of fat-soluble antioxidants from and to the skin surface, 2) the pro- and antiinflammatory skin properties and 3) the innate antimicrobial activity of the skin [8–10]. Although the composition, biosynthesis, secretion and function of cutaneous lipids are well known from extensive and eloquent work done in the 1970s, little is known about their role in controlling microbial infection and colonization. Recent work suggests that lipids have more of a direct role than was previously thought in innate immune defence against epidermal bacterial infections.

In this chapter, we will briefly review the structure and composition of the skin and tear film. We will then review their role in innate immune defences, describe the antimicrobial lipids found in these tissues and secretions, and present their antimicrobial activities. We then suggest potential mechanisms for lipid antimicrobial activity on Gram-negative and Gram-positive bacteria and yeasts. We will weave the newly described antimicrobial activity of lipids into that of other innate immune mechanisms of the skin and tear film, suggesting areas of synergistic activity. Finally, we present the concept that isolated or purified lipids could serve as pharmaceuticals to improve therapies to treat and control a wide variety of cutaneous infections and inflammatory disorders.

Throughout the chapter, fatty acids will be designated by their common name, with the fatty acid structure in parentheses [11]. The formulae show the number of carbon atoms followed by the number of double bonds. For example, lauric acid (C12 : 0) is 12 carbon atoms long with zero double bonds.

5.2 Innate Immune Mechanisms in Skin

5.2.1 The Extensive Cutaneous Microbial Burden

The skin is commonly inhabited by a variety of microbial species [12]. Their role in skin health and disease is currently being reassessed [13]. Commonly isolated commensal microorganisms and opportunistic microorganisms include Gram-positive cocci, Gram-positive bacilli, some Gram-negative bacilli and fungi (Table 5.1). Concentrations can vary from 10^2 to 10^6 colony-forming units (CFU)/cm^2 depending upon the anatomical location [14–16]. Moist areas of the forehead and armpit can contain $\sim 10^5$ to 10^7 CFU/cm^2 whereas dry areas of the back and forearm can contain $\sim 10^2$ CFU/cm^2.

The skin is occasionally infected with pathogens (Table 5.2) capable of causing a variety of cutaneous lesions, including papules and nodules, erythematous plaques, vesicles and bullae, and ulcers [17]. Cuts, burns, puncture wounds, abrasions, surgical-incision sites, placement sites of intravenous catheters and transdermal devices are all potential sites of skin infection.

The Gram-positive cocci, *Staphylococcus* spp. and *Streptococcus* spp., contaminate cuts, burns, puncture wounds and abrasions. They cause impetigo, furunculosis, abscesses, blisters and erysipelas [17–19]. In extreme cases they cause cellulitis, necrotizing fasciitis and scalded-skin syndrome.

The Gram-positive bacilli, *Corynebacterium* spp. and *Brevibacterium* spp., induce sepsis in immunocompromised patients, infect patients with implants and cause pitted keratolysis,

Table 5.1 *Bacterial commensals commonly associated with normal human skin.*

Genus	Species
Gram-positive cocci	
Staphylococcus	*S. aureus, S. auricularis, S. capitis, S. cohnii, S. epidermidis, S. hominis, S. saccharolyticus, S. simulans, S. warneri, S. xylosus, S. saprophyticus* and *S. haemolyticus*
Micrococcus	*M. lylae, M. nishinomiyaensis, M. kristinae, M. sedentarius, M. roseus, M. varians* and *M. luteus*
Gram-positive bacilli	
Corynebacterium	*C. jeikeium, C. urealyticum, C. minutissimum, C. xerosis, C. lipophilicus* and *C. striatum*
Propionibacterium	*P. acnes, P. avidum* and *P. granulosum*
Brevibacterium	*B. epidermidis*
Dermabacter	*D. hominis*
Gram-negative bacilli	
Acinetobacter	*A. johnsonii, A. lwoffi* and *A. calcoaceticus anitratus*

Summarized from [12,14,15,107].

Table 5.2 *Bacterial skin infections in normal individuals, immunocompromised individuals and travelers, and associated with specific occupations.*

Microorganisms	Infections	References
Staphylococcus spp. including *S. aureus* and *MRSA*	Contamination of cuts, burns, puncture wounds and abrasions; colonization and aggravation of lesions such as those caused by atopic dermatitis; impetigo, furunculosis, abscesses, blisters, scalded-skin syndrome and toxic-shock syndrome; catheter- and implant-associated infections, urinary tract infections and peritonitis	[12, 17–19]
Streptococcus spp.	Contamination of cuts, burns, puncture wounds (including tattoos and piercings) and abrasions; also causes impetigo, erysipelas, angular cheilitis, paronychia, ecthyma, cellulitis, lymphangitis, lymphadenitis, necrotizing fasciitis; secondary infections of other rashes (eczema, scabies, chickenpox); cutaneous manifestations associated with streptococcal infections in other parts of the body (scarlet fever following *Streptococcus pyogenes* infection of the throat)	[12, 17–19]
Corynebacterium and *Brevibacterium* spp.	Sepsis and cutaneous infections in immunocompromised patients, patients with implants and device-related nosocomial infections, pitted keratolysis, erythrasma, trichobacteriosis, dermatophytosis complex, breast abcesses, endocarditis, peritonitis, pneumonia, bacteraemia, urinary tract infections, arthritis, osteomyelitis; bacteremia and sepsis in an AIDS patient via contaminated catheter	[12, 19–21, 109]
Bacillus anthracis	Occupational exposure to infected ruminants results in cutaneous infection with pruritic papules, ulcers or black eschars	[19, 108]
Propionibacterium acnes	Acne vulgaris, a common disorder of the pilosebaceous follicles	[22]
Brucellosis spp.	Occupational exposure to infected ruminants results in cutaneous infection with urticaria-like papules, papulonodular erysipelas, erythema nodosum-like lesions and purpura	[19]
Erysipelothrix rhusiopathiae	Occupational exposure to fish results in cutaneous infection with red, well-demarcated lesions	[19]

Table 5.2 *Bacterial skin infections in normal individuals, immunocompromised individuals and travelers, and associated with specific occupations.* (Continued)

Microorganisms	Infections	References
Acinetobacter baumannii	Skin and soft-tissue infection associated with war trauma results in cellulitis with overlying vesicles progressing to necrotizing infection with haemorrhagic and nonhaemorrhagic bullae	[24]
Pseudomonas spp. including *P. aeruginosa*	Super-hydrated skin (trench foot), paronychia, stasis ulcers, burns, ulcers, abscesses, cellulitis and erysipelas-like infections	[12, 23, 110]
Proteus mirabilis	Swamp foot	[12]
Haemophilus influenzae	Cellulitis of the face in young children (especially young males) and abscesses	[12]
Pasteurella multocida	Associated with bites from animals (especially cats)	[12]
Aeromonas hydrophila	Exposure to fresh/salt water	[12]
Neisseria gonorrhoeae	Cutaneous infection follows minor trauma and genital contact	[12]
Mycobacterium spp.	Lesions vary and include erythematous papules or pustules that form a verrucous or violaceous plaque or nodule (*M. marinum, M. kansasii* and *M. scrofulaceum*); painful or painless subcutaneous nodules in the cervical, submandibular, submaxillary or preauricular region that may ulcerate with discharge of a serosanguineous material (*M. avium-intracellulare* complex); painless subcutaneous swellings on legs and forearms that enlarge to form firm nodules which may ulcerate and become necrotic (*M. ulcerans*); also causes cutaneous tuberculosis. *M. marinum* is frequently associated with aquatic activities/occupations	[12, 17, 19, 25]
Borrelia burgdorferi	Lyme disease – erythema chronicum migrans	[17, 19]
Treponema pallidum	Syphilis – painless chancre at the site of infection	[19]
Micrococcus sedentarius	Pitted keratolysis	[12]

erythrasma and trichobacteriosis [19–21]. *P. acnes* is involved in the pathogenesis of acne vulgaris, a common disorder of the pilosebaceous units [22]. Exposure to fish results in *E. rhusiopathiae* cutaneous infections with red, well-demarcated lesions [19]. Individuals in developing countries handling meat, hair or hides are at risk of contracting cutaneous *B. anthracis* infections consisting of pruritic papules, ulcers or black eschars [19].

The Gram-negative bacilli are often found as transient organisms [23]. They have a tendency to flourish when the Gram-positive flora is suppressed by antibiotics. Gram-negative bacilli can be site-specific, particularly the enterics found near the perineum. These coliform-like organisms include *Achromobacter*, *Alcaligenes*, *Escherichia* and *Serratia*. Some noncoliform-like Gram-negative bacilli are opportunistic in nature. These microorganisms include *Pseudomonas*, *Proteus* and *Klebsiella*. *A. baumannii* infections are associated with war-trauma wounds and induce cellulitis with overlying vesicles progressing to necrotizing infection with haemorrhagic and nonhaemorrhagic bullae [24].

Atypical mycobacteria are ubiquitous in the environment and cause a variety of cutaneous infections, especially in immunocompromised patients [25]. These include the slow-growing mycobacteria *M. marinum*, *M. kansasii* and *M. avium-intracellulare* complex and the rapidly-growing mycobacteria *M. fortuitum*, *M. chelonei* and *M. abscessus*. Lesions vary and include erythematous papules or pustules that form a verrucous or violaceous plaque or nodule (*M. marinum*, *M. kansasii* and *M. scrofulaceum*); painful or painless subcutaneous nodules in the cervical, submandibular, submaxillary or preauricular region that may ulcerate with discharge of a serosanguineous material (*M. avium-intracellulare*); and painless subcutaneous swellings on legs and forearms that enlarge to form firm nodules which may ulcerate and become necrotic (*M. ulcerans*) [17, 19, 25].

B. burgdorferi is among the *Borrelia* spp. that cause Lyme disease [26]. Local infection of the skin is followed by haematogenous spread, invasion of endothelial cells and infection of many tissues with lymphocytic and histiocytic inflammatory infiltrates. *B. burgdorferi* replicates in skin [17, 19] and causes a very low-density spirochetemia, erythema migrans, myalgias, arthralgias, arthritis, carditis, conjunctivitis, neurologic involvement and acrodermatitis.

Treponema pallidum ssp. *pallidum* causes syphilis. Transmission occurs through direct contact with active lesions; *T. pallidum* penetrates the intact mucous membrane of the genital tract or through abraded epithelium. From 0 to 90 days, a chancre develops at the site of infection [19].

The conjunctivae, lacrimal apparatus and eyelids are also commonly inhabited by a variety of microbial species. Gram-positive cocci including *Streptococcus pneumoniae*, *S. pyogenes*, *S. aureus* and *S. epidermidis*, Gram-positive bacilli including *Corynebacterium* spp. and *Propionibacterium* spp., Gram-negative cocci including *Moraxella* spp. and *Neisseria* spp., Gram-negative bacilli including *Enterobacteriaceae*, *Haemophilus influenzae* and *Haemophilus parainfluenzae* and fungi including *Candida* spp. are all found [27].

All of these microbial species in and on the skin are kept under control by localized innate immune responses utilizing a vast array of constitutively produced proteins and peptides [6, 28, 29]. In addition, cutaneous cells respond to injury and exposure to microorganisms or their microbial antigens by producing cytokines, chemokines and growth factors; all of which are involved in inflammation, angiogenesis and reepithelialization [30].

5.2.2   Cutaneous Innate Immune Mechanisms

Antimicrobial peptides are produced by many tissues and cell types [31–34]. They are a particularly important component of cutaneous innate immunity [35–37]. These peptides limit colonization of commensal microorganisms, opportunistic microorganisms and occasionally transient pathogenic microorganisms; act on host cells to stimulate cytokine production, cell migration, proliferation, maturation and extracellular matrix synthesis; activate different immune and inflammatory cells; and enhance wound healing by promoting keratinocyte migration and proliferation (Table 5.3). They include the cathelicidins, defensins, dermcidin, psoriasin, adrenomedullin, lactoferrin, MUC1 and lysozyme [28,38,39]. Novel peptides have also been isolated and include S100A15 and histone H4 [40,41].

The cathelicidins, defensins, dermcidin, psoriasin, adrenomedullin, lactoferrin, MUC1 and lysozyme have different modes of killing and often exhibit synergistic antimicrobial activity, particularly against *S. aureus* [42–44]. For example, dermcidin has time-dependent

Table 5.3 *Innate and adaptive immune functions of antimicrobial peptides in cutaneous tissues.*

Peptide	Innate and adaptive immune function
Defensins (HBD2)	• HBD2, HBD3 and HBD4 induce keratinocytes to produce cytokines IL-6, IL-10 and IL-18 and chemokines CXCL10 (IP-10), CCL2 (MCP-1), CCL20 (MIP-3α) and CCL5 (RANTES) [111,112] • HBD2 acts as a chemotaxin for mast cells [113] • HBD2 stimulates mast cells to mobilize intracellular Ca(2+) and release histamine or generate PGD(2) [114] • HBD3 and HBD4 induces mast cell degranulation, prostaglandin D2 production, intracellular Ca2+ mobilization and chemotaxis [115]
Cathelicidins (LL-37)	• LL-37 enhances the production of IL-8 in neutrophils and induces both mRNA expression and protein release of human neutrophil peptides (HNPs) 1–3 [116] • LL-37 induces mast cell chemotaxis [117] • LL-37 stimulates mast cells to mobilize intracellular Ca(2+) and release histamine or generate PGD(2)[114] • LL-37 induces keratinocytes to produce IL-18 [111]
Dermcidin	• Induces keratinocytes to produce proinflammatory cytokines (TNF-α) and the chemokines CXCL8 (IL-8) and CCL20 (MIP-3α) [118]
Adrenomedullin	• Active against pathogenic and commensal strains of bacteria, but not yeast [119]
Lysozyme	• Expression is localized to the cytoplasm of epidermal cells in the skin, pilosebaceous follicles and eccrine glands [28]
Psoriasin	• Stimulates normal keratinocytes [120] • Activates neutrophils to produce IL-6, CXCL8 (IL-8), TNF-α, macrophage inflammatory protein-1alpha (MIP-1alpha)/CCL3, MIP-1beta/CCL4 and CCL20 (MIP-3α) [121] • Induces phosphorylation of mitogen-activated protein kinase p38 and extracellular signal-regulated kinase (ERK), but not c-Jun N-terminal kinase (JNK) [121] • Enhances mRNA expression and induces the extracellular release of human neutrophil peptides (HNPs) 1–3 [121]
RNase7	• Effective against Gram-positive and Gram-negative bacteria [6]

bactericidal activity, which is followed by bacterial membrane depolarization. Dermcidin-derived peptides do not induce pore formation in bacterial membranes, in contrast to the mode of action of LL-37. HBD1, HBD2, HBD3, LL-37 and lysozyme in various combinations also have synergistic or additive antibacterial effects against *S. aureus* that are enhanced in an acidic environment.

The amino acid composition, amphipathicity, cationic charge and size allow antimicrobial peptides to attach to and insert into well-defined membrane bilayers, forming pores through 'barrel-stave', 'toroid pore' or 'carpet' mechanisms. However, there is mounting evidence that antimicrobial peptides can also 1) alter cytoplasmic membrane septum formation, 2) inhibit cell-wall synthesis, 3) inhibit nucleic-acid synthesis, 4) inhibit protein synthesis or 5) inhibit enzymatic activity [28, 31].

In normal human skin, cathelicidins (LL-37) and human beta-defensins are negligible, but they accumulate in chronic facial skin inflammatory diseases such as psoriasis and rosacea but not in atopic dermatitis [44, 45]. Several antimicrobial peptides are affected, including cathelicidin, HBD2 and HBD3, which are lower in lesional skin of atopics compared with other inflammatory skin diseases, and dermcidin, which is decreased in sweat [46]. LL-37 and HBD2 are present in the superficial epidermis of patients with psoriasis but not in acute and chronic lesions of patients with atopic dermatitis. LL-37 is present in high levels in the facial skin of patients with rosacea and the proteolytically processed forms of these peptides are different from those present in normal individuals. These findings suggest a role of cathelicidin in skin inflammatory responses, possibly an exacerbated innate immune response in the pathogenesis of these diseases [47].

The induction of antimicrobial peptides can in some cases be body-site specific. A good example is the production of antimicrobial peptides and proteins by human ceruminous glandular cells to protect the surface of the external auditory meatus of the human external auditory canal [39]. The ceruminous glands in the skin of the human external auditory canal are modified apocrine glands, which, together with sebaceous glands, produce the cerumen: the ear wax. Cerumen plays an important role in the protection of the ear canal against physical damage and microbial invasion. Numerous antimicrobial proteins and peptides are present in the ceruminous glandular cells: beta-defensin-1, beta-defensin-2, cathelicidin, lysozyme, lactoferrin, MUC1 and secretory component of IgA.

The induction of antimicrobial peptides, in other cases, can be at select epithelial interfaces, particularly at sites of injury and infection [48]. Large increases in the expression of cathelicidin LL-37 and CRAMP in human and murine skin respectively occur after sterile incisions, or in mice following infection by Group A *Streptococcus*. The appearance of cathelicidins in skin is due to both synthesis within epidermal keratinocytes and deposition from granulocytes that migrate to the site of injury.

Antimicrobial peptides and proteins clearly limit colonization of commensal and pathogenic microorganisms, but there are other factors that also provide initial, nonspecific defence of surface epithelia and are involved in skin-barrier development and survival. These include antimicrobial lipids and fatty acids at the skin surface [49, 50].

5.3   Types and Locations of Lipids of the Skin and Tear Film

The skin is a complex and multilayered tissue consisting of a hypodermis (innermost layer), dermis (inner layer) and epidermis (outermost layer, with the stratum corneum).

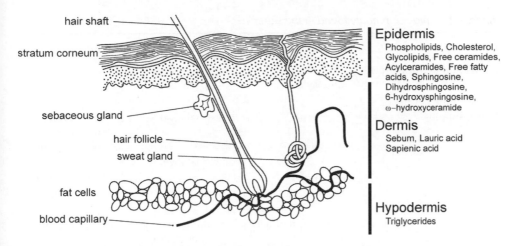

Figure 5.1 *Conceptualization of the architecture of the skin, showing the location and production of lipids on the epidermal, dermal and hypodermal layers.*

These layers have unique architectures (Figure 5.1), contain lipids and fatty acids (Table 5.4), and have a variety of physiological and immunological functions.

5.3.1 Hypodermal and Dermal Lipids

The hypodermis contains a layer of adipose tissue and areolar tissue. Here, lipids are found within adipocytes as fats, which serve to insulate and cushion the body and regulate body temperature.

Over the hypodermis is the dermis, a layer of connective tissue containing a vascular network and sensory nerve endings. The dermal reticular layer contains dense and irregular connective tissue with glands and hair follicles. The dermal papillary layer contains loose connective tissue, the vascular network and the sensory nerves. Fibroblasts in the dermis produce collagen, elastin and ground substance of connective tissue. Nonkeratinized cells are involved in innate and adaptive immune protection.

5.3.2 Epidermal Lipids

The outermost layer is the epidermis. It is more complex with respect to its lipid composition and lipid metabolism [51–56]. The epidermal stratum germinativum layer is separated from the dermis by a thin layer of basement membrane and contains the germinal cells necessary for the regeneration and healing of the overlying epidermal layers. The germinal cells have desmosomes on their outer membranes, which appear as 'prickles' of the prickle-cell layer in the stratum spinosum.

As cells mature and differentiate, they accumulate keratin and dense basophilic keratohyalin granules, forming the stratum granulosum layer. Simultaneously, protein-rich keratohyalin granules and lipid-rich lamellar granules accumulate. The latter lamellar granules contain phospholipids, cholesterol and glycolipids. Hydrolytic enzymes also accumulate in granules, which ultimately convert the phospholipids and glycolipids into ceramide and

Table 5.4 *Lipids commonly found in the skin.*

Lipid	Site of production	Location	Reference
Lipids in the hypodermis			
Fats	Adipocytes	Adipose tissue	
Lipids in the dermis			
Phospholipids		Dermal layer	(Phil Wertz, unpublished observation)
Cholesterol		Dermal layer	(Phil Wertz, unpublished observation)
Lipids in the epidermis			
Phospholipids	Lamellar granules	Stratum granulosum	[51–53, 56, 122]
Cholesterol	Lamellar granules	Stratum granulosum	[51–53, 56, 122]
Glycolipids	Lamellar granules	Stratum granulosum	[51–53, 56, 122]
Ceramides	Keratinocytes	Stratum corneum	[1, 52, 59]
Covalently bound ceramides	Keratinocytes	Stratum corneum	[1, 52, 59]
Acylceramides	Keratinocytes	Stratum corneum	[1, 52, 59]
Cholesterol	Keratinocytes	Stratum corneum	[1, 52, 59]
Fatty acids (free)	Keratinocytes	Stratum corneum	[1, 6, 52, 59]
Sphingosine	Keratinocytes	Stratum corneum	[6, 57, 123]
Dihydrosphingosine	Keratinocytes	Stratum corneum	[57, 123]
6-hydroxy-sphingosine	Keratinocytes	Stratum corneum	[57, 123]
ω-hydroxyceramide	Keratinocytes	Cornified layer	[61–63]
Secreted lipids			
Lauric acid (C12 : 0)	Sebum	Sebaceous glands	
Sapienic acid (C16 : 1Δ6)	Sebum	Sebaceous glands	

fatty acids. The cells in the stratum granulosum layer are pushed to the surface, where they begin to die, forming the stratum corneum.

Between the stratum granulosum and stratum corneum layers, the contents of lamellar granules are extruded into the intercellular spaces. Here, enzymes hydrolyse sphingomyelin and glucosylceramides to ceramide, and phospholipids are converted into saturated fatty acids. Enzymes also hydrolyse phosphoglycerides and cholesterol to monounsaturated cholesterol esters.

At the skin surface is the stratum corneum, a complex, lipid-rich region containing ceramides, acylceramides, cholesterol and free fatty acids, and at least two different ceramidases [52, 57–59]. The ceramides consist of normal fatty acids and α-hydroxyacids or ω-hydroxyacids, amide-linked to sphingosines, dihydrosphingosines, phytosphingosines and 6-hydroxysphingosines. They are the source of epidermal long-chain bases [57, 60].

The acylceramides have 30- through 34-carbon-long ω-hydroxyacids amide-linked to long-chain bases and bearing linoleic acid ester-linked to the ω-hydroxyl group. Acylceramides are derived from acylglucosylceramides associated with the lamellar granules and are important for the physical organization of lipids in the stratum corneum. The free fatty acids are saturated, straight-chains 20 to 28 carbons in length.

In the stratum-corneum layer, ω-hydroxyceramide molecules become covalently attached to the outer surfaces of cornified cells. This layer is very hydrophobic and likely forms a lipid-permeability barrier [61–63]. The ω-hydroxyceramides are derived from acylglucosylceramide precursors and contain 30- through 34-carbon hydroxyacids amide-linked to sphingosine bases. This forms a chemically-bound lipid envelope around the corneocytes, giving it a typical bilayer appearance [63].

5.3.3 Secreted Lipids

Squalene, wax monoesters, triglycerides and lesser amounts of cholesterol and cholesterol esters are secreted from sebaceous glands to the skin surface. For example, lauric acid (C12 : 0) and sapienic acid (C16 : 1Δ6) are found in 3.5 and 36.0% of the composition of free fatty acids associated with human hair fat [64]. Secreted triglycerides undergo hydrolysis to lauric acid and sapienic acid by acid lipases from the lamellar granules and to a lesser extent by commensal bacterial lipases. Lauric acid, sapienic acid and sphingoid bases have antibacterial activity. The epidermal acid lipase may also have a role in hydrolysis of the sebaceous triglycerides. This could provide a mechanism for accelerating hydrolysis in the event of a challenge that would make the skin surface more vulnerable to colonization by bacteria other than the normal flora. Anything that damages the normal barrier provided by the stratum corneum could trigger a defensive response.

5.3.4 Tear-Film Lipids

Tears are an extremely complex mixture which provide a mechanical and antimicrobial barrier (Table 5.5) at the ocular surface [65–69]. The aqueous component of tears is produced by lacrimal glands and contains electrolytes, water and a large variety of proteins, peptides and glycoproteins [70]. A mucin component is secreted by epithelial tissues at the mucosal surface and contains glycoproteins that serve as antioxidants, lubricants and inhibitors of bacterial adherence. The lipid component of tears, of interest to this chapter, is produced by the meibomian glands and contains fatty, oily substances called meibum [71]. These glands are specialized sebaceous glands that secrete their product at the edges of the eyelids. They are sebaceous follicles and are not associated with hairs.

Meibum is a complex mixture of polar and nonpolar lipids. The polar-lipid fraction contains several phospholipids (e.g. phosphatidylcholine, phosphatidylethanolamine, sphingomyelin and a few unidentified phospholipids), ceramides and cerebrosides that form a surfactant-like layer interphase between the aqueous-tear component and the thicker nonpolar-lipid component of the tear film [72]. The nonpolar-lipid fraction maintains the structural integrity of the tear film and protects the ocular surface from losing water [69, 71, 73, 74]. As in sebum from the skin surface, the major nonpolar lipids include wax esters, triglycerides, sterols, sterol esters, diacylglycerols and fatty acids ranging from 12 to 18 carbons in length [69, 72]. Some of the wax esters contain unusually long fatty alcohol components compared to sebaceous wax esters [75, 76]. However, unlike sebum, meibum

Table 5.5 *Antimicrobial components and lipids commonly found in the ocular-surface epithelium and Meibomian glands.*

Component	Site of production	Function	Reference
Ocular epithelium			
MIP-3α	Scraped corneal epithelium and primary cultured conjunctival cells	Defence of the ocular surface	[70, 126]
Tβ4	Scraped corneal epithelium and primary cultured conjunctival cells	Defence of the ocular surface	[70]
HBD1, 2, 3	Ocular surface epithelium	Defence of the ocular surface	[70, 127, 128]
LL-37	Human corneal and conjunctival epithelial cells	Defence of the ocular surface	[70, 129]
CXCL-1	Cornea	Defence of the ocular surface	[70, 130]
Mucin layer			
Glycoproteins	Epithelial tissues	Antioxidation, lubrication and inhibition of bacterial adherence	
Lipid layer			
Myristic fatty acid	Meibomian gland	Structural integrity of the tear film and defence of the ocular surface	[71]
Palmitic fatty acid	Meibomian gland	Structural integrity of the tear film and defence of the ocular surface	[71]
Stearic fatty acid	Meibomian gland	Structural integrity of the tear film and defence of the ocular surface	[71]
Oleic fatty acid	Meibomian gland	Structural integrity of the tear film and defence of the ocular surface	[71]
Myristamide fatty acid amide	Meibomian gland	Structural integrity of the tear film and defence of the ocular surface	[71]
Palmitamide fatty acid amide	Meibomian gland	Structural integrity of the tear film and defence of the ocular surface	[71]
Stearamide fatty acid amide	Meibomian gland	Structural integrity of the tear film and defence of the ocular surface	[71]
Erucamide fatty acid amide	Meibomian gland	Structural integrity of the tear film and defence of the ocular surface	[71]
Oleamide fatty acid amide	Meibomian gland	Structural integrity of the tear film and defence of the ocular surface	[71]

does not contain squalene [69]. Recently, (O-acyl)-ω-hydroxyacids have been identified [76]. These unusual lipids contain ω-hydroxyacids with 30, 32 and 34 carbons and one double bond, and oleic acid is the main fatty acid ester-linked to the ω-hydroxyl group. It is thought that the bulk of the lipid film is composed of nonpolar components, while the polar lipids and (O-acyl)-ω-hydroxyacids interface the lipid film with the underlying aqueous phase.

The lipid film itself provides a physical barrier that helps to prevent microorganisms from reaching the surface of the eye. The fatty acid fraction is capable of providing antimicrobial activity. The other nonpolar lipids are mostly not antimicrobial, but the diglycerides may contribute to this function. At least some of the polar lipids and (O-acyl)-ω-hydroxyacids are potentially antimicrobial but have not yet been tested. Oleamide has been reported as an antimicrobial lipid found in meibum [71]; however, more recent work indicates that this is a contaminant [69].

5.4 Functions of Lipids

5.4.1 Cutaneous Lipids as Permeability Barriers

Cutaneous lipids are necessary for skin-barrier development and survival. Elovl4 is part of a family of fatty acid elongases. Removing very-long-chain fatty acids by disruption of Elovl4 results in a smaller, weaker stratum corneum that loses its ability to prevent water loss in a desiccating environment [3]. Mice lacking a functional Elovl4 (e.g. *Elovl4*$^{270x/270x}$ and *Elovl4*$^{-/-}$ mice) have a significant reduction in free fatty acids longer than 26 carbons in their skin. They dehydrate very quickly and die perinatally. Similarly, the depletion of very-long-chain fatty acids containing ceramide in Elovl4-deficient mice also leads to impaired stratum corneum barrier function [77]. Together these studies show that Elovl4 is essential for lipid biosynthesis and very-long-chain fatty acids are essential for skin-permeability barrier function and neonatal survival.

5.4.2 Cutaneous Lipids as Innate Immune Mechanisms

The ability of free fatty acids to kill microorganisms has been known for well over a century. For a review, see the article by Bayliss [78]. Fatty acids are among the most actively studied groups of lipids [50, 79, 80]. They are antimicrobial against Gram-positive bacteria, Gram-negative bacteria and *C. albicans* but exceptions occur among them. In the 1940s, Burtenshaw found that cutaneous lipid extracts, likely containing fatty acids, killed *S. aureus in vitro* [81]. More recently, other lipids in the stratum corneum have been found to have antimicrobial activities and include free sphingosines, dihydrosphingosines and 6-hydroxysphingosines [49, 82–85].

5.5 Antimicrobial Activity of Lipids and Their Mechanisms of Killing

Lipids are necessary for microbial growth, microbial membrane integrity and microbial energy storage. The glycerol moiety and fatty acid components of phosphoglycerides are used for microbial catabolism. Of course, all microorganisms have polar lipids in their cytoplasmic membranes (Gram-positive and Gram-negative bacteria) and the inner leaflet of their outer membrane (Gram-negative bacteria) [86]. The cytoplasmic membrane is an enclosed lipid bilayer surrounding the cytoplasm and serves as a permeability barrier. It contains enzyme and transporter proteins for protein export and DNA replication. Depending upon the bacterial species, the cell membrane will contain phosphatidylglycerol, phosphatidylethanolamine, lyso-cardiolipin, cardiolipin and in some species high levels of phosphatidylserine [87]. Often, parts of these membranes are released into the extracellular environments as small membrane vesicles that can fuse with other prokaryotic or eukaryotic membrane surfaces [88]. Microorganisms also contain lipid inclusions and microcompartments of varying shapes and compositions. Triacylglycerol inclusions and neutral storage-lipid inclusions are just two examples [89, 90].

What differentiates the growth-promoting effects of lipids from the antimicrobial effects of lipids for bacteria is not known. However, a number of recent studies have provided some insight. First, the importance of lipids in innate immunity can be inferred from case reports of individuals with lipid deficiencies who have skin infections. Second, the importance of lipids in innate immunity can be tested by determining the minimal inhibitory concentrations of lipids against pathogenic bacteria capable of causing skin infections.

5.5.1   Lipid Deficiencies

In atopic dermatitis, individuals have diminished ceramide concentrations in the stratum corneum [91]. Levels of sphingosine and covalently bound ω-hydroxyceramides in the cornified envelope are reduced [92, 93]. Patients with atopic dermatitis are also colonized by *S. aureus* [93]. Since free ceramides and covalently bound ω-hydroxyceramides are likely sources of long-chain bases, levels of ceramides in the stratum corneum will result in lower levels of free sphingosine [91]. Similarly, decreased levels of sapienic acid (C16 : 1Δ6) in the skin may be associated with vulnerability of the stratum corneum of patients with atopic dermatitis to colonization by *S. aureus* [94]. Here, sapienic acid may be involved in the defence mechanism in healthy skin and this deficit could potentially trigger increased susceptibility of the skin to colonization by *S. aureus*.

The importance of lipids in innate immunity can also be shown using animal models; such studies suggest the existence of a regulated, lipid-based antimicrobial effector pathway. C57BL/6 mice with a recessive germ-line mutation in the allele of the stearoyl coenzyme A desaturase 1 gene (Scd1) are susceptible to *S. pyogenes* and *S. aureus* skin infections [95]. They produce lower amounts of sebum and cannot synthesize the monounsaturated fatty acids palmitoleate (C16 : 1) and oleate (C18 : 1). Interestingly, intradermal administration of monounsaturated fatty acids to *S. aureus*-infected mice partially rescues the phenotype and indicates that an additional component of the sebum may be required to improve bacterial clearance.

5.5.2   Lipid Antimicrobial Activity and Mechanism of Action

Nonpolar epidermal lipids, sebaceous-gland lipids and sebaceous-gland lipid byproducts all have varying degrees of antimicrobial activity [49, 50]. Although antimicrobial activity is both lipid-specific and microorganism-specific, there appear to be some interesting trends

Table 5.6   *Reported minimum inhibitory concentrations of sphingolipids and fatty acids for Gram-positive bacteria, Gram-negative bacteria and yeasts.*

Microorganism	Sphingosine	Dihydro-sphingosine	Phyto-sphingosine	Sapienic acid	Lauric acid
Gram-positive cocci					
S. aureus	2[a]	8–60[a]	200[b] 2[a]	30[a]	1[c]
S. epidermidis					4[c]
MRSA	4–5[a]	8–60[a]	2–3[a]		400[e]
M. luteus			200[b]		
Gram-positive bacilli					
P. acnes			200[b]		4[c]
C. xerosis			160[b]		
Gram-negative bacilli					
E. coli	42[a]	>50[a]	400[b]		
P. aeruginosa			100[b]		
Mycobacteria					
11 species including M. avium					6–25[d]
Yeasts					
C. albicans	6–18[a]	100[a]	6 12[a]		

[a][49].
[b]Creative development (Cosmetics) Ltd. UK (http://www.creative-developments.co.uk/papers/APDs%201999.htm). Complete growth inhibition within one hour.
[c][99].
[d][124].
[e][125].

when minimum inhibitory concentrations and ultrastructural damage are examined and compared (Table 5.6).

5.5.2.1   Sphingolipids

Sphingosine and phytosphingosine have potent antimicrobial activity. Dihydrosphingosine, although active, is somewhat less potent in its antimicrobial activity. For example, free sphingosine bases are antimicrobial for Gram-positive bacteria [84, 85]. Sphinganine, sphingosine, dimethylsphingosine, phytosphingosine and stearylamine are antimicrobial for *C. albicans*, while dihydrosphingosine is not [82].

Sphingosine disrupts the bacterial cell [82]. Sphinganine-treated *S. aureus* has multiple lesions in the cell wall, evaginations in the plasma membrane and a loss of ribosomes in the cytoplasm. The cell-wall lesions may be sequelae of the affected plasma membrane.

Sphingosine and dihydrosphingosine also interfere with cell-wall synthesis [82]. Glycerol monolaurate interferes with signal-transduction pathways, inhibits the synthesis of beta-lactamase, toxins and exo-proteins of *S. aureus*, and inhibits induction of vancomycin resistance, but not of erythromycin-inducible macrolide resistance [96–98].

5.5.2.2 *Fatty acids*

The fatty acids in cutaneous lipids, and of interest to this chapter, are lauric acid and sapienic acid [49, 80, 99]. Lauric acid is antimicrobial for *P. acnes*, *S. aureus*, *C. albicans* and *S. epidermidis* but not for Group A *Streptococcus* or Group B *Streptococcus* [99]. Sapienic acid is antimicrobial for *S. aureus*.

Although not tested directly, many of the lipids found in meibum have reported antimicrobial activity. Oleamide has a structural similarity to oleic acid [100]. Oleic acid has antimicrobial activity, particularly against *Streptococcus* spp. [50], and it is interesting to speculate that oleamide would also have similar antimicrobial activity. Lipids also contain myristic, palmitic, stearic and oleic free fatty acids and myristamide, palmitamide, stearamide and erucamide fatty acid amides [71]. Myristic fatty acid (C14 : 0) is antimicrobial against some Gram-negative bacteria [50].

Fatty acids do not disrupt the integrity of the bacterial or fungal cell. Often there are no visible effects on bacterial cell walls in either scanning electron microscopy or thin sections examined by transmission electron microscopy. Rather, the site of action appears to be the plasma membrane, which is often partially dissolved or missing. For example, Group B *Streptococcus* treated with 10 mM monocaprin for 30 minutes are killed by disintegration of the cell membrane, leaving the bacterial cell wall intact [101]. The plasma membrane and electron-transparent granules are gone. Interestingly, there are no visible effects of monocaprin on the bacterial cell wall directly. No changes can be seen in either scanning electron microscopy or thin sections examined by transmission electron microscopy. Similarly, *C. albicans* treated with capric acid (C10 : 0) has a disrupted or disintegrated plasma membrane with a disorganized and shrunken cytoplasm [80]. Again, no visible changes are seen in either the shape or the size of the cell wall. Whether there is a general fluidizing effect resulting in leakage of cellular contents or a more specific interaction with membrane components is not yet known.

5.6 Synergy of Cutaneous Lipids and Other Innate Immune Molecules

Antimicrobial lipids act synergistically with other permeability enhancers and cationic antimicrobial peptides [102, 103]. Ethanol is a permeability enhancer that increases fluidity of the cell membrane. It likely enhances the partitioning of lipids into the bacterial membrane or facilitates diffusion of lipids into the microbial cytoplasm. For example, 10^8 CFU/ml methicillin-resistant *S. aureus* are resistant to killing by solutions of sapienic acid (100 μg/ml) or 15% ethanol for over 35 minutes. However, methicillin-resistant *S. aureus* are completely killed in a solution containing both sapienic acid (100 μg/ml) and 15% ethanol in as little as five minutes [49].

Synergy also occurs among sphingolipids and LL-37. LL-37 can permeate membranes and induce pore formation/membrane disintegration and inhibit bacterial macromolecular synthesis, especially RNA and protein synthesis, without binding to microbial DNA or RNA [42]. Minimal inhibitory concentrations are significantly lower for *E. coli*, *S. aureus*, methicillin-resistant *S. aureus* and *C. albicans* when a subinhibitory concentration of LL-37 is added to sphingosine [103].

Synergy also occurs among histone H4, a peptide extracted from human sebocytes, and free fatty acids from human sebum [41]. Recombinant histone H4 has antimicrobial activity against *S. aureus* and *P. acnes*. However, histone H4 enhances the antimicrobial action of free fatty acids in human sebum.

5.7 Lipids as Therapeutic Agents

The high antimicrobial activity and low toxicity of lipids suggest that they may have applications as therapies for the treatment and control of a wide variety of skin infections and inflammatory disorders. For example, phytosphingosine is an ideal candidate for treating acne vulgaris [104, 105]. Phytosphingosine is antimicrobial for *P. acnes in vitro* and downregulates the proinflammatory chemokines IL-8, CXCL2 and endothelin-1 in primary human keratinocytes. It reduces the release of both lactate dehydrogenase and interleukin 1α in response to sodium dodecyl sulfate, is antiinflammatory when tested in an organotypic skin model and enhances the resolution of acne when applied topically in combination with benzoyl peroxide.

The high antimicrobial activity of lipids also suggests that they may have applications in personal hygiene as deodorants. A variety of commensal microorganisms break down axillary secretions containing dehydroepiandrosterone, androsterone sulfates, androstenol sulfates and glycosides to byproducts, which contribute to body malodour [106]. Therefore, use of antimicrobial lipids in antiperspirants would inhibit the growth of these microorganisms and significantly reduce the amount of body malodour.

Fatty acids in general have applications as therapies for the treatment and control of a wide variety of skin infections by cutaneous pathogens. The list is long [49,80,101]. For example, caprylic acid (C8 : 0), myristic acid (C14 : 0), oleic acid (C18 : 1), monocaprylin (C8 : 0), monocaprin (C10 : 0), monolaurin (C12 : 0), monomyristin (C14 : 0), monopalmitolein (C16 : 1) and monoolein (C18 : 1) are all antimicrobial for Group A *Streptococcus*, Group B *Streptococcus* and *S. aureus*. Capric acid (C10 : 0) is antimicrobial for Group A and Group B *Streptococcus*; lauric acid (C12:0), palmitoleic acid (C16:1) and sapienic acid (C16:1Δ6) are antimicrobial for *S. aureus*; and capric acid and lauric acid are antimicrobial for *C. albicans*.

Lauric acid has promise as a potential therapeutic for the treatment of acne [99]. Minimal inhibitory concentrations of lauric acid are over 15 times lower than those of benzoyl peroxide, indicating that it has stronger antimicrobial properties than those of benzoyl peroxide. In addition, lauric acid is not cytotoxic *in vitro* to human sebocytes or *in vivo* after intradermal injection and epicutaneous application on mouse ears. Interestingly, application of lauric acid effectively decreases the number of *P. acnes* colonized on mouse ears, relieving both *P. acnes*-induced swelling and granulomatous inflammation.

5.8 Conclusions

The skin contains a complex mixture of lipids that contribute to localized innate immunity. These lipids have antimicrobial activity when isolated from tissues and tested against

Gram-positive bacteria, Gram-negative bacteria and yeasts. The spectrum of antimicrobial activity of lipids is currently under investigation by a variety of research groups for a variety of pathogens. The site of lipid antimicrobial activity appears to be the microbial plasma membrane but this needs to be confirmed. Ultimately, lipids may serve as therapeutics to treat a wide variety of cutaneous diseases and inflammatory disorders. Further investigation will provide a variety of new therapeutic avenues for treatment and prevention of cutaneous infection and inflammation. Similarly, the tear film contains a complex mixture of lipids that may also have antimicrobial activity. Like skin lipids, they may function individually or synergistically with other tear-film components to provide an antimicrobial barrier at the ocular surface.

Acknowledgments

We thank Patricia J. Conrad for assistance with the figure. This work was supported by funds from R01 DE014390 from the National Institute of Dental and Craniofacial Research, National Institutes of Health.

References

1. Proksch, E., Brandner, J.M. and Jensen, J.M. (2008) The skin: an indispensable barrier. *Exp. Dermatol.*, **17**, 1063–1072.
2. Lee, S.H., Jeong, S.K. and Ahn, S.K. (2006) An update of the defensive barrier function of skin. *Yonsei Med. J.*, **47**, 293–306.
3. Cameron, D.J., Tong, Z., Yang, Z. *et al.* (2007) Essential role of Elovl4 in very long chain fatty acid synthesis, skin permeability barrier function, and neonatal survival. *Int. J. Biol. Sci.*, **3**, 111–119.
4. Schmid-Wendtner, M.H. and Korting, H.C. (2006) The pH of the skin surface and its impact on the barrier function. *Skin Pharmacol. Physiol.*, **19**, 296–302.
5. Levin, J. and Maibach, H. (2008) Human skin buffering capacity: an overview. *Skin Res. Technol.*, **14**, 121–126.
6. Elias, P.M. (2007) The skin barrier as an innate immune element. *Semin. Immunopathol.*, **29**, 3–14.
7. Jungersted, J.M., Hellgren, L.I., Jemec, G.B. and Agner, T. (2008) Lipids and skin barrier function: a clinical perspective. *Contact Dermatitis*, **58**, 255–262.
8. Smith, K.R. and Thiboutot, D.M. (2008) Thematic review series: skin lipids. Sebaceous gland lipids: friend or foe? *J. Lipid Res.*, **49**, 271–281.
9. Zouboulis, C.C., Baron, J.M., Bohm, M. *et al.* (2008) Frontiers in sebaceous gland biology and pathology. *Exp. Dermatol.*, **17**, 542–551.
10. Zouboulis, C.C. (2004) Acne and sebaceous gland function. *Clin. Dermatol.*, **22**, 360–366.
11. Robinson, P.G. (1982) Common names and abbreviated formulae for fatty acids. *J. Lipid Res.*, **23**, 1251–1253.
12. Noble, W.C. (ed.) (1992) *The skin microflora and microbial skin disease*, Cambridge University Press, Cambridge, UK.
13. Cogen, A.L., Nizet, V. and Gallo, R.L. (2008) Skin microbiota: a source of disease or defence? *Brit. J. Dermatol.*, **158**, 442–455.
14. Tannock, G.W. (1995) *Normal Microflora: An Introduction to Microbes Inhabiting the Human Body*, Chapman & Hall, London, UK.

15. Tannock, G.W. (ed.) (1999) *Medical Importance of the Normal Microflora*, Kluwer Academic Publishers, London, UK.
16. Fredricks, D.N. (2001) Microbial ecology of human skin in health and disease. *J. Investig. Dermatol. Symp. Proc.*, **6**, 167–169.
17. Hochedez, P. and Caumes, E. (2008) Common skin infections in travelers. *J. Travel Med.*, **15**, 252–262.
18. Bernard, P. (2008) Management of common bacterial infections of the skin. *Curr. Opin. Infect. Dis.*, **21**, 122–128.
19. Harries, M.J. and Lear, J.T. (2004) Occupational skin infections. *Occup. Med. (Lond)*, **54**, 441–449.
20. Bayston, R. and Higgins, J. (1986) Biochemical and cultural characteristics of 'JK' coryneforms. *J. Clin. Pathol.*, **39**, 654–660.
21. Blaise, G., Nikkels, A.F., Hermanns-Le, T. *et al.* (2008) Corynebacterium-associated skin infections. *Int. J. Dermatol.*, **47**, 884–890.
22. Coenye, T., Honraet, K., Rossel, B. and Nelis, H.J. (2008) Biofilms in skin infections: *Propionibacterium acnes* and acne vulgaris. *Infect. Disord. Drug Targets*, **8**, 156–159.
23. Noble, W.C. and Savin, J.A. (1971) 'Gram-negative' infections of the skin. *Brit. J. Dermatol.*, **85**, 286–289.
24. Sebeny, P.J., Riddle, M.S. and Petersen, K. (2008) *Acinetobacter baumannii* skin and soft-tissue infection associated with war trauma. *Clin. Infect. Dis.*, **47**, 444–449.
25. Bhambri, S., Bhambri, A. and Del Rosso, J.Q. (2009) Atypical mycobacterial cutaneous infections. *Dermatol. Clin.*, **27**, 63–73.
26. Connolly, S.E. and Benach, J.L. (2005) The versatile roles of antibodies in Borrelia infections. *Nat. Rev. Microbiol.*, **3**, 411–420.
27. Hull, M.W. and Chow, A.W. (2007) Indigenous microflora and innate immunity of the head and neck. *Infect. Dis. Clin. North Am.*, **21**, 265–282.
28. Radek, K. and Gallo, R. (2007) Antimicrobial peptides: natural effectors of the innate immune system. *Semin. Immunopathol.*, **29**, 27–43.
29. Elias, P.M. (2005) Stratum corneum defensive functions: an integrated view. *J. Invest. Dermatol.*, **125**, 183–200.
30. Schauber, J. and Gallo, R.L. (2008) Antimicrobial peptides and the skin immune defense system. *J. Allergy Clin. Immunol.*, **122**, 261–266.
31. Brogden, K.A. (2005) Antimicrobial peptides: pore formers or metabolic inhibitors in bacteria? *Nat. Rev. Microbiol.*, **3**, 238–250.
32. Brown, K.L. and Hancock, R.E. (2006) Cationic host defense (antimicrobial) peptides. *Curr. Opin. Immunol.*, **18**, 24–30.
33. Ganz, T. (2004) Antimicrobial polypeptides. *J. Leukoc. Biol.*, **75**, 34–38.
34. Lehrer, R.I. (2004) Primate defensins. *Nat. Rev. Microbiol.*, **2**, 727–738.
35. Yamasaki, K. and Gallo, R.L. (2008) Antimicrobial peptides in human skin disease. *Eur. J. Dermatol.*, **18**, 11–21.
36. Braff, M.H. and Gallo, R.L. (2006) Antimicrobial peptides: an essential component of the skin defensive barrier. *Curr. Top. Microbiol. Immunol.*, **306**, 91–110.
37. Braff, M.H., Bardan, A., Nizet, V. and Gallo, R.L. (2005) Cutaneous defense mechanisms by antimicrobial peptides. *J. Invest. Dermatol.*, **125**, 9–13.
38. Niyonsaba, F. and Ogawa, H. (2005) Protective roles of the skin against infection: implication of naturally occurring human antimicrobial agents beta-defensins, cathelicidin LL-37 and lysozyme. *J. Dermatol. Sci.*, **40**, 157–168.
39. Stoeckelhuber, M., Matthias, C., Andratschke, M. *et al.* (2006) Human ceruminous gland: ultrastructure and histochemical analysis of antimicrobial and cytoskeletal components. *Anat. Rec. A. Discov. Mol. Cell Evol. Biol.*, **288**, 877–884.
40. Buchau, A.S., Hassan, M., Kukova, G. *et al.* (2007) S100A15, an antimicrobial protein of the skin: regulation by *E. coli* through Toll-like receptor 4. *J. Invest. Dermatol.*, **127**, 2596–2604.
41. Lee, D.Y., Huang, C.M., Nakatsuji, T. *et al.* (2009) Histone H4 Is a major component of the antimicrobial action of human sebocytes. *J. Invest. Dermatol.*, **129**, 2489–2496.

42. Senyurek, I., Paulmann, M., Sinnberg, T. *et al.* (2009) Dermcidin-derived peptides show a different mode of action than the cathelicidin LL-37 against *Staphylococcus aureus*. *Antimicrob. Agents Chemother.*, **53**, 2499–2509.
43. Chen, X., Niyonsaba, F., Ushio, H. *et al.* (2005) Synergistic effect of antibacterial agents human beta-defensins, cathelicidin LL-37 and lysozyme against *Staphylococcus aureus* and *Escherichia coli*. *J. Dermatol. Sci.*, **40**, 123–132.
44. Ong, P.Y., Ohtake, T., Brandt, C. *et al.* (2002) Endogenous antimicrobial peptides and skin infections in atopic dermatitis. *N. Engl. J. Med.*, **347**, 1151–1160.
45. Yamasaki, K., Di Nardo, A., Bardan, A. *et al.* (2007) Increased serine protease activity and cathelicidin promotes skin inflammation in rosacea. *Nat. Med.*, **13**, 975–980.
46. Hata, T.R. and Gallo, R.L. (2008) Antimicrobial peptides, skin infections, and atopic dermatitis. *Semin. Cutan. Med. Surg.*, **27**, 144–150.
47. Yamasaki, K. and Gallo, R.L. (2009) The molecular pathology of rosacea. *J. Dermatol. Sci.*, **55**, 77–81.
48. Dorschner, R.A., Pestonjamasp, V.K., Tamakuwala, S. *et al.* (2001) Cutaneous injury induces the release of cathelicidin anti-microbial peptides active against group A Streptococcus. *J. Invest. Dermatol.*, **117**, 91–97.
49. Drake, D.R., Brogden, K.A., Dawson, D.V. and Wertz, P.W. (2008) Thematic review series: skin lipids. Antimicrobial lipids at the skin surface. *J. Lipid Res.*, **49**, 4–11.
50. Thormar, H. and Hilmarsson, H. (2007) The role of microbicidal lipids in host defense against pathogens and their potential as therapeutic agents. *Chem. Phys. Lipids*, **150**, 1–11.
51. Freinkel, R.K. and Traczyk, T.N. (1985) Lipid composition and acid hydrolase content of lamellar granules of fetal rat epidermis. *J. Invest. Dermatol.*, **85**, 295–298.
52. Gray, G.M. and Yardley, H.J. (1975) Different populations of pig epidermal cells: isolation and lipid composition. *J. Lipid Res.*, **16**, 441–447.
53. Landmann, L. (1988) The epidermal permeability barrier. *Anat Embryol (Berl)*, **178**, 1–13.
54. Madison, K.C., Sando, G.N., Howard, E.J. *et al.* (1998) Lamellar granule biogenesis: a role for ceramide glucosyltransferase, lysosomal enzyme transport, and the Golgi. *J. Investig. Dermatol. Symp. Proc.*, **3**, 80–86.
55. Squier, C.A., Wertz, P.W. and Cox, P. (1991) Thin-layer chromatographic analyses of lipids in different layers of porcine epidermis and oral epithelium. *Arch. Oral Biol.*, **36**, 647–653.
56. Wertz, P.W., Downing, D.T., Freinkel, R.K. and Traczyk, T.N. (1984) Sphingolipids of the stratum corneum and lamellar granules of fetal rat epidermis. *J. Invest. Dermatol.*, **83**, 193–195.
57. Wertz, P.W. and Downing, D.T. (1990) Ceramidase activity in porcine epidermis. *FEBS Lett.*, **268**, 110–112.
58. Yada, Y., Higuchi, K. and Imokawa, G. (1995) Purification and biochemical characterization of membrane-bound epidermal ceramidases from guinea pig skin. *J. Biol. Chem.*, **270**, 12677–12684.
59. Law, S., Wertz, P.W., Swartzendruber, D.C. and Squier, C.A. (1995) Regional variation in content, composition and organization of porcine epithelial barrier lipids revealed by thin-layer chromatography and transmission electron microscopy. *Arch. Oral Biol.*, **40**, 1085–1091.
60. Ponec, M., Weerheim, A., Lankhorst, P. and Wertz, P. (2003) New acylceramide in native and reconstructed epidermis. *J. Invest. Dermatol.*, **120**, 581–588.
61. Wertz, P.W. and Downing, D.T. (1987) Covalently bound omega-hydroxyacylsphingosine in the stratum corneum. *Biochim. Biophys. Acta*, **917**, 108–111.
62. Wertz, P.W. and Downing, D.T. (1989) Free sphingosines in porcine epidermis. *Biochim. Biophys. Acta*, **1002**, 213–217.
63. Swartzendruber, D.C., Wertz, P.W., Madison, K.C. and Downing, D.T. (1987) Evidence that the corneocyte has a chemically bound lipid envelope. *J. Invest. Dermatol.*, **88**, 709–713.
64. Weitkamp, A.W., Smiljanic, A.M. and Rothman, S. (1947) The free fatty acids of human hair fat. *J. Am. Chem. Soc.*, **69**, 1936–1939.
65. Ohashi, Y., Dogru, M. and Tsubota, K. (2006) Laboratory findings in tear fluid analysis. *Clin. Chim. Acta*, **369**, 17–28.
66. Davidson, H.J. and Kuonen, V.J. (2004) The tear film and ocular mucins. *Vet. Ophthalmol.*, **7**, 71–77.

67. Tiffany, J.M. (2008) The normal tear film. *Dev. Ophthalmol.*, **41**, 1–20.
68. Bron, A.J., Tiffany, J.M., Gouveia, S.M. *et al.* (2004) Functional aspects of the tear film lipid layer. *Exp. Eye Res.*, **78**, 347–360.
69. Butovich, I.A. (2008) On the lipid composition of human meibum and tears: comparative analysis of nonpolar lipids. *Invest. Ophthalmol. Vis. Sci.*, **49**, 3779–3789.
70. Huang, L.C., Jean, D., Proske, R.J. *et al.* (2007) Ocular surface expression and in vitro activity of antimicrobial peptides. *Curr. Eye Res.*, **32**, 595–609.
71. Nichols, K.K., Ham, B.M., Nichols, J.J. *et al.* (2007) Identification of fatty acids and fatty acid amides in human meibomian gland secretions. *Invest. Ophthalmol. Vis. Sci.*, **48**, 34–39.
72. Shine, W.E. and McCulley, J.P. (2003) Polar lipids in human meibomian gland secretions. *Curr. Eye Res.*, **26**, 89–94.
73. Nicolaides, N., Santos, E.C., Papadakis, K. *et al.* (1984) The occurrence of long chain alpha, omega-diols in the lipids of steer and human meibomian glands. *Lipids*, **19**, 990–993.
74. Nicolaides, N., Kaitaranta, J.K., Rawdah, T.N. *et al.* (1981) Meibomian gland studies: comparison of steer and human lipids. *Invest. Ophthalmol. Vis. Sci.*, **20**, 522–536.
75. Butovich, I.A., Uchiyama, E. and McCulley, J.P. (2007) Lipids of human meibum: mass-spectrometric analysis and structural elucidation. *J. Lipid Res.*, **48**, 2220–2235.
76. Butovich, I.A. (2009) Cholesteryl esters as a depot for very long chain fatty acids in human meibum. *J. Lipid Res.*, **50**, 501–513.
77. Li, W., Sandhoff, R., Kono, M. *et al.* (2007) Depletion of ceramides with very long chain fatty acids causes defective skin permeability barrier function, and neonatal lethality in ELOVL4 deficient mice. *Int. J. Biol. Sci.*, **3**, 120–128.
78. Bayliss, M. (1936) Effect of the chemical constitution of soaps upon their germicidal properties. *J. Bacteriol.*, **31**, 489–504.
79. Kabara, J.J. and Vrable, R. (1977) Antimicrobial lipids: natural and synthetic fatty acids and monoglycerides. *Lipids*, **12**, 753–759.
80. Bergsson, G., Arnfinnsson, J., Steingrimsson, O. and Thormar, H. (2001) In vitro killing of *Candida albicans* by fatty acids and monoglycerides. *Antimicrob. Agents Chemother.*, **45**, 3209–3212.
81. Burtenshaw, J.M. (1942) The mechanisms of self disinfection of the human skin and its appendages. *J. Hyg. (Lond.)*, **42**, 184–209.
82. Bibel, D.J., Aly, R., Shah, S. and Shinefield, H.R. (1993) Sphingosines: antimicrobial barriers of the skin. *Acta Derm. Venereol.*, **73**, 407–411.
83. Bibel, D.J., Aly, R. and Shinefield, H.R. (1992) Antimicrobial activity of sphingosines. *J. Invest. Dermatol.*, **98**, 269–273.
84. Bibel, D.J., Aly, R. and Shinefield, H.R. (1995) Topical sphingolipids in antisepsis and antifungal therapy. *Clin. Exp. Dermatol.*, **20**, 395–400.
85. Payne, C.D., Ray, T.L. and Downing, D.T. (1996) Cholesterol sulfate protects *Candida albicans* from inhibition by sphingosine in vitro. *J. Invest. Dermatol.*, **106**, 549–552.
86. Brogden, K.A. (2009) Cytopathology of pathogenic prokaryotes, in *Ultrastructural Pathology: The Cellular Basis of Disease*, 2nd edn (ed. N. Cheville), Blackwell Publishers, Ames, IA, USA, pp. 425–524.
87. Verkley, A.J., Ververgaert, P.H., Prins, R.A. and van Golde, L.M. (1975) Lipid-phase transitions of the strictly anaerobic bacteria *Veillonella parvula* and *Anaerovibrio lipolytica*. *J. Bacteriol.*, **124**, 1522–1528.
88. Beveridge, T.J. (1999) Structures of gram-negative cell walls and their derived membrane vesicles. *J. Bacteriol.*, **181**, 4725–4733.
89. Alvarez, H.M. and Steinbuchel, A. (2002) Triacylglycerols in prokaryotic microorganisms. *Appl. Microbiol. Biotechnol.*, **60**, 367–376.
90. Kalscheuer, R., Stoveken, T., Malkus, U. *et al.* (2007) Analysis of storage lipid accumulation in *Alcanivorax borkumensis*: evidence for alternative triacylglycerol biosynthesis routes in bacteria. *J. Bacteriol.*, **189**, 918–928.
91. Imokawa, G. (2001) Lipid abnormalities in atopic dermatitis. *J. Am. Acad. Dermato.l*, **45**, S29–32.

92. Macheleidt, O., Kaiser, H.W. and Sandhoff, K. (2002) Deficiency of epidermal protein-bound omega-hydroxyceramides in atopic dermatitis. *J. Invest. Dermatol.*, **119**, 166–173.

93. Arikawa, J., Ishibashi, M., Kawashima, M. *et al.* (2002) Decreased levels of sphingosine, a natural antimicrobial agent, may be associated with vulnerability of the stratum corneum from patients with atopic dermatitis to colonization by *Staphylococcus aureus*. *J. Invest. Dermatol.*, **119**, 433–439.

94. Takigawa, H., Nakagawa, H., Kuzukawa, M. *et al.* (2005) Deficient production of hexadecenoic acid in the skin is associated in part with the vulnerability of atopic dermatitis patients to colonization by *Staphylococcus aureus*. *Dermatology*, **211**, 240–248.

95. Georgel, P., Crozat, K., Lauth, X. *et al.* (2005) A toll-like receptor 2-responsive lipid effector pathway protects mammals against skin infections with gram-positive bacteria. *Infect. Immun.*, **73**, 4512–4521.

96. Ruzin, A. and Novick, R.P. (2000) Equivalence of lauric acid and glycerol monolaurate as inhibitors of signal transduction in *Staphylococcus aureus*. *J. Bacteriol.*, **182**, 2668–2671.

97. Ruzin, A. and Novick, R.P. (1998) Glycerol monolaurate inhibits induction of vancomycin resistance in *Enterococcus faecalis*. *J. Bacteriol.*, **180**, 182–185.

98. Projan, S.J., Brown-Skrobot, S., Schlievert, P.M. *et al.* (1994) Glycerol monolaurate inhibits the production of beta-lactamase, toxic shock toxin-1, and other staphylococcal exoproteins by interfering with signal transduction. *J. Bacteriol.*, **176**, 4204–4209.

99. Nakatsuji, T., Kao, M.C., Fang, J.Y. *et al.* (2009) Antimicrobial property of lauric acid against *Propionibacterium acnes*: its therapeutic potential for inflammatory acne vulgaris. *J. Invest. Dermatol.*, **129**, 2480–2488.

100. Nojima, H., Ohba, Y. and Kita, Y. (2007) Oleamide derivatives are prototypical anti-metastasis drugs that act by inhibiting Connexin 26. *Curr. Drug Saf.*, **2**, 204–211.

101. Bergsson, G., Arnfinnsson, J., Steingrimsson, O. and Thormar, H. (2001) Killing of Gram-positive cocci by fatty acids and monoglycerides. *APMIS*, **109**, 670–678.

102. Robertson, E., Brogden, K., Wertz, P. *et al.* (2005) Antimicrobial activity of human skin lipids, Abstr. 3452. 83rd General Session IADR meeting, Baltimore, MD, USA.

103. Robertson, E., Burnell, K., Qian, F. *et al.* (2006) Synergistic activity of human skin lipids and LL-37, Abstr. 2113. 84th General Session IADR meeting, Orlando, FL, USA.

104. Klee, S.K., Farwick, M. and Lersch, P. (2007) The effect of sphingolipids as a new therapeutic option for acne treatment. *Basic Clin. Dermatol*, **40**, 155–156.

105. Pavicic, T., Wollenweber, U., Farwick, M. and Korting, H.C. (2007) Anti-microbial and -inflammatory activity and efficacy of phytosphingosine: an in vitro and in vivo study addressing acne vulgaris. *Int. J. Cosmet. Sci.*, **29**, 181–190.

106. Labows, J., Reilly, J., Leyden, J. and Preti, G. (1999) Axillary odor determination, formulation, and control, in *Antiperspirants and Deodorants*, 2nd edn (ed. K. Laden), Marcel Dekker, New York, NY, USA, pp. 59–82.

107. Granato, P.A. (2003) Pathogenic and indigenous microorganisms of humans, in *Manual of Clinical Microbiology. 1*, (eds P.R. Murray, E.J. Baron, J.H. Jorgensen *et al.*), ASM Press, Washington, DC, USA, pp. 44–54.

108. Oncul, O., Ozsoy, M.F., Gul, H.C. *et al.* (2002) Cutaneous anthrax in Turkey: a review of 32 cases. *Scand. J. Infect. Dis.*, **34**, 413–416.

109. Janda, W.M., Tipirneni, P. and Novak, R.M. (2003) *Brevibacterium casei* bacteremia and line sepsis in a patient with AIDS. *J. Infect.*, **46**, 61–64.

110. Hojyo-Tomoka, M.T., Marples, R.R. and Kligman, A.M. (1973) Pseudomonas infection in superhydrated skin. *Arch. Dermatol.*, **107**, 723–727.

111. Niyonsaba, F., Ushio, H., Nagaoka, I. *et al.* (2005) The human beta-defensins (-1, -2, -3, -4) and cathelicidin LL-37 induce IL-18 secretion through p38 and ERK MAPK activation in primary human keratinocytes. *J. Immunol.*, **175**, 1776–1784.

112. Niyonsaba, F., Ushio, H., Nakano, N. *et al.* (2007) Antimicrobial peptides human beta-defensins stimulate epidermal keratinocyte migration, proliferation and production of proinflammatory cytokines and chemokines. *J. Invest. Dermatol.*, **127**, 594–604.

113. Niyonsaba, F., Iwabuchi, K., Matsuda, H. *et al.* (2002) Epithelial cell-derived human beta-defensin-2 acts as a chemotaxin for mast cells through a pertussis toxin-sensitive and phospholipase C-dependent pathway. *Int. Immunol.*, **14**, 421–426.

114. Niyonsaba, F., Someya, A., Hirata, M. *et al.* (2001) Evaluation of the effects of peptide antibiotics human beta-defensins-1/-2 and LL-37 on histamine release and prostaglandin D(2) production from mast cells. *Eur. J. Immunol.*, **31**, 1066–1075.
115. Chen, X., Niyonsaba, F., Ushio, H. *et al.* (2007) Antimicrobial peptides human beta-defensin (hBD)-3 and hBD-4 activate mast cells and increase skin vascular permeability. *Eur. J. Immunol.*, **37**, 434–444.
116. Zheng, Y., Niyonsaba, F., Ushio, H. *et al.* (2007) Cathelicidin LL-37 induces the generation of reactive oxygen species and release of human alpha-defensins from neutrophils. *Brit. J. Dermatol.*, **157**, 1124–1131.
117. Niyonsaba, F., Iwabuchi, K., Someya, A. *et al.* (2002) A cathelicidin family of human antibacterial peptide LL-37 induces mast cell chemotaxis. *Immunology*, **106**, 20–26.
118. Niyonsaba, F., Suzuki, A., Ushio, H. *et al.* (2009) The human antimicrobial peptide dermcidin activates normal human keratinocytes. *Brit. J. Dermatol.*, **160**, 243–249.
119. Allaker, R.P. and Kapas, S. (2003) Adrenomedullin and mucosal defence: interaction between host and microorganism. *Regul. Pept.*, **112**, 147–152.
120. Niyonsaba, F., Hattori, F., Maeyama, K. *et al.* (2008) Induction of a microbicidal protein psoriasin (S100A7), and its stimulatory effects on normal human keratinocytes. *J. Dermatol. Sci.*, **52**, 216–219.
121. Zheng, Y., Niyonsaba, F., Ushio, H. *et al.* (2008) Microbicidal protein psoriasin is a multifunctional modulator of neutrophil activation. *Immunology*, **124**, 357–367.
122. Grayson, S., Johnson-Winegar, A.G., Wintroub, B.U. *et al.* (1985) Lamellar body-enriched fractions from neonatal mice: preparative techniques and partial characterization. *J. Invest. Dermatol.*, **85**, 289–294.
123. Stewart, M.E. and Downing, D.T. (1995) Free sphingosines of human skin include 6-hydroxysphingosine and unusually long-chain dihydrosphingosines. *J. Invest. Dermatol.*, **105**, 613–618.
124. Saito, H., Tomioka, H. and Yoneyama, T. (1984) Growth of group IV mycobacteria on medium containing various saturated and unsaturated fatty acids. *Antimicrob. Agents Chemother.*, **26**, 164–169.
125. Kitahara, T., Koyama, N., Matsuda, J. *et al.* (2004) Antimicrobial activity of saturated fatty acids and fatty amines against methicillin-resistant *Staphylococcus aureus*. *Biol. Pharm. Bull.*, **27**, 1321–1326.
126. Shirane, J., Nakayama, T., Nagakubo, D. *et al.* (2004) Corneal epithelial cells and stromal keratocytes efficently produce CC chemokine-ligand 20 (CCL20) and attract cells expressing its receptor CCR6 in mouse herpetic stromal keratitis. *Curr. Eye Res.*, **28**, 297–306.
127. McDermott, A.M., Redfern, R.L. and Zhang, B. (2001) Human beta-defensin 2 is up-regulated during re-epithelialization of the cornea. *Curr. Eye Res.*, **22**, 64–67.
128. McDermott, A.M., Redfern, R.L., Zhang, B. *et al.* (2003) Defensin expression by the cornea: multiple signalling pathways mediate IL-1beta stimulation of hBD-2 expression by human corneal epithelial cells. *Invest. Ophthalmol. Vis. Sci.*, **44**, 1859–1865.
129. Gordon, Y.J., Huang, L.C., Romanowski, E.G. *et al.* (2005) Human cathelicidin (LL-37), a multifunctional peptide, is expressed by ocular surface epithelia and has potent antibacterial and antiviral activity. *Curr. Eye Res.*, **30**, 385–394.
130. Spandau, U.H., Toksoy, A., Verhaart, S. *et al.* (2003) High expression of chemokines Gro-alpha (CXCL-1), IL-8 (CXCL-8), and MCP-1 (CCL-2) in inflamed human corneas in vivo. *Arch. Ophthalmol.*, **121**, 825–831.

6

Antimicrobial Lipids and Innate Immunity

Halldor Thormar

Faculty of Life and Environmental Sciences, University of Iceland, Reykjavik, Iceland

Lipids and Essential Oils as Antimicrobial Agents   Halldor Thormar
© 2011 John Wiley & Sons, Ltd

6.1   Introduction

Innate immunity is an inborn host defence mechanism which responds rapidly when a host organism is invaded by another organism, especially a microbial pathogen [1]. The aim is to destroy and eliminate the pathogen before it can cause a widespread infection. An innate immune system is found in all plants and animals and is therefore phylogenetically ancient. Its major role is to protect the host in a general way from infection by pathogenic bacteria, viruses, fungi and protozoa by production of antimicrobial molecules which either kill the pathogens or inhibit their multiplication. In vertebrates, it also constitutes the first line of host defence against invading pathogens by causing an immediate activation of phagocytic cells, which engulf and destroy the pathogen, and by initiating an inflammatory response. The primary innate immune response is later followed by the secondary response of the adaptive or acquired immune system, which can recognize an infinite number of molecular patterns (antigenic determinants) and confers a specific, long-lasting protection to the host against the invading pathogen. In comparison, the innate immune system is much less specific, with a restricted number of pathogen-associated recognition molecules and no memory of prior exposures. The adaptive immunity is mediated either by antibodies or by T lymphocytes. Similarly, innate immunity in animals is mediated both by cells and by noncellular factors. Cells of the innate immune response include nonphagocytic leukocytes, such as natural killer cells (NK cells), mast cells, eosinophils and basophils, and also phagocytic leukocytes, such as neutrophils, macrophages and dendritic cells. The best-known noncellular components of the innate immune system are proteins, such as lysozyme, lactoferrin and lactoperoxidase, as well as proteins of the alternative complement pathway, and antimicrobial peptides, such as defensins and cathelicidins. Cytokines, which are low-molecular-weight proteins, such as tumour necrosis factor-alpha, interleukins and interferons, also play a role as regulators in the innate as well as the adaptive immune system. The cells of the innate immune system carry receptors which recognize a large variety of molecular patterns on bacteria, viruses, fungi and protozoa, called pathogen-associated molecular patterns (PAMPs). These microbial PAMPs are not shared by any molecular structures on host cells and thus distinguish the pathogen from the host. The PAMPs are conserved microbial structures which are absolutely essential for the survival or pathogenicity of the microorganisms. Any change in these structures, for example by mutation, would therefore be fatal to the pathogen or eliminate its pathogenic effect [1–4]. This greatly reduces the possibility of the pathogen developing resistance to the innate immune response. The pattern-recognition receptors (PRRs) on cells of the innate immune system have broad specificity and each PRR can recognize and bind to PAMPs of many related pathogens, for example to peptidoglycans on the cell wall of Gram-positive bacteria or to lipopolysaccharides in the outer membrane of Gram-negative bacteria. Prominent among the PRRs are the Toll-like receptors (TLRs), which are present as transmembrane proteins on macrophages, dendritic cells and epithelial cells, and were first disovered in the fruit fly *Drosphila melanogaster* [5–7]. At least 12 different TLRs are known in mammals and each of these recognizes a subset of PAMPs on a large variety of pathogens.The recognition and binding of TLRs to PAMPs on invading pathogens initiates a signaling pathway which activates a number of cytokine genes. The cytokines trigger an inflammatory response, which results in accumulation of leukocytes and blood plasma in tissues at the infection site. Inflammation is an early response to infection and is critical to host defence, due to the influx of leukocytes, complement proteins, lysozyme and antimicrobial peptides [8]. The best-known antimicrobial peptides, defensins and cathelicidins, are secreted in response to pathogen-activated TLRs on leukocytes and epithelial cells in the skin and mucosal membranes. However, some defensins are secreted without PRR activation and are constitutively produced on epithelial surfaces lining the respiratory, gastrointestinal and genitourinary tracts. They therefore mount an immediate innate immune response upon pathogen invasion in the skin and mucosa. Defensins are cationic peptides ranging in length from about 20 to 50 amino acid residues. They are believed to kill pathogenic bacteria by penetrating the bacterial membranes, forming pores and leading to irreparable autolysis. Host cells seem to be protected from this effect by a difference in the electric charge of bacterial and host membranes. Thus, bacterial surfaces are negatively charged and are therefore a high-affinity target for the cationic, positively-charged peptides. On the other hand, the outer membranes of eukaryotic host cells carry no net electric charge and therefore interact much less with antimicrobial peptides.

The idea that antibacterial lipids may play a role in nonspecific host defence against bacteria goes back to the early 1900s. Thus Klotz in 1905 [9] showed that neutral fats, fatty acids and soaps – sodium or potassium salts of fatty acids – are present in inflammatory foci. In 1907 Noguchi [10] found that alcohol extracts from various organs contained alkali salts of oleic acid and that they were antibacterial. He concluded that antibacterial fatty acids and their salts in blood and organs might play a role in the natural host defence mechanism. Landsteiner and Ehrlich in 1908 [11] confirmed the antibacterial effect of lipids extracted from leukocytes and lymphoid tissues, and stated that these are mostly due to oleic acid. Like Noguchi, they suggested that lipids found in tissue extracts are a part of the complement system which causes lysis of bacteria. Lamar in 1911 [12] extracted fatty acids from human lung with resolving lobar pneumonia and suggested that the decrease in the number and virulence of pneumococci present in the exudate depended in part upon the presence of antibacterial fatty acids and their salts. This assumption was further strengthened by the finding that pneumococci treated *in vitro* with salts of fatty acids became gradually less virulent. The quantity of antibacterial substances in inflammatory foci could be considerable and was assumed to destroy the infecting bacteria and thus contribute to recovery from the infection. Flexner, in the introduction to Lamar's paper [12], stated that recovery from bacterial infections had not been fully explained by the activities of antibodies and phagocytes and suggested that antibacterial chemical substances, such as fatty acids and their salts, might contribute to the recovery from local bacterial infections.

In the following decades of the twentieth century, little attention was paid to the possible function of lipids in the natural defence of the body against bacteria. It was not until 1941 that Burtenshaw's elegant experiments [13] showed that self-disinfection of the skin was, at least partly, due to antimicrobial fatty acids. His work was confirmed by several studies of other workers and created some interest in the role of antimicrobial lipids in host defence. In particular, their presence in the skin and mucosal membranes and in human milk was actively studied from this point of view for a number of years. However, few studies have been carried out in the last few decades on the role of antimicrobial lipids in innate immunity, although recent studies indicate that they may be one of the factors in the complex multifactorial innate immune response.

6.2   The Role of Human Milk Lipids in Innate Immunity

6.2.1   Breast-Feeding Protects Infants against Infection

Observations indicating that mortality is lower in breast-fed than in nonbreast-fed infants go back to the eighteenth and nineteenth centuries [14]. However, in the early studies the high mortality rate in bottle-fed babies may have been caused, at least partly, by contaminated feedings due to poor hygiene [15]. In a more recent study, a lower morbidity rate in breast-fed compared with bottle-fed infants was reported and the health advantage of breast-feeding was still significant after controlling for socioeconomic factors [16]. In another study [15], including a large number of infants, morbidity requiring hospitalization was clearly shown to be associated with artificial feeding, after a number of other factors that might affect morbidity had been controlled for, such as educational level, maternal age, family size, day-care exposure and birth weight. The advantage of breast-feeding was most obvious during the first two months of life, where morbidity was 16-fold higher in bottle-fed than in breast-fed infants. The difference decreased with time and was fourfold during the first four months and twofold for the first year of life, even though many of the breast-feeders were partially receiving artificial feeding during the later months. It was concluded that breast-feeding protects infants more effectively against serious than against mild illnesses, both intestinal and respiratory, and that the protection increases proportionally with the extent and duration of breast-feeding, at least up to a certain age [15]. In a similar study [17], it was shown that breast-feeding during the first three months of life significantly reduced the incidence of infections requiring hospitalization. Numerous other studies have suggested that breast-fed infants are more resistant to a variety of infections, particularly intestinal and respiratory, than bottle-fed infants, especially during the early months of life [18, 19]. Thus, many observations made over a number of years have shown that breast-feeding has a profound effect on the colonization of bacteria in the gastrointestinal tract and throats in the newborn [18]. The intestinal tract is less often heavily colonized with *Escherichia coli* and other harmful bacteria in breast-fed than in bottle-fed babies and a protective effect of breast milk against enteric infections has been suggested by several observations. For example, epidemics of enterocolitis among newborns caused by pathogenic strains of *E. coli* were brought under control by administration of unprocessed breast milk [20, 21]. Similarly, a study in Guatemala showed that diarrhoea was uncommon in breast-fed infants in spite of a common exposure to enteropathogenic *E. coli*, shigella and *Salmonella*. On the other hand, diarrhoea appeared after weaning [22]. A study done in California on a middle-class population showed a significantly lower incidence of acute gastroenteritis in breast-fed compared with bottle-fed infants, only 1 of 107 hospitalized babies being breast-fed at the time of admission [23]. There is a general agreement that morbidity and mortality from gastroenteritis is much lower in breast-fed compared to bottle-fed infants in Western industrialized societies, as well as in underdeveloped countries with poor sanitation. However, babies in Third World countries may benefit more from breast-feeding, and should optimally be fed mother's milk up to the age of three years, compared with at least six months in Europe and North America [19].

A number of studies have shown that breast-fed infants are less susceptible to respiratory infections than bottle-fed babies [15, 18]. Thus, in a study conducted in Britain and controlled for factors such as social class, it was found that hospital admissions of infants with respiratory syncytial virus (RSV) infection were significantly fewer in breast-fed than in bottle-fed babies [24]. In another well-controlled British study it was shown that breast-feeding reduced the number of hospitalizations by half, compared with bottle-fed infants with RSV lower respiratory infection [25]. These, and a number of other studies reviewed by Cunningham [19], strongly support the conclusion that breast-feeding is particularly protective against serious disease of the lower respiratory tract, such as pneumonia caused by RSV, which is one of the most important respiratory pathogens in infancy. The advantages of breast-feeding are especially important where the climate or living conditions are unfavourable. They are most evident during the first six months of life but may be significant up to two years of age. Not only are respiratory infections less frequent in breast-fed babies, but the infections that occur are likely to be less severe [19].

A beneficial effect of breast-feeding on otitis media was indicated by a number of studies published in the 1950s [18] and has been confirmed in later studies [26–28]. In addition, there are several reports showing a protective effect of breast-feeding in other bacterial infections, at least during the first six months of life, for example in bacteremia and meningitis caused by invasive *Haemophilus influenzae* infection [19, 29, 30]. In a more recent study of the protective role of breast-feeding against febrile urinary tract infections it was shown that ongoing exclusive breast-feeding significantly reduced the risk of urinary tract infection [31]. The protection was strongest directly after birth, but decreased until seven months of age.

6.2.2 Factors in Human Milk which Protect against Infection

Soon after the discovery at the end of the nineteenth century of the immune system of humans and other mammals, and its many functions in host defence against infectious agents, Ehrlich and his coworkers showed that immunity against tetanus toxin in mice could be transferred to nonimmune pups by milk from immunized dams [14]. This indicated that the mammary gland is an immunological organ and laid the foundation for the understanding of the immune factors in human milk and their protective functions. In the first half of the twentieth century, specific antibodies, protective against microbial pathogens, were found in human milk. However, it was not until after 1950 that their molecular nature was determined. They were found to consist mostly of a special type of immunoglobulin A protein, secretory IgA (sIgA), produced in the mammary gland and directed primarily against enteric and respiratory pathogens. Lysozyme, an enzyme with lytic activity against bacteria, was described by Alexander Fleming in 1922 [32] and subsequently shown to be present in human milk. It was the first nonimmunoglobulin protein found to have antibacterial activities. Its hydrolytic activity causes breakdown of peptidoglycans in the cell walls of susceptible bacteria. Later a number of other antimicrobial proteins were found in milk, and these are now considered a part of the innate immune system [18, 33–37]. Lactoferrin has a strong bacteriostatic effect due to its binding of iron, which is essential for the growth of certain bacteria such as *E. coli*. Like sIgA, lysozyme and lactoferrin are stable in the gastrointestinal tract because of their resistance to digestive enzymes. Secretory IgA, lysozyme and lactoferrin are found in much higher concentrations in human milk than in cows' milk [37]. They remain in high concentration in human milk throughout lactation [34]. Lactoperoxidase, which is found in much lower amounts in human than in cows' milk, has been found to kill *E. coli*, *Salmonella typhimurium* and *Pseudomonas aeruginosa* in the presence of H_2O_2. The bactericidal activity was greatest at pH 5 and below and may contribute to the prevention of enteric infections in neonates [38]. C3 and C4 complements are also found in human milk but in lower amounts than in bovine milk and are not detectable after the first month of lactation [34].

Antimicrobial peptides have recently been demonstrated in human milk [39, 40]. The milk, particularly colostrum, was found to contain high concentrations of several types of defensin molecules, both alpha- and beta-defensins, which were produced by milk leukocytes expressing defensin mRNA [39]. It has also been shown that cathelicidin antimicrobial peptides are present in human milk and are secreted as mature peptides by cells in the milk [40]. The milk cathelicidin peptides were shown to be active against both Gram-positive and Gram-negative bacteria in human milk environment. Murine cathelicidin peptide was identified in mouse breast tissues, suggesting that it is also secreted into the milk by cells of the mammary gland.

In addition to proteins and peptides, many different oligosaccharides are found in human milk, some of which have structural similarities to epithelial cell receptors. They may inhibit mucosal attachment of certain enteric and respiratory bacteria and thus reduce the risk of infection [14, 36, 37, 41, 42]. A 2004 study provides evidence that human milk oligosaccharides are protective against infant diarrhea caused by either bacteria or viruses [43].

Cellular components of the immune system have also been studied in milk. Leukocytes in human milk occur in the highest concentrations in the colostrum during the first days of lactation. They are mostly neutrophils and macrophages but there are also some lymphocytes [14]. The phagocytes in human milk seem to be immunologically active, as judged by their morphology and motility [37].

How does this multitude of immune factors in breast milk, adaptive and innate, protect the suckling infant from infections of microbial pathogens? The sIgG of human milk is known to contain antibodies against enteric and respiratory pathogens that are prevalent in the mother–infant environment [37]. Antibody-producing cells migrate from the lymph nodes in the mother's intestines to the breast, where the specific antibodies are secreted into the milk. A similar migration of lymphocytes seems to take place from the respiratory system to the mammary gland, where they produce antibodies against respiratory pathogens. Because of their resistance to digestive enzymes and to acid conditions, sIgG molecules travel unharmed through the gastrointestinal tract of nursing babies. Specific sIgG antibodies have been shown to reduce the number of coliform bacteria in the stools [44]. How they protect against respiratory infection, such as RSV infection, is not known, but it is known that human milk contains neutralizing sIgA antibodies against viruses, for example against RSV [24, 45]. They might therefore protect against severe infection by coating the pharyngeal entrance to the lower respiratory tract, due to inhalation of the milk and regurgitation through the nose by the nursing infant [24, 35]. Likewise, it is not known exactly how breast-feeding might protect the infant against urinary tract infection [31], but urinary excretion of sIgA with antibody activities against *E. coli* has been demonstrated in infants fed human milk. The concentration of sIgA in the urine correlates with the amount of milk ingested [46].

Like sIgG, lysozyme and lactoferrin are stable proteins which are transported to the lower intestinal tract. Lysozyme is found in a significant amount in the faeces of breast-fed infants [18], where it may cause a nonspecific killing of bacteria and thus act in concert with sIgA [35]. Lactoferrin is also found in the stools of breast-fed low-birth-weight infants and is excreted in large amounts in the urine [47]. Lactoferrin may help to protect the urinary tract against infection by *E. coli*, since it has been shown that it has a powerful bacteriostatic effect on *E. coli*, particularly in combination with specific antibodies [48]. It is not known whether the presence of sIgA and lactoferrin in the urine of infants is due to gastrointestinal absorption from the milk and subsequent renal filtration via the blood. It has also been suggested that a local production of these immune factors is promoted in the kidneys by breast-feeding [49]. However, a close correlation between lactoferrin fragments in the stools and urine supports the notion that the urinary lactoferrin originates in the gastrointestinal tract [47], rather than being produced in the kidneys.

Substantial amounts of oligosaccharides have been found to be excreted into the urine of breast-fed infants during the first month of life [41]. There was a close correlation between an infant's urinary oligosaccharide profiles and those found in the mother's milk. A fraction of neutral oligosaccharides from breast milk was found to inhibit the adhesion of a strain of *E. coli* to uroepithelial cells. This *E. coli* strain had been isolated from the urine of an infant with urinary tract infection. It has also been shown that oligosaccharides from human milk bind to *E. coli* isolated from patients with urinary tract infection [50]. These studies indicate that antibacterial oligosaccharides in the urine of breast-feeding infants are derived directly from the milk and support the observation that breast-feeding has a protective effect against urinary tract infections [31].

The function of activated phagocytic cells in breast milk is not fully understood, although they may play a part in protecting the maternal gland against infection [18]. Colostral phagocytes have been shown to kill enteropathogenic *E. coli*, which is an important etiological agent of acute diarrhoea in newborns [51]. Since intact colostral leukocytes seem to be able to reach the lower intestinal tract of infants, their phagocytic activity may represent an additional mechanism in the control of acute enteric infections during the first weeks of life.

In conclusion, numerous studies have shown that breast milk provides protection against infectious agents early in life by transferring passive immunity to neonates until their development of a functional immune system, either innate or adaptive. Many different agents of the adaptive and innate immune system, present in human milk, interact in defending infants against invading pathogens [37]. Although not as well supported by experimental data, breast-feeding also appears to reduce the risk of chronic diseases appearing later in life [14, 19, 36]. The protective effect of breast-feeding and the immunological factors involved have recently been thoroughly reviewed [52–54].

Table 6.1 gives an overview of the major noncellular components of the innate immune system found in human milk.

6.2.3 Antmicrobial Lipids in Human Milk and Their Possible Protective Function

The antimicrobial activity of milk lipids has been known for almost half a century. Sabin and Fieldsteel showed in 1962 [55] that human milk reduced the infectivity of several viruses, such as herpes simplex virus (HSV) and poliovirus. They demonstrated that the activity against HSV, and a number of other enveloped viruses, was different from that against poliovirus, which does not contain a lipid envelope. Thus, the antipoliomyelytic activity was completely eliminated by heating the milk at 100 °C for 30 minutes, whereas the antiherpetic activity was not affected by this treatment. Furthermore, the antiherpes

Table 6.1 Noncellular innate immune factors in human milk.

Antimicrobial factors[a]	Chemical compound	Primary function	Susceptible pathogens[b]	Ref.[c]
Lysozyme	Protein	Degradation of[d] peptidoglycans	Gram-positive bacteria E. coli, Salmonella[e]	[35,37,54]
Lactoferrin	Protein	Iron-binding	E. coli	[14,37,48]
Lactoperoxidase	Protein	Oxidation by H_2O_2	E. coli, Salmonella	[38]
Complements	Protein	Lysis, opsonization[f]	Gram-+/−[g]	[4,37]
Defensins	Cationic peptides	Membrane disruption[h]	Gram-+/−[g]	[39]
Cathelicidins	Cationic peptides	Membrane disruption[h]	E.coli, staphylococci, streptococci	[40]
Oligosaccharides	—	Receptor analogues[i]	E. coli, Salmonella	[37,42,50]
Triglycerides[j]	Lipids	Membrane disruption	E.coli, Salmonella enveloped viruses such as RSV	[59,64,79]

[a] The major antimicrobial factors in human milk.
[b] Examples of susceptible pathogenic microbes.
[c] References.
[d] Damage of bacterial cell walls, particularly of Gram-positive bacteria, leading to cell lysis.
[e] Lysozyme in conjunction with lactoferrin.
[f] Complements of the alternative pathway bind to many different PAMPs on bacterial surfaces, for example to lipopolysaccharides on Gram-negative bacteria and to peptidoglycans on Gram-positive bacteria. The binding leads to lysis or to phagocytosis of the opsonized bacteria.
[g] Gram-positive and Gram-negative bacteria.
[h] Leads to killing of Gram-positive and Gram-negative bacteria, protozoa, fungi and enveloped viruses.
[i] Receptor analogues inhibit mucosal attachment of certain enteric and respiratory bacteria.
[j] Antimicrobial fatty acids are released from milk triglycerides by hydrolysis caused by lipases in the milk and in the gastrointestinal tract.

virus activity was found in the cream fraction of the milk, whereas protein fractions, which contained the activity against poliovirus, had no effect on herpes virus or the other enveloped viruses. It was concluded that human milk contains antiviral substances which are not proteins and are unrelated to specific antibodies. These findings were confirmed by several workers. Thus, Fieldsteel [56] found high antiviral activity against Japanese B encephalitis virus and Friend and Rauscher leukaemia viruses in 43 samples of human milk collected at various times postpartum. He found the activity to be associated with the cream fraction and to be remarkably heat-stable. Falkler *et al.* [57] observed antiviral activity against yet another enveloped virus, dengue virus, in 34 colostrum and milk samples. The activity was found only in the lipid fraction and remained constant over a period of 10 months. No antibody activity was found in the immunoglobulin of the milk samples. Sarkar *et al.* [58] found that mouse mammary tumour virus lost infectivity and the virus particles were degraded by incubation with human milk at 37 °C for 18 hours. The cream fraction was more active than whole milk in destroying the virus. Electron microscopy of negatively stained virus particles showed that the milk primarily affected the lipid envelope of the virus, making the particles permeable and degrading their structure and infectivity. Although the activity was shown to be associated with the fat fraction of the milk, the destructive agent was not identified and it was even suggested that the RNAase present in human milk might play a role.

In 1971, György [59] suggested that a fatty acid present in human milk was active against a strain of virulent staphylococci in mice. Previously, he and his coworkers had shown that repeated injections of a mixture of human milk and a low dose of the bacterium protected mice against a subsequent challenge with a larger dose of bacteria. The 'antistaphylococcal factor' in human milk was thermostable and fractionation of the milk showed that it was an unsaturated fatty acid, similar to but not identical with linoleic acid. However, this study did not explain the nature of the fatty acid protection of the mice. A few years later, Welsh and coworkers [60] showed that the antiviral activity of human milk was associated with free fatty acids and monoglycerides. In this study, the effect of human milk on a number of viruses was examined, including HSV, influenza virus and Semliki Forest virus, which all contain a lipid envelope, and adenovirus and SV40, which are non-enveloped. All the enveloped viruses were inactivated by incubation with milk samples at 37 °C for 30 minutes, whereas there was no significant effect on adenovirus and SV40. Cows' milk and a synthetic milk formula did not show activity against any of the viruses. The nonspecific antiviral activity was heat-stable and associated with the cream fraction. Semliki Forest virus was selected for experiments to further locate the activity. The study of the milk lipids showed that free fatty acid and monoglyceride fractions had the highest antiviral activity, whereas the diglyceride fraction of the milk fat was not active. Oleic acid and its monoglyceride were found to be highly active, as was linoleic acid. Since together they constitute about 50% of the fatty acids in the triglycerides of human milk [59, 61], oleic and linoleic acids thus seemed to account for a large part of the antiviral activity of the milk. Notably, in this study some of the human milk samples showed no antiviral activity in their fat fraction. This was found to be related to a low lipase activity in the milk and thus to a high content of nonhydrolysed triglycerides and a corresponding low content of antiviral free fatty acids and monoglycerides. The milk lipase was therefore a prerequisite for releasing the antiviral lipids of the milk, although the lipase is not antiviral itself. No lipase activity was found in the cows' milk, in agreement with an earlier report [59].

In a series of studies on the antimicrobial activity of human milk, Isaacs and coworkers [62–66] confirmed and extended the earlier studies by Welsh *et al.* [60]. They found that fresh human milk does not inactivate enveloped viruses, such as vesicular stomatitis virus and HSV. After storage in a refrigerator at 4 °C for two to five days, or even in a freezer at −20 °C for 15 to 50 days, most milk samples, but not all, became highly antiviral and reduced the virus titre by more than 1000-fold after incubation at 37°C for 30 minutes. The antiviral activity was located in the cream fraction. When milk was separated into cream and supernatant fractions before storage, it was found that a factor in the supernatant was required for the appearance of antiviral activity in the lipid fraction [64]. It was considered most likely that this factor was a lipase, since endogenous lipases were known which hydrolysed triglycerides to monoglycerides and free fatty acids [59,60]. The enzyme activities of bile salt-stimulated lipase and lipoprotein lipase in milk samples from various donors were therefore measured and compared to the antiviral activities in the milk after storage at 4°C. It was found that samples with low levels of lipoprotein lipase did not become antiviral upon storage, in contrast to those with high levels of this enzyme. On the other hand, no relation was found between bile salt-stimulated lipase and antiviral activity. It was concluded that the appearance of antiviral activity in human milk upon storage was caused by hydrolysis of triglycerides to free fatty acids by lipoprotein lipase in the milk [62, 64]. This was confirmed by heating the milk before storage and by addition of lipase inhibitors. In both cases, antiviral activity failed to appear upon storage of the milk. Additionally, when purified lipoprotein lipase was added to milk samples, they became antiviral in a shorter time than in the absence of added enzyme [66].

To answer the question of whether human milk becomes antiviral in the suckling infant, samples were obtained from low-birth-weight neonates fed human milk through a nasogastric tube [62, 64–66]. The milk fed to the infants was either from their mothers or from donors and was without antimicrobial activity. At one and three hours after feeding, samples of the stomach contents were removed by aspiration and tested for antimicrobial activities. All of the one-hour samples were found to be highly active against the enveloped viruses tested, such as vesicular stomatitis virus, HSV and human immunodeficiency virus (HIV). The three-hour samples were either not antiviral or had lower activities than the corresponding one-hour samples. The variability of antimicrobial activity at three hours probably reflects various rates of digestion by the infants, since emptying of the stomach may be complete around that time [62]. Surprisingly, one- and three-hour samples from infants fed pasteurized instead of fresh human milk did not inactivate viruses. This indicated that the inactivation of viruses in the samples was not caused by gastric secretions in the stomach contents. This conclusion was also supported by the finding that the antiviral activity was located in the lipid fraction of the one-hour stomach samples, which contains monoglycerides and free fatty acids [65]. Most likely, the release of monoglycerides and free fatty acids in the stomach was due to a rapid hydrolysis of milk triglycerides by salivary and gastric lipases, and not by lipoprotein lipase in the milk, which is not stable in the gastric environment [64].

In addition to enveloped viruses, *Staphylococcus epidermidis*, *E. coli* and *Salmonella enteritidis* were killed by stomach contents from milk-fed infants, although the activity varied greatly among the samples [65, 66].

The possible protective function of antimicrobial lipids in human milk has to be considered in context with the complex multifunctional immune factors present in the milk, described earlier. One can only speculate about the function of endogenous lipases in human milk. It can be hypothesized that small amounts of milk stay on the oral and pharyngeal mucosa of babies after feeding, long enough to allow the milk lipoprotein lipase and the lingual lipase in the saliva to hydrolyse milk triglycerides and release antimicrobial monoglycerides and free fatty acids on the mucosal surfaces. The antimicrobial lipids might then act in concert with other antimicrobial factors in the milk to kill pathogens in the oropharynx before they enter the lower respiratory tract by aspiration, as suggested earlier for RSV. The antiviral effect of milk lipids would be most relevant, because many viruses which cause lower respiratory illness are enveloped and susceptible to a rapid inactivation by lipids.

It appears more likely that milk lipids have a protective function against infectious agents in the gastrointestinal than in the respiratory tract, but evidence for this is still scarce. An excellent review of the early work on lipids as resistance factors in human milk was given by Kabara in 1980 [67]. A later study of the relative risk of low-fat consumption for acute gastrointestinal illness in children over one year of age showed that those fed only low-fat milk were five times more likely to develop intestinal infections than children fed whole milk [68]. The effect of milk lipids on intestinal infections has also been studied experimentally in rats. The experiments showed that feeding with high milk-fat diets reduced intestinal colonization of *Listeria monocytogenes* in orally infected rats, but not colonization of *S. enteritidis*, which is a Gram-negative bacterium and much less susceptible to the antibacterial activity of lipids [69].

The only direct evidence for the presence of antimicrobial lipids in the gastrointestinal tract of humans is the results from studies of gastric contents from neonates, collected one and three hours after feeding, as previously described [62,64–66]. In addition to enveloped viruses, the antimicrobial lipids in the stomach contents also killed *E. coli* and *S. enteritidis*. These Gram-negative enteric bacteria have been found to be resistant to free fatty acids and monoglycerides at neutral pH, probably due to a layer of hydrophilic lipopolysaccharides on the surface of their outer membrane [70]. Both *E. coli* and *Salmonella* become susceptible to the bactericidal effect of lipids at pH <5 [71]. It was previously suggested that acid environment may affect the lipopolysaccharide layer in the outer membrane of these Gram-negative bacteria and make the membrane more permeable to the antibacterial lipid molecules, allowing them to penetrate into the inner membrane (cell membrane), which is the primary site for their antibacterial action [72]. This is in agreement with the observation that enteric bacteria are killed by milk lipids in the acid environment of the stomach [66]. Antibacterial lipids may thus contribute to the prevention of enteric infections in neonates, together with other milk factors previously described. Fatty acids and monoglycerides released by hydrolysis of milk fat by digestive lipases are present not only in the stomach but also at other sites in the digestive tract. Thus, gastric lipase is still active after entering the duodenum, where it promotes the activity of pancreatic lipase, the major lipolytic enzyme involved in the digestion of dietary triglycerides [73]. Their action, in concert with bile salt-stimulated lipase, results in a complete digestion of triglycerides in the intestinal tract, with free glycerol and fatty acids as the final products [74]. Although most lipid is absorbed, a small amount is excreted in the stools of newborn infants [67]. Antimicrobial lipids may therefore be present in most parts of the gastrointestinal tract of neonates.

Most viruses which are pathogenic in the intestines of children, for example rotavirus, are non-enveloped and are therefore not affected by antiviral lipids. However, a human enteric coronavirus, which is an enveloped virus, can cause necrotizing enterocolitis in infants [62,75]. It is therefore possible that antiviral lipids could protect against this infection.

Giardia lamblia is a pathogenic protozoan parasite, especially prevalent in children, which colonizes the upper small intestine and can cause severe diarrhoea and failure to thrive. In some cases the symptoms disappear spontaneously within a few days, whereas in others they may persist for a long time, even though circulating or secretory antibodies are present. It has been suggested that the course of the infection may be influenced by innate immune factors in the intestines [76]. Free fatty acids, released by hydrolysis of milk triglycerides, have been shown to kill *Giardia lamblia* [76,77], and giardiacidal fatty acids may therefore have a role in preventing or controlling human giardiasis [78].

In spite of the potent antimicrobial activity of hydrolysed milk lipids, mostly demonstrated *in vitro*, their clinical significance in the respiratory and gastrointestinal tracts of infants is uncertain. Notably, it is not only fats in human milk but also those in infant formulas that are hydrolysed by lipases in the stomach of infants to release antimicrobial fatty acids and monoglycerides [65]. Although formula feeds do not contain the variety of antimicrobial factors naturally found in breast milk, they nevertheless have a potential for antimicrobial activity due to their contents of lipids, similar to those in human milk. This activity could possibly be enhanced by addition of medium-chain fatty acids or mono-glycerides to synthetic formulas [79, 80], or even by addition of lipases which hydrolyse formula triglycerides before they enter the gastrointestinal tract [81].

6.3   Antimicrobial Lipids in the Pulmonary Mucosa

The airway surface of the respiratory tract is continuously exposed to the external environment and the numerous microbes in the inhaled air. In spite of this, the mucous membranes of the lower respiratory tract are normally free of microbial colonization and are cleared of external particles by very effective cleaning mechanisms, both physical, chemical and biological. The viscous mucus layer, covering the conducting airway epithelium, contains mucin glycoproteins and acts as a physical barrier where microorganisms and other particles are removed by mucocilary clearance. Below the viscous layer is a thin layer of liquid which covers the epithelium. The liquid layer contains a variety of antimicrobial proteins and peptides and other factors, which, in addition to phagocytic and epithelial cells, constitute the innate immune system of the respiratory tract [82, 83].

6.3.1   The Innate Immune System of the Respiratory Tract

The serous layer, which covers the airway epithelium, contains a wide variety of antimicrobial molecules that are constitutively expressed and therefore always present. As first demonstrated by Fleming [32], the human respiratory secretions contain lysozyme, which, as previously described, is a potent antimicrobial enzyme that hydrolyses the peptidoglycan layer in the cell walls of bacteria, for example pneumococci (*Streptococcus pneumoniae*), and causes bacteriolysis. It is found in a high concentration in nasal secretions and in lower concentrations in bronchoalveolar lavage. Another antimicrobial protein, present in respiratory mucous membranes, is lactoferrin, which is found in concentrations similar to those of lysozyme. Peroxidase, secretory leukoproteinase inhibitor, secretory phospholipase A2 and the broad-spectrum antimicrobial peptides β-defensin and hCAP-18/LL-37 cathelicidin are found in lower concentrations, along with a variety of other potential antimicrobial factors [82–84]. The specific ability to kill different bacterial species varies among the factors and they can work additively, antagonistically or synergistically; that is, the activity of one factor enhances the activity of another. Thus, lysozyme and lactoferrin show synergistic activity against bacteria. The multiple effector mechanisms, used by the various factors to kill microbes or inhibit their growth, greatly increase their protective power.

The constitutively expressed antimicrobial factors are produced by cells in submucosal glands as well as by secretory cells in the epithelial layer and by leukocytes. These chemical factors, in addition to resident macrophages, provide a first line of defence against the constant exposure of the respiratory tract to microorganisms in the environment [83]. Furthermore, invading pathogens promptly induce an innate immune response in the epithelial cells of the airway and in alveolar macrophages, resulting in synthesis of additional chemical factors with antimicrobial activities as well as in recruitment of phagocytic and inflammatory cells. The inducible response has been studied in organ and cell culture systems [82]. The results indicate that surface receptors on airway epithelial cells recognize and respond directly to molecular patterns on pathogens (PAMPs), first to induce the production of antimicrobial factors and second to initiate a sequence of events leading to an inflammatory response. TLRs are expressed on human airway epithelial cells and have been shown to recognize and bind to PAMPs, on both Gram-negative and Gram-positive bacteria [82].

6.3.2   Antimicrobial Fatty Acids in the Alveoli of Animal and Human Lungs

The question of how antimicrobial lipids might fit into the complex multifactorial spectrum of factors constituting the innate immune system of the respiratory tract has been most actively addressed by Coonrod and his coworkers [85–89]. In their studies, they were most concerned with the mechanism of killing of pneumococci in the alveoli of the lungs. In previous *in vivo* studies, alveolar macrophages had been shown to have a major role in the killing of inhaled *Staphylococcus aureus*. However, preliminary studies of the activity of rat alveolar macrophages *in vitro* showed that their bactericidal activity was greatly enhanced by, or even dependent on, the addition of extracellular material derived from the alveolar lining in rats [90]. These results were confirmed in a study by Juers *et al.* [91], who tested alveolar-lining material obtained by bronchoalveolar lavage, either from exsanguinated rats or from human subjects by either whole-lung lavage under general anesthesia or limited lavage from a segment of the lung [91]. Their studies showed that a factor enhancing the bactericidal capacity of rat alveolar macrophages was present in both rat and human alveolar lining material and suggested that the factor was located in a lipid fraction of the alveolar material [91].

Coonrod *et al.* [85, 86] studied the effect of alveolar lining material on bacteria, in order to determine whether or not the extracellular material had a direct bactericidal activity. For the study they used bronchoalveolar lavage fluid (BALF) collected from the lungs of exsanguinated rats. After removal of the leukocytes and concentration of the fluid by positive-pressure filtration it was incubated with pneumococci at 37 °C. Samples were harvested at intervals and the number of viable bacteria counted as colony-forming units. The results showed that the BALF, which contained the extracellular alveolar material, was highly bactericidal and killed more than 80% of the pneumococci within 45 minutes. Microscopic examination revealed a massive bacteriolysis. Similar studies showed that several species of streptococci and bacilli were also killed by active preparations of lavage fluid. These bacteria developed increased membrane permeability but most of them did not undergo lysis. The study suggested that killing of pneumococci, and possibly other bacteria, in the alveoli of rats might occur independent of phagocytes [85]. The role of extracellular killing of pneumococci in the alveoli was further established by histologic studies of the lungs of rats after inhalation of bacteria radiolabelled with iron (^{59}Fe). There was little association with macrophages during clearance of the pneumococci, in contrast to radiolabelled staphylococci which were used as control. The early killing of inhaled pneumococci, unlike staphylococci, therefore appeared to take place outside of macrophages [92]. Preliminary experiments indicated that the extracellular antipneumococcal activity was heat- and trypsin-resistant and was located in the surfactant fraction of the rat BALF. This fraction, which was obtained as a pellet by centrifuging lavage fluid at 55 000 G for 20 minutes, contained water-insoluble lamellar bodies and membraneous lipid and consisted mainly of phospholipids, but also of neutral lipids including triglycerides. Further studies showed that the bactericidal activity could be quantitatively recovered from the surfactant fraction by chloroform extraction and that the chloroform extract contained neutral lipids, whereas the phospholipids eluted in methanol. Fractionation by thin-layer chromatography showed that most of the bactericidal activity was confined to the free fatty acids fraction of neutral lipids. A further analysis revealed that they were mainly a mixture of saturated and unsaturated long-chain fatty acids, of which palmitoleic, oleic and linoleic acids were the most abundant and accounted for most of the antibacterial activity. Small amounts of more highly unsaturated fatty acids with longer chain lengths, such as arachidonic acid, were also present. It was concluded that the bactericidal and bacteriolytic activities of rat bronchoalveolar extracellular material were mainly due to unsaturated fatty acids [87].

The source of the BALF is complex and the fluid is a mixture of airway and alveolar components. The origin of the fatty acids in the fluid is therefore uncertain. A high lipase activity has been demonstrated in the BALF from rats, mostly due to lipoprotein lipase. The lipase is probably produced by alveolar macrophages, since it has been found that rat alveolar macrophages secrete lipoprotein lipase in culture [93]. It is therefore possible that fatty acids are released from lipoproteins or triglycerides in the alveolar lining material by lipoprotein lipase hydrolysis or other lipase activity. However, despite the presence of antimicrobial fatty acids in bronchoalveolar secretions, their role in the lung defences against microbes in the rat model is unknown. Any role of fatty acids in direct killing of bacteria or other microorganisms *in vivo* would most likely be confined to the alveoli rather than to the conducting airways. The reason for this is that mucoproteins in the airway mucus bind free fatty acids and inhibit their antibacterial activity [88].

The role of lipids in the lung defences of host animals other than rats is even less clear. The amounts of free fatty acids and lipoprotein lipase in BALF surfactants have been found to vary greatly among different species, such as guinea pigs, rabbits and dogs, and are in all cases much lower than in rats [89, 93]. The reason for this is unclear, but may be either that the level of fatty acids in the BALF does not accurately reflect their amount on alveolar surfaces or that their role in the lung defence against pathogens varies among different species. The latter interpretation seems to be more likely, since experiments using radiolabelling of inhaled pneumococci with ^{59}Fe showed that extracellular killing of the bacteria in alveoli, compared with killing by macrophages, was more rapid in rats than in rabbits. In rabbits, in contrast to rats, alveolar macrophages were the most active in clearance of the inhaled pneumococci [94].

The possible role of free fatty acids in the defence of pulmonary mucosa in humans is of particular interest. Both the amount of free fatty acids, as a percentage of surfactant lipids, and the lipase activity were much lower in BALF from humans than from rats [89, 93]. Finley and Ladman [95] found low amounts of free fatty acids in pulmonary surfactants from the lungs of cigarette smokers, whereas a higher percentage was reported in the surfactant lipids from BALF of patients with various lung diseases [96]. In contrast to these findings, Jonsson *et al.* [97] found no detectable fatty acids in alveolar lining material from BALF of healthy humans and, unlike in the results of Juers *et al.* [91], human alveolar lining material did not stimulate the uptake of bacteria by pulmonary alveolar macrophages. The differences in the results of these workers may be due to differences in the human subjects used in the studies, being either patients with bronchopulmonary disease or healthy individuals.

As emphasized by Coonrod [98,99], alveolar macrophages are of primary importance in protection of the lung against microbes. Not only do they kill microorganisms intracellularly by phagaocytosis but they also secrete antibacterial factors including lysozyme, peptides and iron-binding proteins, in addition to lipase, which may release antimicrobial fatty acids from lipids in the broncholaveolar lining fluid. The exact role of any of these factors in lung defences is unknown, but free fatty acids may be one of the participants. Clearly, more work is needed to further elucidate the role of free fatty acids in the innate pulmonary host defence in animals and humans.

6.4 Antimicrobial Skin Lipids

6.4.1 The Self-Disinfecting Power of the Human Skin

It has been known since the beginning of the twentieth century that the human skin has an extraordinary power to rid its surface of harmful bacteria coming from the environment and to maintain its endogenous bacterial flora. The ability to destroy exogenous microbes has been referred to as the 'self-disinfecting power of the skin' [100]. Early experiments showed that the rate of killing of bacteria varied on different surface areas of the skin, for example on the fingertips versus the palms of the hands versus the forearms. Of these three areas, the killing rate was highest on the palmar skin and lowest on the forearm. Also, the susceptibility of various bacteria to the disinfectant power of the skin differed considerably. For example, *S. aureus* was more resistant than *Streptococcus pyogenes*. In these experiments, suspensions of bacteria were spread over two equal sites on each surface area of the skin and allowed to evaporate. A sample was harvested immediately from one site of each pair and from the other site one or two hours later. The relative disinfectant power was calculated as the ratio of the number of organisms recovered from the first sample to the number recovered from the second sample, called the 'fall in viability ratio' [101]. The question of the nature of the disinfectant power of the skin was addressed in some of the early work. Drying was by some workers considered to contribute to destruction of microorganisms on the skin, and moisture was found to partially inhibit the disinfecting process [102]. There was also some evidence that the self-disinfecting power of different surface areas of the skin depended on their degree of acidity [103]. The acid secretion of the sweat glands was considered by some workers to be of importance as a disinfecting agent. Burtenshaw [101] counted the sweat ducts in the finger, palmar and forearm skin of cadavers and found that the ducts were most numerous in the palms, fewer in the fingers and very scarce in the skin of the forearms. Their numbers were therefore in the same order as the power of self-disinfection in these three areas. This seemed to support the view that sweat was of importance as a skin disinfectant.

6.4.2 The Role of Skin Lipids in the Self-Disinfection of Human Skin

The first study to show that skin lipids are antibacterial and may play a role in the self-disinfection of human skin was done by Burtenshaw in 1942 [13]. In his experiments, intended to throw light on the mechanism of self-disinfection, he applied a cylinder containing either saline, ether or alcohol to his forearm or palm and scraped the submerged area of the skin with a microscope slide, until the detached epithelium formed an opalescent suspension in the fluid. The saline suspensions were tested directly against haemolytic streptococci, whereas the ether and alcohol extracts were evaporated to dryness and the residues were resuspended in saline. A saline suspension of the test bacteria was added to each suspension of skin scrapings, adjusted to either slightly alkaline or to acidic pH, and incubated for 45 to 150 minutes. A sample was then removed from each suspension, diluted 50- to 100-fold in saline and a small amount tested on blood agar for surviving bacteria. In the saline suspensions all of the antibacterial activity was found in the skin scrapings and none in the saline supernatants, showing that the disinfecting agent was not water-soluble. Tests on the ether extracts indicated, on the other hand, that the agent was ether-soluble. Notably, it was observed that scraping of a skin area with ether greatly reduced the self-disinfecting power of the skin upon subsequent inoculation of streptococci on its surface, as indicated by the rate of their disappearance. Ether extracts of hair, nails and cerumen were highly active against the streptococci and in some experiments also against *S. aureus* and *S. epidermidis*. Fractionation of the ether extracts and testing of the various fractions against streptococci demonstrated that all of the bactericidal activity was in the fraction containing fatty and hydroxy acids. Further studies of this fraction showed that it contained oleic acid and other long-chain fatty acids and their soaps. In nearly all experiments, the disinfecting extracts and fractions thereof were far more bactericidal at an acid than at a more alkaline pH, leading to the conclusion that the less dissociated fatty acid molecules were more active against bacteria than the soap molecules. As had been pointed out by Osterhout [104], a less dissociated molecule may penetrate more easily into an organism. The experiments therefore suggested that the disinfecting effect of acidity on the skin was, at least partly, due to release of undissociated fatty acids from their soaps at low pH.

These studies were partly confirmed and extended by Ricketts and his coworkers [105], who studied self-disinfection (self-sterilization) of the skin of young adult volunteers. Multiple sites on the forearm were contaminated with various bacteria and the sites were then covered with either an occlusive nylon dressing, which allowed a normal evaporation of water from the skin, or a polythene dressing, which was much less permeable to water vapour and kept the skin moist. At intervals, each site was sampled by swabbing and the swabs were tested for bacteria on suitable culture media. The results showed that *S. pyogenes*, *Bacterium coli*, *Pseudomonas pyocyanea* and *Chromobacterium prodigiosum* disappeared from the normally dry skin under the nylon dressing within one day, wheras *S. aureus* survived for two to three days. The same was true for skin naturally contaminated with *S. aureus*. However, *B. coli*, *P. pyocyanea* and *C. prodigiosum* survived much longer in sites under polythene dressing, where the skin was moist, than on the dry skin. For these Gram-negative bacteria, drying of the skin seemed to play a role in the rate of killing, whereas the rate of killing of the Gram-positive cocci was not affected by the dryness of the skin.

A series of experiments were carried out in which the skin lipids of one forearm were largely removed by acetone extraction for three minutes, while the other forearm was not treated with acetone. Multiple sites on each arm were then contaminated with *S. pyogenes* and the rates of self-sterilization on the arms were compared. The results of swabbing one to two hours later showed that the acetone-treated sites yielded significantly higher counts of *S. pyogenes*, indicating that bactericidal skin lipids had been removed by acetone. In order to determine the nature of these lipids, Ricketts and his coworkers analysed the composition of the acetone extracts. As before, the lipids were extracted from the surface of the skin of human volunteers by immersion of the hand and forearm in acetone. An average amount of 57 mg lipids was recovered from the skin of each individual by extraction for three minutes. A fractionation of the total skin lipids showed that 35 to 40% consisted of free fatty acids and the remaining 60% was neutral fats. Of the free fatty acids about half, or 20% of the total lipids, was found to be saturated. The saturated fatty acid fraction contained a mixture of at least seven components, of which stearic acid (C18 : 0) seemed to be the most abundant. Of the unsaturated 20% fraction, oleic acid (C18 : 1) was the most prominent.

The saturated and unsaturated fatty acid fractions from the skin extracts, as well as oleic and stearic acids, were tested for bactericidal activities against *S. pyogenes*, *S. aureus*, *B. coli*, *P. pyocyanea* and micrococci, which are normally present in skin microflora. Low concentrations of unsaturated fatty acid fractions and of oleic acid killed *S. pyogenes* in 10 minutes, whereas much higher concentrations of saturated fatty acids were needed to kill this bacterium. *S. aureus* was less susceptible than *S. pyogenes* to the bactericidal effect of the unsaturated fatty acid fraction and to oleic acid. The other bacteria tested were much less affected by the fatty acids, even at pH 5.5.

This study showed that the removal of surface lipids reduced the self-sterilizing power of the skin against *S. pyogenes* and to a lesser degree against *S. aureus*, but had no effect on resident micrococci or the Gram-negative *B. coli* and *P. pyocyanea*. The sensitivity of *S. pyogenes* and *S. aureus* to fatty acids extracted from the skin surface was found to run parallel with their rate of disappearance from normal skin. It was concluded that for *S. pyogenes* the major factor in self-sterilization was the bactericidal effect of unsaturated fatty acids, particularly oleic acid, whereas drying of the skin was the major factor in the killing of micrococci and the Gram-negative bacilli. For *S. aureus* both factors seemed to have an effect. A practical aspect of this study is illustrated by the success of nylon dressings, which permit drying of a wound and the surrounding skin, with a resulting decrease in bacterial contamination.

This work was confirmed in a similar study by Aly *et al.* [106], who demonstrated that *S. pyogenes*, *S. aureus* and *Candida albicans*, which were artificially applied to human forearm skin, disappeared over time, although in the case of *S. aureus* and *Candida* the rate of bacterial reduction varied greatly among individual volunteers. The number of bacteria was not reduced on acetone-extracted skin, whereas when the acetone extracts were replaced on the skin, reduction was restored. Natural skin lipids therefore seem to play an important role in elimination of these pathogenic organisms from the skin, all of which were isolated from human cutaneus infections. In contrast, the growth on the skin of *P. aeruginosa*, a Gram-negative bacterium, was not inhibited by acetone extracts.

6.4.3 Sebum Lipids as a Source of Antimicrobial Fatty Acids in the Skin

As outlined above, it has been known for decades that fatty acids play a role in the self-disinfection of the skin and it has been assumed that they are derived from sebaceous lipids. Sebaceous glands, which are found in the skin of most mammals, including humans, secrete a complex mixture of lipids, the sebum, into the hair follicles. The sebum is transported through the follicular canal to the surface of the skin. The secretion of the sebacious glands consists mostly of debris of dead lipid-producing cells, the cebocytes, which disintegrate to form the sebum. It does not contain phospholipids from the cell membranes but is mostly composed of triglycerides, wax esters, squalene and some cholesterol [107, 108]. Triglycerides produced by the sebacious glands of humans are unique in that they contain numerous fatty acids of extraordinary variety not found in internal tissues [107]. Thus, they contain fatty acids with unusual patterns of unsaturation, for example the palmitoleic acid isomer C16 : 1Δ6 (sapienic acid), which makes up a considerable part of human skin lipids. This fatty acid is equally antibacterial as the more common isomer C16 : 1Δ9, which is found in other tissues. In addition to fatty acids with unusual positions of double bonds, fatty acids synthesized by sebacious glands have odd-numbered as well as even-numbered carbon chains, branched chains and a very wide range of chain lengths, extending far beyond 24 carbon atoms [107]. During the passage from the sebacious gland through the follicular canal to the skin surface, the triglycerides of the sebum are hydrolysed into free fatty acids and glycerol. This was clearly demonstrated in a study in which the lipids of sebacious glands were compared with those of skin-surface sebum and hair clippings from human skin. The study showed that the main difference in the compositions of lipids from these three locations was in the triglyceride and free fatty acid content. The percentage triglyceride was highest in the glands and lowest on the hair clippings, whereas the opposite was true for free fatty acids [109]. The hydrolysis of sebum triglycerides is caused, at least partly, by a lipase of the bacterium *Propionibacterium acnes* in the sebaceous follicles. Thus, it was shown that suppression of *P. acnes* by oral tetracycline treatment greatly reduced the percentage of free fatty acid on the skin surface [110] and a strong correlation was found between the number of *P. acnes* and the percentage of free fatty acids on the skin [111]. The role of skin bacteria in the hydrolysis of sebum triglycerides was supported by another study, which showed a relative decrease in free fatty acids and increase in triglycerides on the skin of patients with acne after receiving topical antibacterial therapy [112]. On the other hand, it is possible that some free fatty acids are produced directly in the sebacious glands, because such production has been demonstrated in cultured human sebocytes [113].

The free fatty acid composition of human skin sebum was analysed by Wille and Kydonieus [109] and the antibacterial spectrum of the fatty acids was studied. Total sebum lipids from the skin surface effectively reduced the number of viable *S. aureus* and *Streptococcus salivarius* by 4 to 5 log$_{10}$, but had no effect on most Gram-negative bacteria studied, including *E. coli* and *P. aeruginosa*. Only the fatty acid fraction of the sebum lipids showed bactericidal activity. Further fractionation into saturated and monounsaturated fatty acids showed that both fractions were equally effective against *S. aureus*. Of the saturated fatty acids, lauric acid (C12 : 0), a minor component of the skin lipids, was the most active. Sapienic acid (C16 : 1Δ6) was the predominant monounsaturated fatty acid, and the next most abundant unsaturated acid was an unusual oleic acid isomer, C18 : 1Δ8. Both C16 : 1Δ6 and its common isomer C16 : 1Δ9 were antibacterial against *S. salivarius*, *S. pyogenes* and a human-wound isolate of *Corynebacterium* sp., the minimum inhibitory concentration being 10 to 20 µg/ml. They were not active against several different Gram-negative bacteria, including *P. acnes*, or against *C. albicans*. Notably, a methyl-branched derivative, methylpalmitoleic acid, was as effective as palmitoleic acid against *S. salivarius*. The antimicrobial activity of the oleic acid isomer C18 : 1Δ8 was not tested in this study. Although the palmitoleic isomers did not inhibit the growth of *C. albicans*, they were effective in preventing the adherence of pathogenic yeast cells to mammalian stratum corneum and may therefore have a prophylactic potential [109].

This study demonstrated that free fatty acids in human sebum are strongly antibacterial and may to a great extent account for the self-disinfecting activity of the skin surface. Thus, it largely confirmed and extended the older studies of Burtenshaw and Ricketts and their coworkers [103, 105]. Some of the fatty acids with the strongest antibacterial activity seem to be unique and exclusively present in the sebum. This supports the notion that the antimicrobial activity of sebum is an intrinsic property of sebacious glands [109] which could be added to the multitude of important functions attributed to this skin organelle [114, 115]. Further studies of a possible relationship between the rate of skin infections and sebum excretion would be of interest, for example in young children with reduced excretion of sebum.

Several studies of the effect of fatty acids on *S. aureus* have thrown a light on their role in the innate immune response [116–124]. About 20% of the human population consists of persistent carriers of *S. aureus* and about 60% of intermittent carriers, with about 30% carrying the bacterium at any time. People most often harbour the bacterium in the nose, particularly in the nostrils, and from there it can spread to the surface of the skin and even to other areas of the body [116]. *S. aureus* and the usually nonpathogenic *S. epidermidis* mostly reside on the skin as harmless commensals. However, under certain conditions *S. aureus* can cause a variety of skin infections, such as pimples, boils, abscesses and cellulitis, but also life-threatening infections of other organs, for example pneumonia, endocarditis and wound infections, and systemic conditions such as toxic shock syndrome due to secretion of a toxin. Recent studies support the notion that bactericidal lipids, as part of the innate immune response, are instrumental in preventing colonization of *S. aureus* from leading to skin infections and that they are also effective in overcoming infection. On the other hand, strains of *S. aureus* which develop resistance to bactericidal lipids demonstrate increased pathogenicity.

Studies of patients with atopic dermatitis (AD) show that their skin is more often colonized with *S. aureus* than that of nonatopics, with more than 90% of patients being colonized [117, 118]. Takigawa and coworkers [119] analysed the sebum lipids in the skin of AD patients and found that there was significantly less of the antibacterial palmitoleic acid isomer sapienic acid (C16 : 1Δ6) in their sebum, compared with healthy controls. This was reflected in a significantly lower level of free C16 : 1Δ6 on the skin of patients, and the number of *S. aureus* colonizing the skin correlated with the concentration of the fatty acid on the skin surface. Takigawa and coworkers [119] carried out a small clinical trial to study whether application of a lotion containing C16 : 1Δ6 had an effect on the colonization of *S. aureus* in nonlesional skin of AD patients. In six of eight patients the number of colonies was not significantly increased in the treated site, in contrast to a distinct increase in the corresponding control site. It was suggested that the deficit in free C16 : 1Δ6 in AD skin is associated with the increased susceptibility to colonization by *S. aureus* and that this fatty acid may therefore be involved in the defence mechanism against the bacteria.

S. aureus is a highly adaptable pathogen which easily develops antibiotic resistance. Bacterial resistance to the innate defences of human skin leads to colonization and various types of skin infection. A 2007 study shows that the expression of the *S. aureus* surface-protein IsdA protects the bacterium against the bactericidal action of human-skin fatty acids, particularly C16 : 1Δ6, and makes it more virulent as a skin pathogen [120]. The surface of IsdA-producing bacterial cells is less hydrophobic than the surface of cells not producing the protein, reducing the adhesion of fatty acids to the surface of the bacteria

and thus making them more resistant. This study also shows that treatment with purified
C16 : 1Δ6 leads to a significant reduction in the bacterial load and improvement of clinical
signs in a mouse model of *S. aureus* infection [120].

Another study analysed the response of strains of *S. aureus* to the presence of linoleic
(C18 : 2) and oleic acid (C18 : 1) and revealed an upregulation of many genes, including
genes encoding virulence determinants and cell wall-associated proteins [121]. Among
the proteins showing increased expression was a cell wall-anchored protein which was
shown to cause an increase in the resistance of *S. aureus* to the bactericidal action of the
fatty acids and increased virulence in a mouse model infected with *S. aureus*. A decreased
hydrophobicity of the cell wall was also observed as a result of adaptation of *S. aureus*
to the fatty acids. As discussed above, this causes increased resistance of the bacterium
to fatty acids and increased virulence. Another adaptive change was an increase in the
biosynthesis of staphyloxanthin, which is a carotenoid inserted into the cell membrane of
S. aureus, giving the bacterium its gold colour. It has been suggested that an increase in
staphyloxanthin stabilises the cell membrane and counteracts the bactericidal increase in
fluidity caused by insertion of fatty acids [122].

Analysis of homogenates from staphylococcal abscesses, generated experimentally in
mice, revealed a pool of antistaphylococcal lipids within the abscesses which contained
long-chain unsaturated fatty acids and monoglycerides, mostly monopalmitolein (C16 :
1) and monopalmitin (C16 : 0) – 69% and 21%, respectively – and less of some other
monoglycerides [123, 124]. The data showed that 2-monoglycerides in the homogenates
had the highest specific activity against *S. aureus* and differed in their activities against
different strains of the bacterium. The survival of *S. aureus* within abscesses was found to
be controlled by the bactericidal monoglycerides. Different strains were killed at different
rates, depending on their sensitivity to the monoglycerides. These studies therefore showed
that the elimination of staphylococci within experimentally produced abscesses was caused
by bactericidal lipids.

It has been reported that glycerol monolaurate and lauric acid inhibit the post-exponential
production of toxic shock syndrome toxin-1 and other staphylococcal exoproteins at lipid
concentrations that do not affect bacterial growth [125, 126]. The lipids seem to act at
the cell membrane by interfering with the signal transduction mechanism, which leads to
transcription of exotoxin genes. Inhibition of toxin production in the anthrax bacillus by
subbacteriostatic concentrations of glycerol monolaurate has also been reported [127]. As
in *S. aureus*, the inhibitory effect was shown to occur at the transcriptional level. Although
apparently not relevant to the question of the role of lipids in innate immunity, this effect
of toxin inhibition may have therapeutic implications.

A 2009 study [128] has confirmed the antibacterial activity of lauric acid against *S.
aureus* and *S. epidermidis*, and has furthermore found it to be active against the Gram-
negative bacterium *P. acnes*. These authors suggested using lauric acid as a treatment of
acne vulgaris, in contrast to the opinion of Wille and Kydonieus [109], who considered
lauric acid too toxic for topical skin treatment.

6.4.4   Stratum Corneum as a Source of Antimicrobial Lipids in the Skin

As discussed in the previous sections, the antimicrobial activity of skin-surface lipids was for
years ascribed to free fatty acids derived from sebum triglycerides, and their importance in

cutaneous antimicrobial defence has been confirmed in recent studies. However, the fact that the skin on some regions of the body, for example the abdomen, is devoid of sebacious glands without being particularly prone to infections, along with other observations, led Miller *et al.* [129] to study the human stratum corneum as another source of antimicrobial lipids in the skin. In their study they used fresh skin specimens from the abdomens of cadavers and analysed the lipids extracted from the stratum corneum. The lipid composition of stratum corneum was different from that of sebum, mainly in that it contained glycosphingolipids and phospholipids and no squalene and wax esters. The lipid fractions were tested for killing of *S. aureus*, either *in vitro* or on acetone-wiped forearm skin of volunteers. In both studies, free fatty acids showed the highest bactericidal activity against *S. aureus*, but phopholipids and glycosphingolipids were also active. Thus, stratum corneum lipids were shown to exhibit substantial antibacterial activity, not only of free fatty acids but also of phopholipids and glycosphingolipids, which are most likely derived from membrane bilayers of epidermal keratinocytes.

Experiments using hairless mice indicate that free fatty acids in the stratum corneum are derived from phopholipase-mediated hydrolysis of phospholipids in the outer epidermis during cornification and shedding of superficial corneocytes from the skin surface. This phospholipid-to-free fatty acids hydrolysis was inhibited by topical application of phospholipase inhibitors, which resulted in accumulation of phospholipids and depletion of fatty acids in the stratum corneum, causing an increase in skin-surface pH and functional abnormalities [130, 131]. Normal acidification of the stratum corneum inhibits pathogenic bacteria, as shown by the fact that alkalinization of the skin surface by urea in diaper rash can lead to infection by a variety of bacteria and yeast [131, 132].

The antimicrobial activities of a number of representative phospholipids and sphingolipids were screened *in vitro* against a variety of bacteria. Only sphingosines, which are long-chain unsaturated amino alcohols that act as building blocks of stratum corneum sphingolipids, were found to have a broad spectrum activity [133]. They were bactericidal against *S. aureus*, *S. pyogenes*, *P. acnes* and *C. albicans* at slightly acid pH, but ineffective against *E. coli*. Sphingosines are also fungistatic against certain dermatophytes [134]. Because free sphingosines are present in the stratum corneum and other epidermal layers [135], they may contribute to the cutaneous antimicrobial barrier. The antibacterial and antifungal activities of sphingosines were tested *in vivo* in human volunteers and in guinea pigs [136]. In the human study, which involved preventive topical application of sphingosine in ethanol, up to 3 $\log_{10}$ reduction in viable *S. aureus* and *C. albicans* was observed, compared with vehicle and untreated controls. In the animal model, there was a significant reduction in fungal infection of the skin as a result of sphingosine therapy. Thus, simple sphingolipids may be of importance as antimicrobial agents of the cutaneous barrier. This is supported by a study which showed an association between reduced levels of sphingosine in the stratum corneum of patients with atopic dermatitis and increased colonization by *S. aureus* [137]. This suggests that the increased colonization resulted from a deficiency of sphingosine as a natural antimicrobial agent.

The stratum corneum forms a protective barrier at the surface of the skin, where acidic pH, drying effect, normal bacterial flora and antimicrobial lipids and peptides, all linked together, contribute to a varying degree to the protective functions [138]. Together with the sebacious glands [108, 114, 115], the stratum corneum is believed to play the most important role in the antimicrobial defence of the skin.

6.4.5 Antimicrobial Lipids as a Part of the Innate Immune Defence

A recent study indicates that synthesis of antimicrobial lipids can occur as a specific response to bacterial invasion of the skin. This supports the notion that lipids play an active role in host defence against invading pathogens and are truly a part of the innate immune system [139]. In this study, a failure to clear an experimental infection by *S. aureus* and *S. pyogenes* in the skin of mice with innate immunodeficiency was linked to a mutation in an enzyme necessary for the synthesis of palmitoleic and oleic acids. The synthetic pathway is Toll-like receptor 2 (TLR2)-responsive and it is of interest that TLR2 and -6 have been found in human sebocytes. This suggests that antimicrobial lipids produced by sebocytes play a role in the innate immune defence of the skin and that there is an inducible, TLR-responsive, lipid-based antimicrobial effector pathway in the skin. As was described previously, TLRs in vertebrates recognize specific PAMPs on invading pathogens and activate various types of innate immune response aimed at destroying the pathogen. If such pathways are functional in production of antimicrobial skin lipids, clinical application of natural lipids may be feasible for treatment and/or prevention of skin infections.

In a recent study, incubation of sebocyte cultures with lauric, oleic or palmitic acid was found to greatly enhance the expression of the antimicrobial peptide beta-defensin (hBD)-2 by the sebocytes [140]. The upregulated expression of hBD-2 by the sebum fatty acids was specific, since an increase in synthesis of other antimicrobial human-skin peptides was not observed. Thus, sebum fatty acids may not only show direct antibacterial activity on the skin surface but also induce synthesis of antibacterial peptides by the sebaceous glands. Antimicrobial fatty acids and peptides could thus be linked together to enhance the innate immune defence of the skin.

6.5   Conclusions

Antimicrobial lipids, particularly fatty acids, are present in the skin and in mucosal membranes of the lungs, as well as in the gastrointestinal tract, where they are commonly found in infants as breakdown products of breast-milk triglycerides. Several studies indicate that antimicrobial lipids play a role in the host defence against pathogenic microorganisms in skin and mucosa, but the question of whether this role is merely incidental or if lipids have a biological mission as protectors against pathogens has not been unequivocally answered. However, recent studies support the notion that they have a specific function in the innate immune response, possibly in connection with antimicrobial peptides. Thus, lipids may be a factor in the complex multifunctional innate immune system.

References

1. Medzhitov, R. and Janeway, Jr, C.A. (2000) Innate immune recognition, mechanisms and pathways. *Immunol. Rev.*, **173**, 89–97.
2. Medzhitov, R. and Janeway, Jr, C.A. (1997) Innate immunity: impact on the adaptive immune response. *Curr. Opin. Immunol.*, **9**, 4–9.
3. Janeway, Jr, C.A. and Medzhitov, R. (2002) Innate immune recognition. *Annu. Rev. Immunol.*, **20**, 197–216.

4. Medzhitov, R. and Janeway, Jr, C.A. (2002) Decoding the patterns of self and nonself by the innate immune system. *Science*, **286**, 298–300.
5. Aderem, A. and Ulevitch, R.J. (2000) Toll-like receptors in the induction of the innate immune response. *Nature*, **406**, 782–787.
6. Takeda, K. and Akira, S. (2005) Toll-like receptors in innate immunity. *Int. Immunol.*, **17**, 1–14.
7. Akira, S., Uematsu, S. and Takeuchi, O. (2006) Pathogen recognition and innate immunity. *Cell*, **124**, 783–801.
8. Zasloff, M. (2002) Antimicrobial peptides of multicellular organisms. *Nature*, **415**, 389–395.
9. Klotz, O. (1905) Studies upon calcareous degeneration. I. The process of pathological calcification. *J. Exp. Med.*, **7**, 633–675.
10. Noguchi, H. (1907) Über gewisse chemische Komplementsubstanzen. *Biochem. Zeitschr.*, **6**, 327–357.
11. Landsteiner, K. und Ehrlich, H. (1908) Ueber bakterizide Wirkungen von Lipoiden und ihre Beziehung zur Komplementwirkung. *Centralbl. f. Bakt.*, **45**, 247–257.
12. Lamar, R.V. (1911) Chemo-immunological studies on localized infections. First paper: action on the pneumococcus and its experimental infections of combined sodium oleate and antipneumococcus serum. *J. Exp. Med.*, **13**, 1–23.
13. Burtenshaw, J.M.L. (1942) The mechanism of self-disinfection of the human skin and its appendages. *J. Hyg.*, **42**, 184–210.
14. Goldman, A.S. (2001) The immunological system in human milk: the past – a pathway to the future. *Adv. Nutr. Res.*, **10**, 15–37.
15. Cunningham, A.S. (1979) Morbidity in breast-fed and artificially fed infants. II. *J. Pediatr.*, **95**, 685–689.
16. Cunningham, A.S. (1977) Morbidity in breast-fed and artificially fed infants. *J. Pediatr.*, **90**, 726–729.
17. Fallot, M. E., Boyd, 3rd, J. L. and Oski, F. A. (1980) Breast-feeding reduces incidence of hospital admissions for infection in infants. *Pediatrics*, **65**, 1121–1124.
18. Hanson, L.A. and Winberg, J. (1972) Breast milk and defence against infection in the newborn. *Arch. Dis. Child.*, **47**, 845–848.
19. Cunningham, A.S., Jelliffe, D.B. and Jelliffe, E.F.P. (1991) Breast-feeding and health in the 1980s: a global epidemiologic review. *J. Pediatr.*, **118**, 659–666.
20. Svirsky-Gross, S. (1958) Pathogenic strains of coli (O,111) among prematures and the use of human milk in controlling the outbreak of diarrhea. *Ann. Paediatrici.*, **190**, 109–115.
21. Tassovatz, B. (1961) Human milk and its action of protection against intestinal infections in the newborn (French). *Sem. Hop.*, **37**, 285–288.
22. Mata, L.J. and Urrutia, J.J. (1971) Intestinal colonization of breast-fed children in a rural area of low socioeconomic level. *Ann. NY Acad. Sci.*, **176**, 93–109.
23. Larsen, Jr, S.A. and Homer, D.R. (1978) Relation of breast versus bottle feeding to hospitalization for gastroenteritis in a middle-class US population. *J. Pediatr.*, **92**, 417–418.
24. Downham, M.A.P.S., Scott, R., Sims, D.G. *et al.* (1976) Breast-feeding protects against respiratory syncytial virus infection. *Brit. Med. J.*, **2**, 274–276.
25. Pullan, C.R., Toms, G.L., Martin, A.J. *et al.* (1980) Breast-feeding and respiratory syncytial virus infection. *Brit. Med. J.*, **281**, 1034–1036.
26. Saarinen, U.M. (1982) Prolonged breast feeding as prophylaxis for recurrent otitis media. *Acta Paediatr. Scand.*, **71**, 567–571.
27. Sassen, M.L., Brand, R. and Grote, J.J. (1994) Breast-feeding and acute otitis media. *Am. J. Otolaryngol.*, **15**, 351–357.
28. Duffy, L.C., Faden, H., Wasielewski, R. *et al.* (1997) Exclusive breastfeeding protects against bacterial colonization and day care exposure to otitis media. *Pediatrics*, **100**, E7.
29. Cochi, S.L., Fleming, D.W., Hightower, A.W. *et al.* (1986) Primary invasive *Haemophilus influenzae* type b disease: a population-based assessment of risk factors. *J. Pediatr.*, **108**, 887–896.
30. Takala, A.K., Eskola, J., Palmgren, J. *et al.* (1989) Risk factors of invasive *Haemophilus influenzae* type b disease among children in Finland. *J. Pediatr.*, **115**, 694–701.
31. Mårild, S., Hansson, S., Jodal, U. *et al.* (2004) Protective effect of breastfeeding against urinary tract infection. *Acta Paediatr.*, **93**, 164–168.

32. Fleming, A. (1922) On a remarkable bacteriolytic element found in tissues and secretions. *Proc. Roy. Soc. Ser. B*, **93**, 306–317.
33. Goldman, A.S. and Smith, C.W. (1973) Host resistance factors in human milk. *J. Pediatr.*, **82**, 1082–1090.
34. McClelland, D.B.L., McGrath, J. and Samson, R.R. (1978) Antimicrobial factors in human milk. Studies of concentration and transfer to the infant during the early stages of lactation. *Acta Paediatr. Scand. Suppl.*, **271**, 1–20.
35. Welsh, J.K. and May, J.T. (1979) Anti-infective properties of breast milk. *J. Pediatr.*, **94**, 1–9.
36. Cunningham, A.S. (1987) Breast-feeding and health. *J. Pediatr.*, **110**, 658–659.
37. Goldman, A.S. (1993) The immune system of human milk: antimicrobial, antiinflammatory and immunomodulating properties. *Pediatr. Infect. Dis. J.*, **12**, 664–671.
38. Reiter, B., Marshall, V.M.E., Björck, L. and Rosén, C.G. (1976) Nonspecific bactericidal activity of the lactoperoxidases-thiocyanate-hydrogen peroxide system of milk against *Escherichia coli* and some gram-negative pathogens. *Infect. Immun*, **13**, 800–807.
39. Armogida, S.A., Yannaras, N.M., Melton, A.L. and Srivastava, M.D. (2004) Identification and quantification of innate immune system mediators in human breast milk. *Allergy Asthma Proc.*, **25**, 297–304.
40. Murakami, M., Dorschner, R.A., Stern, L.J. *et al.* (2005) Expression and secretion of cathelicidin antimicrobial peptides in murine mammary glands and human milk. *Pediatr. Res.*, **57**, 10–15.
41. Coppa, G.V., Gabrielli, O., Giorgi, P. *et al.* (1990) Preliminary study of breastfeeding and bacterial adhesion to uroepithelial cells. *Lancet*, **335**, 569–571.
42. Coppa, G.V., Zampini, L., Galeazzi T. *et al.* (2006) Human milk oligosaccharides inhibit the adhesion to Caco-2-cells of diarrheal pathogens: *Escherichia coli*, *Vibrio cholerae*, and *Salmonella fyris*. *Pediatr. Res.*, **59**, 377–382.
43. Morrow, A.L., Ruiz-Palacios, G.M., Altaye, M. *et al.* (2004) Human milk oligosaccharides are associated with protection against diarrhea in breast-fed infants. *J. Pediatr.*, **145**, 297–303.
44. Michael, J.G., Ringenback, R. and Hottenstein, S. (1971) The antimicrobial activity of human colostral antibody in the newborn. *J. Infect. Dis.*, **124**, 445–448.
45. Toms, G.L., Gardner, P.S., Pullan, C.R. *et al.* (1980) Secretion of respiratory virus inhibitors and antibody in human milk throughout lactation. *J. Med. Virol.*, **5**, 351–360.
46. Goldblum, R.M., Schanler, R.J., Garza, C. and Goldman, A.S. (1989) Human milk feeding enhances the urinary excretion of immunologic factors in low birth weight infants. *Pediatr. Res.*, **25**, 184–188.
47. Goldman, A.S., Garza, C., Schanler, R.J. and Goldblum, R.M (1990) Molecular forms of lactoferrin in stool and urine from infants fed human milk. *Pediatr. Res.*, **27**, 252–255.
48. Bullen, J.J., Rogers, H.J. and Leigh, L. (1972) Iron-binding proteins in milk and resistance to *Escherichia coli* infection in infants. *Br. Med. J*, **1**, 69–75.
49. Prentice, A. (1987) Breast feeding increases concentrations of IgA in infants' urine. *Arch. Dis. Child.*, **62**, 792–795.
50. Rosenstein, I.J., Stoll, M.S., Mizuochi, T. *et al.* (1988) New type of adhesive specificity revealed by oligosaccharide probes in *Escherichia coli* from patients with urinary tract infection. *Lancet*, **2**, 1327–1330.
51. Honorio-Franca, A.C., Carvalho, M.P., Isaac, L. *et al.* (1997) Colostral mononuclear phagocytes are able to kill enteropathogenic *Escherichia coli* opsonized with colostral IgA. *Scand. J. Immunol.*, **46**, 59–66.
52. Lawrence, R.M. and Pane, C.A. (2007) Human breast milk: current concepts of immunology and infectious diseases. *Curr. Probl. Pediatr. Adolesc. Health Care*, **37**, 7–36.
53. Chirico, G., Marzollo, R., Cortinovis, S. *et al.* (2008) Antiinfective properties of human milk. *J. Nutr.*, **138**, 1801S–1806S.
54. Lönnerdal, B. (2003) Nutritional and physiologic significance of human milk proteins. *Am. J. Clin. Nutr.*, **77**, 1537S–1543S.
55. Sabin, A.B. and Fieldsteel, A.H. (1962) Antipoliomyelytic activity of human and bovine colostrum and milk. *Pediatr.*, **29**, 105–115.
56. Fieldsteel, A.H. (1974) Nonspecific antiviral substances in human milk active against arbovirus and murine leukemia virus. *Cancer Res*, **34**, 712–715.

57. Falkler, Jr, W.A., Diwan, A.R. and Halstead, S.B. (1975) A lipid inhibitor of dengue virus in human colostrum and milk: with a note on the absence of anti-dengue secretory antibody. *Arch. Virol.*, **47**, 3–10.
58. Sarkar, N.H., Charney, J., Dion, A.S. and Moore, D.H. (1973) Effect of human milk on the mouse mammary tumor virus. *Cancer Res*, **33**, 626–629.
59. György, P. (1971) The uniqueness of human milk. Biochemical aspects. *Am. J. Clin. Nutr.*, **24**, 970–975.
60. Welsh, J.K., Skurrie, I.J. and May, J.T. (1978) Use of Semliki Forest virus to identify lipid-mediated antiviral activity and anti-alphavirus immunoglobulin A in human milk. *Infect. Immun.*, **19**, 395–401.
61. Lammi-Keefe, C.J. and Jensen, R.G. (1984) Lipids in human milk: a review. 2. Composition and fat-soluble vitamins. *J. Pediatr. Gastroenterol. Nutr.*, **3**, 172–198.
62. Isaacs, C.E., Thormar, H. and Pessolano, T. (1986) Membrane-disruptive effect of human milk: inactivation of enveloped viruses. *J. Infect. Dis.*, **154**, 966–971.
63. Thormar, H., Isaacs, C.E., Brown, H.R. *et al.* (1987) Inactivation of enveloped viruses and killing of cells by fatty acids and monoglycerides. *Antimicrob. Agents Chemother.*, **31**, 27–31.
64. Isaacs, C.E. and Thormar, H. (1990) Human milk lipids inactivate enveloped viruses, in *Breast-feeding, Nutrition, Infection and Infant Growth in Developed and Emerging Countries*, (eds S.A. Atkinson, L.A. Hanson and R.K. Chandra), ARTS Biomedical Publishers and Distributors, St. John's, Newfoundland, Canada, pp. 161–174.
65. Isaacs, C.E., Kasyap, S., Heird, W.C. and Thormar, H. (1990) Antiviral and antibacterial lipids in human milk and infant formula feeds. *Arch. Dis. Child.*, **65**, 861–864.
66. C. E. Isaacs and H. Thormar (1991) The role of milk-derived antimicrobial lipids as antiviral and antibacterial agents, in *Immunology of Milk and the Neonate* (eds J. Mestecki and P.L. Ogra), Plenum Press, New York, NY, USA, pp. 159–165.
67. Kabara, J.J. (1980) Lipids as host-resistance factors of human milk. *Nutr. Rev.*, **38**, 65–73.
68. Koopman, J.S., Turkish, V.J., Monto, A.S. *et al.* (1984) Milk fat and gastrointestinal illness. *Am. J. Public Health*, **74**, 1371–1373.
69. Sprong, R.C., Hulstein, M.F. and Van Der Meer, R. (1999) High intake of milk fat inhibits colonization of Listeria but not of Salmonella in rats. *J. Nutr.*, **129**, 1382–1389.
70. Bergsson, G., Steingrímsson, Ó. and Thormar, H. (2002) Bactericidal effects of fatty acids and monoglycerides on *Helicobacter pylori*. *Int. J. Antimicrob. Agents*, **20**, 258–262.
71. Thormar, H., Hilmarsson, H. and Bergsson, G. (2006) Stable concentrated emulsions of the 1-monoglyceride of capric acid (monocaprin) with microbicidal activities against the food-borne bacteria *Campylobacter jejuni*, *Salmonella* spp., and *Escherichia coli*. *Appl. Environ. Microbiol.*, **72**, 522–526.
72. Shibasaki, I. and Kato, N. (1978) Combined effects on antibacterial activity of fatty acids and their esters against Gram-negative bacteria, in *Symposium on the Pharmacological Effect of Lipids* (ed. J.J. Kabara), The American Oil Chemists' Society, Champaign, IL, USA, pp. 15–24.
73. Miled, N., Canaan, S., Dupuis, L. *et al.* (2000) Digestive lipases: from three-dimensional structure to physiology. *Biochimie*, **82**, 973–986.
74. Bernbäck, S., Bläckberg, L. and Hernell, O. (1990) The complete digestion of human milk triacylglycerol in vitro requires gastric lipase, pancreatic colipase-dependent lipase, and bile salt-stimulated lipase. *J. Clin. Invest.*, **85**, 1221–1226.
75. Resta, S., Luby, J.P., Rosenfeld, C.R. and Siegel, J.D. (1985) Isolation and propagation of a human enteric coronavirus. *Science*, **229**, 978–981.
76. Reiner, D.S., Wang, C.-S. and Gillin, F.D. (1986) Human milk kills *Giardia lamblia* by generating toxic lipolytic products. *J. Infect. Dis.*, **154**, 825–832.
77. Hernell, O., Ward, H., Blackberg, L. and Pereira, M.E. (1986) Killing of *Giardia lamblia* by human milk lipases: an effect mediated by lipolysis of milk lipids. *J. Infect. Dis.*, **153**, 715–720.
78. Gillin, F.D. (1987) *Giardia lamblia*: the role of conjugated and unconjugated bile salts in killing by human milk. *Exp. Parasitol.*, **63**, 74–83.
79. Isaacs, C.E., Litov, R.E. and Thormar, H. (1995) Antimicrobial activity of lipids added to human milk, infant formula, and bovine milk. *J. Nutr. Biochem.*, **6**, 362–366.

80. Hilmarsson, H., Traustason, B.S., Kristmundsdóttir, T. and Thormar, H. (2007) Virucidal activities of medium-and long-chain fatty alcohols and lipids against respiratory syncytial virus and parainfluenza virus type 2: comparison at different pH levels. *Arch. Virol.*, **152**, 2225–2236.
81. Isaacs, C.E., Litov, R.E., Marie, P. and Thormar, H. (1992) Addition of lipases to infant formulas produces antiviral and antibacterial activity. *J. Nutr. Biochem.*, **3**, 304–308.
82. Diamond, G., Legarda, D. and Ryan, L.K. (2000) The innate immune response of the respiratory epithelium. *Immunol. Rev.*, **173**, 27–38.
83. Travis, S.M., Singh, P.K. and Welsh, M.J. (2001) Antimicrobial peptides and proteins in the innate defense of the airway surface. *Curr. Opin. Immunol.*, **13**, 89–95.
84. Hiemstra, P.S. (2007) The role of epithelial β-defensins and cathelicidins in host defense of the lung. *Exp. Lung Res.*, **33**, 537–542.
85. Coonrod, J.D. and Yoneda, K. (1983) Detection and partial characterization of antibacterial factor(s) in alveolar lining material of rats. *J. Clin. Invest.*, **71**, 129–141.
86. Coonrod, J.D., Rehm, S.R. and Yoneda, K. (1983) Pneumococcal killing in the alveolus. *Chest*, **83**, 89S–90S.
87. Coonrod, J.D., Lester, R.L. and Hsu, L.C. (1984) Characterization of the extracellular bactericidal factors of rat alveolar lining material. *J. Clin. Invest.*, **74**, 1269–1279.
88. Coonrod, J.D. (1986) The role of extracellular bactericidal factors in pulmonary host defence. *Sem. Respir. Infect.*, **1**, 118–129.
89. Coonrod, J.D. (1987) Rôle of surfactant free fatty acids in antimicrobial defenses. *Eur. J. Respir. Dis.*, **71** (Suppl. 153), 209–214.
90. LaForce, F.M., Kelly, W.J. and Huber, G.L. (1973) Inactivation of staphylococci by alveolar macrophages with preliminary observations on the importance of alveolar lining material. *Am. Rev. Respir. Dis.*, **108**, 784–790.
91. Juers, J.A., Rogers, R.M., McCurdy, J.B. and Cook, W.W. (1976) Enhancement of bactericidal capacity of alveolar macrophages by human alveolar lining material. *J. Clin. Invest.*, **58**, 271–275.
92. Coonrod, J.D., Marple, S., Holmes, G.P. and Rehm, S.R. (1987) Extracellular killing of inhaled pneumococci in rats. *J. Lab. Clin. Med.*, **110**, 753–766.
93. Coonrod, J.D., Karathanasis, P. and Lin, R. (1989) Lipoprotein lipase: a source of free fatty acids in bronchoalveolar lining fluid. *J. Lab. Clin.Med.*, **113**, 449–457.
94. Coonrod, J.D., Varble, R. and Jarrells, M.C. (1990) Species variation in the mechanism of killing of inhaled pneumococci. *J. Lab. Clin. Med.*, **116**, 354–362.
95. Finley, T.N. and Ladman, A.J. (1972) Low yield of pulmonary surfactant in cigarette smokers. *New Engl. J. Med.*, **286**, 223–227.
96. Ramirez, J., Schwartz, B., Dowell, A.R. and Lee, S.D. (1971) Biochemical composition of human pulmonary washings. *Arch. Intern. Med.*, **127**, 395–400.
97. Jonsson, S., Musher, D.M., Goree, A. and Lawrence, E.C. (1986) Human alveolar lining material and antibacterial defenses. *Am. Rev. Respir. Dis.*, **133**, 136–140.
98. Coonrod, J.D. (1989) Pneumococcal pneumonia. *Semin. Resp. Infect.*, **4**, 4–11.
99. Coonrod, J.D. (1989) Role of leukocytes in lung defenses. *Respiration*, **55** (Suppl. 1), 9–13.
100. Arnold, L., Gustafson, C.J., Hull, T.G. *et al.* (1930) The self-disinfecting power of the skin as a defense against microbic invasion. *Am. J. Epidemiol.*, **11**, 345–361.
101. Burtenshaw, J.M.L. (1938) The mortality of the haemolytic *Streptococcus* on the skin and other surfaces. *J. Hyg.*, **38**, 575–586.
102. Cornbleet, T. and Montgomery, B.E. (1931) Self-sterilizing powers of the skin. *Arch. Derm. Syphilol.*, **23**, 908–919.
103. Burtenshaw, J.M.L. (1945) Self-disinfection of the skin: a short review and some original observations. *Br. Med. Bull.*, **3**, 161–164.
104. Osterhout, W.J.V. (1925) Is living protoplasm permeable to ions? *J. Gen. Physiol.*, **8**, 131–146.
105. Ricketts, C.R., Squire, J.R. and Topley, E. (1951) Human skin lipids with particular reference to the self-sterilising power of the skin. *Clin. Sci.*, **10**, 89–111.
106. Aly, R., Maibach, H.I., Shinefeld, H.R. and Strauss, W.G. (1972) Survival of pathogenic microorganisms on human skin. *J. Invest. Dermatol.*, **58**, 205–210.
107. Nicolaides, N. (1974) Skin lipids: their biochemical uniqueness. *Science*, **186**, 19–26.

108. Stewart, M.E. (1992) Sebacious gland lipids. *Semin. Dermatol.*, **11**, 100–105.
109. Wille, J.J. and Kydonieus, A. (2003) Palmitoleic acid isomer (C16 : 1Δ6) in human skin sebum is effective against Gram-positive bacteria. *Skin Pharmacol. Appl. Skin Physiol.*, **16**, 176–187.
110. Marples, R.R., Downing, D.T. and Kligman, A.M. (1971) Control of free fatty acids in human surface lipids by *Corynebacterium acnes*. *J. Invest. Dermatol.*, **56**, 127–131.
111. Kearney, J.N., Ingham, E., Cunliffe, W.J. and Holland, K.T. (1984) Correlations between human skin bacteria and skin lipids. *Brit. J. Dermatol.*, **110**, 593–599.
112. Thielitz, A., Helmdach, M., Röpke, E-M. and Gollnick, H. (2001) Lipid analysis of follicular casts from cyanoacrylate strips as a new method for studying therapeutic effects of antiacne agents. *Brit. J. Dermatol.*, **145**, 19–27.
113. Zouboulis, C.C., Xia, L., Akamatsu, H. *et al.* (1998) The human sebocyte culture model provides new insights into development and management of seborrhoea and acne. *Dermatology*, **196**, 21–31.
114. Zouboulis, C.C. (2003) Sebacious gland in human skin: the fantastic future of a skin appendage. *J. Invest. Dermatol.*, **120**, xiv–xv.
115. Zouboulis, C.C. (2004) Acne and sebacious gland function. *Clin. Dermatol.*, **22**, 360–366.
116. Kluytmans, J., van Belkum, A. and Verbrugh, H. (1997) Nasal carriage of *Staphylococcus aureus*: epidemiology, underlying mechanisms, and associated risks. *Clin. Microbiol. Rev.*, **10**, 505–520.
117. Goh, C.L., Wong, J.S. and Giam, Y.C. (1997) Skin colonization of *Staphylococcus aureus* in atopic dermatitis patients seen at the National Skin Center, Singapore. *Int. J. Dermatol.*, **36**, 653–657.
118. Abeck, D. and Mempel, M. (1998) Staphylococcus aureus colonization in atopic dermatitis and its therapeutic implications. *Br. J. Dermatol.*, **139** (Suppl 53), 13–16.
119. Takigawa, H., Nakagawa, H., Kuzugawa, M. *et al.* (2005) Deficient production of hexadecenoic acid in the skin is associated in part with the vulnerability of atopic dermatitis patients to colonization by *Staphylococcus aureus*. *Dermatology*, **211**, 240–248.
120. Clarke, S.R., Mohamed, R., Bian, L. *et al.* (2007) The *Staphylococcus aureus* surface protein IsdA mediates resistance to innate defenses of human skin. *Cell Host Microbe*, **1**, 199–212.
121. Kenny, J.G., Ward, D., Josefsson, E. *et al.* (2009) The *Staphylococcus aureus* response to unsaturated long chain free fatty acids: survival mechanisms and virulence implications. *PLoS One*, **4**, e4344.
122. Chamberlain, N.R., Mehrtens, B.G., Xiong, Z. *et al.* (1991) Correlation of carotenoid production, decreased membrane fluidity, and resistance to oleic acid killing in *Staphylococcus aureus* 18Z. *Infect. Immun.*, **59**, 4332–4337.
123. Shryock, T.R. and Kapral, F.A. (1992) The production of bactericidal fatty acids from glycerides in staphylococcal abscesses. *J. Med. Microbiol.*, **36**, 288–292.
124. Engler, H.D. and Kapral, F.A. (1992) The production of a bactericidal monoglyceride in staphylococcal abscesses. *J. Med. Microbiol.*, **37**, 238–244.
125. Projan, S.J., Brown-Skrobot, S., Schlievert, P.M. *et al.* (1994) Glycerol monolaurate inhibits the production of beta-lactamase, toxic shock toxin-1, and other staphylococcal exoproteins by interfering with signal transduction. *J. Bacteriol.*, **176**, 4204–4209.
126. Ruzin, A. and Novick, R.P. (2000) Equivalence of lauric acid and glycerol monolaurate as inhibitors of signal transduction in *Staphylococcus aureus*. *J Bacteriol*, **182**, 2668–2671.
127. Vetter, S.M. and Schlievert, P.M. (2005) Glycerol monolaurate inhibits virulence factor production in *Bacillus anthracis*. *Antimicrob. Agents Chemother.*, **49**, 1302–1305.
128. Nakatsuji, T., Kao, M.C., Fang, J.Y. *et al.* (2009) Antimicrobial property of lauric acid against *Propionibacterium acnes*: its therapeutic potential for inflammatory acne vulgaris. *J. Invest. Dermatol.*, **129**, 2480–2488.
129. Miller, S.J., Aly, R., Shinefield, H.R. and Elias, P.M. (1988) In vitro and in vivo antistaphylococcal activity of human stratum corneum lipids. *Arch. Dermatol.*, **124**, 209–215.
130. Mao-Qiang, M., Feingold, K.R., Jain, M. and Elias, P.M. (1995) Extracellular processing of phospholipids is required for permeability barrier homeostasis. *Lipid Res.*, **36**, 1925–1935.
131. Fluhr, J.W., Kao, J., Jain, M. *et al.* (2001) Generation of free fatty acids from phospholipids regulates stratum corneum acidification and integrity. *J. Invest. Dermatol.*, **117**, 44–51.

132. Brook, I. (1992) Microbiology of secondarily infected diaper dermatitis. *Int. J. Dermatol.*, **31**, 700–702.
133. Bibel, D.J., Aly, R. and Shinefield, H.R. (1992) Antimicrobial activity of sphingosines. *J. Invest. Dermatol.*, **98**, 269–273.
134. Bibel, D.J., Aly, R., Shah, S. and Shinefield, H.R. (1993) Sphingosines: antimicrobial barriers of the skin. *Acta Derm. Venereol.*, **73**, 407–411.
135. Wertz, P.W. and Downing, D.T. (1990) Free sphingosine in human epidermis. *J. Invest. Dermatol.*, **94**, 159–161.
136. Bibel, D.J., Aly, R. and Shinefield, H.R. (1995) Topical sphingolipids in antisepsis and antifungal therapy. *Clin. Exp. Dermatol.*, **20**, 395–400.
137. Arikawa, J., Ishibashi, M., Kawashima, M. *et al.* (2002) Decreased levels of sphingosine, a natural antimicrobial agent, may be associated with vulnerability of the stratum corneum from patients with atopic dermatitis to colonization by *Staphylococcus aureus*. *J. Invest. Dermatol.*, **119**, 433–439.
138. Elias, P.M. and Choi, E.H. (2005) Interactions among stratum corneum defensive functions. *Exp. Dermatol.*, **14**, 719–726.
139. Georgel, P., Crozat, K., Lauth, X. *et al.* (2005) A Toll-like receptor 2-responsive lipid effector pathway protects mammals against skin infections with Gram-positive bacteria. *Infect. Immun.*, **73**, 4512–4521.
140. Nakatsuji, T., Kao, M.C., Zhang, L. *et al.* (2010) Sebum free fatty acids enhance the innate immune defense of human sebocytes by upregulating beta-defensin-2 expression. *J. Invest. Dermatol.*, **130**, 985–994.

7

Lipids as Active Ingredients in Pharmaceuticals, Cosmetics and Health Foods

Thórdís Kristmundsdóttir and Skúli Skúlason

Faculty of Pharmaceutical Sciences, University of Iceland, Reykjavik, Iceland

Lipids and Essential Oils as Antimicrobial Agents   Halldor Thormar
© 2011 John Wiley & Sons, Ltd

7.1 Introduction

Lipids are a group of compounds that have in common the characteristic of being hydrophobic due to the presence of long-chain hydrocarbon residues. They can be either linear, branched or cyclic and the hydrocarbon part can be bound to glycerol, sugars or a phosphate head group. Major lipid groups comprise fatty acids, glycerolipids, glycerophospholipids, sphingolipids and sterols, including cholesterol and cholesteryl esters. For centuries lipids have been used in pharmaceuticals and cosmetics as excipients and in food for nutritional value but at present interest in the use of lipids is more focussed on their potential as active ingredients. It has been known for a long time that some lipids kill microbes. Early reports were on the effects of soaps but research has shown that lipids such as fatty alcohols, fatty acids and their derivatives are potent microbicidal agents and kill enveloped viruses, Gram-positive and Gram-negative bacteria, and fungi. Although the microbicidal properties of lipids have been known for more than a century, this knowledge has so far only been utilized commercially to a very limited extent. For several decades the main focus of study on the activity of lipids was on microbicidal activity and the possible use of lipids in place of traditional antibiotics, which can have undesirable side effects and lead to bacterial and viral resistance. Recent research has indicated the extensive possibilities the microbicidal properties of lipids can offer in treating or preventing infections as well as the ability of lipids to take part in body processes and thereby influence the progress of diseases. Numerous other beneficial properties have been attributed to lipids, ranging from their being antiinflammatory to their affecting cognitive function. Clinical studies carried out in the first decade of the twenty-first century have shown that many of the assumed beneficial properties can in fact be demonstrated in a scientific way.

7.2 Antimicrobial Effects of Lipids

7.2.1 Antibacterial Activity

The antibacterial activities of lipids have been known for over a century and more than three decades ago it was suggested that the clinical use of antimicrobial lipids could be advantageous, but there are few examples of clinical use so far [1, 2]. For the past four decades there has been growing interest in the effects of lipids on microbes and numerous publications have reported research on the microbicidal effects of lipids. The early work on fatty acids and derivatives as antimicrobial agents showed that of the saturated medium-chain fatty acids, ranging in chain length from C6 to C14, C12 had the highest activity against the Gram-positive bacteria tested, but activity against Gram-negative bacteria was very low. For long-chain fatty acids, unsaturation led to increased antibacterial activity, and the monounsaturated fatty acid, palmitoleic acid (C16 : 1), showed the highest activity. For oleic acid (C18 : 1), an additional double bond, as in linoleic acid (C18 : 2), led to increased activity. The *cis*-isomers were active but the *trans*-isomers were inactive. Esterification of fatty acids to monohydric alcohols led to inactive derivatives but esterification to polyhydric alcohols increased biological activity, with monoglycerides being more active than di- or triglycerides [2–5]. The polyunsaturated fatty acid eicosapentaenoic acid (EPA, C20 : 5) has been found to be active against a range of both Gram-positive and Gram-negative bacteria, including methicillin-resistant *Staphylococcus aureus* (MRSA), at micromolar concentrations [6].

As previously stated, Gram-positive bacteria have been found to be more susceptible to the antibacterial effect of lipids than Gram-negative bacteria. Nevertheless, there are exceptions from this, as has been shown by Bergsson, Thormar and coworkers, who reported that *Chlamydia trachomatis, Neisseria gonorrhoea* and *Helicobacter pylori* are extremely susceptible to both free fatty acids and monoglycerides, and that *Campylobacter jejuni* is easily killed by capric acid and monocaprin, its 1-monoglyceride [7–9]. Additionally, Altieri *et al.* investigated the effectiveness of lauric (C12 : 0), myristic (C14 : 0) and palmitic (C16 : 0) acids and their monoglycerides against the Gram-negative bacterias *Escherichia coli, Yersinia enterocolitica* and *Salmonella* spp. A total of 20 ppm of myristic and palmitic acids and their monoglycerides showed a promising bioactivity (60 to 80%) against *E. coli* within 10 to 24 hours and 50 ppm of glycerol monolaurate (monolaurin) inhibited *Y. enterocolitica* and *E. coli* by more than 90% of control in 96 hours. Otherwise, 40 ppm of monolaurin and 30 to 50 ppm of lauric acid reduced *Y. enterocolitica* growth by more than 65% of control. The effect of lauric acid and its monoglyceride against *Salmonella* spp. was moderate, with inhibition of approximately 30% [10].

The exact process of the antibacterial activity of fatty acids is not clear but research has shown that the main target of action is the cell membrane, where the fatty acids disrupt the electron transport chain and oxidative phosphorylation. The antibacterial action could also result from inhibition of enzyme activity and impairment of nutrient uptake, as well as generation of peroxidation and autooxidation degradation products or direct lysis of bacterial cells [11,12]. It has been reported that by applying heat or lowering the pH, *E. coli* and related Gram-negative bacteria are killed by lipids [13,14]. This suggests that heat or low pH removes a permeability barrier in the outer membrane of the bacteria, allowing penetration of the lipid to the inner cell membrane [8].

There is no general conclusion regarding which lipid structure gives the highest microbicidal activity but research has shown that the intensity of the antimicrobial effect is dependent on the chemical structure of the lipid, such as the length of the hydrocarbon chain, the degree of unsaturation and esterification. Of the medium-chain saturated fatty acids, capric acid and lauric acid have been found to be the most active against a range of bacteria.

7.2.2 Antiviral Activity

Thormar *et al.* [15–18] tested the effect of fatty acids with chain lengths ranging from 4 to 20 carbons against enveloped viruses. The results showed that medium-chain saturated fatty acids and monoglycerides and long-chain unsaturated fatty acids have generally the highest virucidal activity and rapidly kill enveloped viruses such as herpes simplex virus (HSV), human immunodeficiency virus type 1 (HIV-1), the lentivirus maedi-visna virus, respiratory syncytial virus (RSV) and parainfluenza and influenza viruses. The activity of unsaturated fatty acids increased with increasing number of double bonds. Monoglycerides were antiviral in lower concentrations than the corresponding fatty acids, but diglycerides were found to be inactive. Only enveloped viruses were inactivated by the lipids [15, 16, 18, 19]. Comparative studies have shown that, with a few exceptions, 1-monoglyceride of the 10-carbon saturated capric acid, monocaprin, is most active against the viruses and bacteria that have been studied; that is, it kills them most rapidly and in the lowest lipid concentration.

Several studies have shown disruption of the viral envelope by virucidal fatty acids, causing disintegration of the viral particles [15, 20]. The virucidal activity of lipids is therefore apparently due to their perturbing effect on the lipid bilayers of the envelope leading to disintegration, as has also been demonstrated for bacterial cell membranes [21]. The exact mechanism of membrane disruption and the reason for the difference in virucidal activities of various fatty acids and monoglycerides are not clear but seem to be related to the length of their carbon chain and the degree of unsaturation, both of which may affect their ability to penetrate the lipid bilayer. Acid environment, pH 4 to 5, generally increases the virucidal activity of lipids, not only of fatty acids but also of monoglycerides and fatty alcohols. It has been suggested that low pH affects viral envelope proteins so that the lipids have easier access [16, 22]. Hilmarsson and coworkers evaluated the effect of pH on the activity of medium- and long-chain fatty alcohols and lipids on RSV and parainfluenza virus type 2, as well as on HSV types 1 and 2, and found that most of the compounds were more active at pH 4.2 than at pH 7 [17, 22]. Bartolotta *et al.* studied the inhibitory activity of saturated fatty acids of variable chain length (C10 to C18) against the multiplication of *Junin virus*. The most active inhibitor was lauric acid and the authors concluded that the action of lauric acid on the virus was due to stimulation of triacylglycerol production in the host cell leading to decreased insertion of viral glycoproteins into the plasma membrane, thereby inhibiting virus maturation and release [23].

7.2.3   Antifungal Activity

Compared to saturated fatty acids, unsaturated fatty acids with double and/or triple bonds are, in general, more potent against fungal pathogens. Undecylenic acid (C11 : 1) has antifungal, antibacterial and antiviral activity and has been used commercially mainly for its antifungal properties [24]. Oleic acid has been found to be fungistatic against a wide spectrum of saprophytic moulds and yeasts. The fatty acid causes a delay of six to eight hours in the germination of fungal spores and is very effective at a low concentration of 0.7% (v/v). The application of this property of oleic acid has found use in food and cosmetics [25].

Cells of the yeast *Candida albicans* appear to be killed by disruption of the cell membrane. The susceptibility of *C. albicans* to several fatty acids and their 1-monoglycerides was tested with a short inactivation time, and ultrathin sections were studied by transmission electron microscopy after treatment with capric acid. The results showed that capric acid, a 10-carbon saturated fatty acid, caused the fastest and most effective killing of all three strains of *C. albicans* tested, leaving the cytoplasm disorganized and shrunken because of a disrupted or disintegrated plasma membrane. Lauric acid, a 12-carbon saturated fatty acid, was the most active at lower concentrations and after a longer incubation time [26]. An evaluation of the microbicidal activity of solutions containing monocaprin against *C. albicans* showed activity against *C. albicans* in 0.156 mg/ml concentrations in planktonic phase and in 2.5 mg/ml in the surface-growing phase. This indicated that monocaprin may have potential as a topical agent against *C. albicans* infection [27].

7.3 Lipids in Pharmaceuticals

Lipids have conventionally been used in pharmaceuticals as solvents for fat-soluble substances, as the oily phase in ointments and creams and as emulsifiers, antioxidants and stabilizers. A more recent use has been as penetration enhancers for drug delivery to the skin and as carriers in liposomes and in micro- and nanoparticles. Encapsulating a drug in a liposome can substantially affect its properties, influence its distribution in the body, decrease its toxic side effects and increase the efficacy of treatment. This has been successfully used for the drugs amphotericin B (fungal infections) and doxorubicin (cancer). Liposomes have been used for dermal delivery as they can increase the residence time of drugs in the stratum corneum and the epidermis, while reducing the systemic absorption of the drug. For the past decade there has also been interest in nanoparticles made from solid lipids (solid lipid nanoparticles) as a novel drug-delivery system, where the lipids used can be highly purified triglycerides, complex glyceride mixtures or even waxes [28, 29].

7.3.1 Lipids in Pharmacopeias

Monographs for 'medium-chain triglycerides', a mixture of triglycerides of saturated fatty acids, mainly of caprylic acid (C8 : 0) and of capric acid (C10 : 0), have been included in several pharmacopeias for a number of years. In the past few years monographs for several other lipids have appeared in pharmacopeias. In the *European Pharmacopoeia* 6th edn 2010 there are monographs for glycerol monocaprylate and glycerol monocaprylocaprate, which is a mixture obtained by direct esterification of glycerol with caprylic and capric acids. Additionally there are monographs for glycerides of omega-3 acids and of esters of propylene glycol and lauric acid. Many fatty acids are generally recognized as safe (GRAS) and are approved by the United States Food and Drug Administration (FDA) and by the European Union (EU) as inactive ingredients in a number of products.

7.3.2 Microbicidal Lipids for Treatment of Mucosal and Skin Infections

A number of studies have suggested that microbicidal lipids, particularly fatty acids, play a role in the natural defence against infections in skin and mucosal membranes [30]. It is logical to assume that topical dosage forms containing microbicidal lipids can be used to treat such infections and thus support the natural defence. With the increased prevalence of drug-resistant microbes, the use of microbicidal lipids could be a good alternative or an addition to conventional antibiotics. Research on the microbicidal activity of lipids has shown that the activity profiles of lipids differ among microbes, that is it varies which lipids are the most active against each bacterium or virus. The *in vitro* studies have shown that medium-chain saturated fatty acids and monoglycerides and long-chain unsaturated fatty acids in most cases have the highest microbicidal activity and rapidly kill enveloped viruses such as HSV, HIV-1, RSV and parainfluenza and influenza viruses [15–17, 19, 22]. Both Gram-positive and Gram-negative bacteria as well as *C. albicans* are also susceptible to long-chain unsaturated and medium-chain saturated fatty acids and monoglycerides, but some of them only at low pH [7–9, 21, 26].

It has been demonstrated that fatty acids can be used to reduce bacterial infection on skin. Clark *et al.* showed that a purified skin fatty acid, *cis*-6-hexadecanoic acid (C16 : 1Δ6), was effective in treating systemic and topical infections in mice by *S. aureus* and stated that the results indicate that natural defence mechanisms can be exploited to combat drug-resistant pathogens [31]. Lukowski *et al.* used emulsions containing lipids extracted from microalgae to prevent dermal colonization of MRSA strains [32].

Knowledge of the microbicidal activity of lipids has stimulated interest in utilizing this information for therapeutic use. In the formulation of dosage forms with lipids as the active ingredient it is important to determine which lipids have the highest microbicidal activity against various pathogens and are therefore the most desirable as active ingredients. Pharmaceutical formulations have been developed as vehicles for microbicidal lipids for the purpose of preventing transmission of pathogens to mucosal membranes or for the treatment of established mucosal or skin infections caused by a virus, a bacterium or a fungus. The lipid that was first used commercially is undecylenic acid, a C11 unsaturated fatty acid with a single double bond at C10, but its microbicidal properties have been studied for more than half a century [24, 33]. Undecylenic acid has antifungal, antibacterial and antiviral properties but has mainly been used commercially as a fungicide. It is FDA approved in over-the-counter medications for skin and nail disorders as well as in antidandruff shampoos and antimicrobial powders and creams [34]. In a double-blind, placebo-controlled clinical trial of undecylenic acid cream it was found to significantly reduce the incidence and duration of viral shedding in recurrent herpes labialis, but the clinical benefits were mainly restricted to patients initiating therapy during the prodrome [35]. Thus, it appeared that early therapy was effective, suggesting that prophylaxis should be evaluated. Following from this it was proposed that undecylenic acid could potentially be effective as a topical microbicide to prevent the spread of sexually transmitted diseases. However, an experiment using mouse and guinea pig models of genital herpes to evaluate the suitability of undecylenic acid as a microbicide showed that significant protection against HSV infection was only provided when undecylenic acid was applied immediately before viral inoculation, indicating that better formulations were needed to extend the duration of protection [33].

In the search for a microbicide that can prevent transmission of HIV-1 and other sexually transmitted diseases there are indications that some monoglycerides could be of value. Monocaprin kills HSV, HIV-1 and *N. gonorrhoea* in large numbers within one minute and *C. trachomatis* in 2.5 minutes, but all of these pathogens infect mucosal membranes, such as genital mucosas. Because monocaprin rapidly kills sexually transmitted bacteria and viruses, hydrogels were prepared which contained monocaprin as the active ingredient [36–38]. These hydrogels contained propylene glycol as a solvent and carbopol and sodium carboxymethyl cellulose as gel-forming ingredients with good adhesion to mucosal membranes. The results indicated that the presence of carbomer 974P (0.65%) in the formulations improves the stability of the monoglyceride but increased amounts of the surfactant polysorbate 20 reduce the microbicidal activity of monocaprin. Monocaprin maintains full microbicidal activity in hydrogels of this composition and they do not show toxic effect in the vaginal mucosa of rabbits [39]. It may be suggested that such hydrogels could be used to prevent or to counteract infections in mucosal membranes, for example sexually transmitted infections by HSV, HIV-1, *C. trachomatis* and *N. gonhorreae*, or to treat skin and mucosal infections by HSV, streptococci, staphylococci and *C. albicans*. The hydrogels were found to prevent infection by HSV-2 in the vaginal mucosa of mice and to reduce skin infection by HSV in hairless mice, without causing irritation or other toxic effects [40]. They did not cause irritation or other discomfort in the vaginal and cervical membranes of young women who participated in a small phase 1 clinical trial and they did not adversely affect the normal bacterial flora [41].

Monolaurin has been considered as a possible microbicide to prevent the transmission of HIV-1. Studies of vaginal transmission of the simian immunodeficiency virus (SIV) in a rhesus monkey model point to opportunities at the earliest stages of infection in which a vaccine or microbicide might be protective. Li *et al.* have shown that monolaurin, which has inhibitory activity against the production of MIP-3 alpha and other proinflammatory cytokines, can inhibit mucosal signalling and the innate and inflammatory response to HIV-1 and SIV *in vitro*, and *in vivo* it can protect rhesus monkeys from acute infection despite repeated intravaginal exposure to high doses of SIV. This new approach, possibly linked to interference with innate host responses that recruit the target cells necessary to establish systemic infection, opens a promising new avenue for the development of effective interventions to block HIV-1 mucosal transmission [42]. Strandberg *et al.* carried out a randomized, double-blind clinical study on the effects of monolaurin on *Lactobacillus*, *Candida* and *Gardnerella vaginalis* vaginal microflora in colonized or infected women (*n* = 36). Women self-administered intravaginal gels containing 0, 0.5 or 5% monolaurate every 12 hours for two days. Vaginal swabs were collected before and immediately after the first gel administration and 12 hours after the final gel administration. *In vitro* monolaurin concentrations of 500 μg/ml killed *Candida*, while a concentration of 10 μg/ml was bactericidal for *G. vaginalis*. Control and monolaurate gels applied vaginally in women did not alter vaginal pH or *Lactobacillus* counts and no adverse events were reported. Control gels reduced *G. vaginalis* counts but not *Candida* counts, whereas monolaurate gels reduced both *Candida* and *G. vaginalis*. The results show that vaginal monolaurin gels in women do not affect *Lactobacillus* negatively but significantly reduce *Candida* and *G. vaginalis* [43].

Lauric acid has shown antibacterial activity against skin infections. A preliminary experiment in which lauric acid was combined with aminoglycosides showed a synergistic effect against MRSA [44]. Lauric acid has been evaluated as a potential treatment against *Propionibacterium acnes*, a Gram-positive bacterium that promotes inflammatory acne. A study assessing the antimicrobial property of lauric acid against *P. acnes* both *in vitro* and *in vivo* reported considerably lower minimum inhibitory concentration (MIC) values than those of benzoyl peroxide, a commonly used over-the-counter drug for acne treatment [45]. In further work, lauric acid, which has a low aqueous solubility, was incorporated into liposomes as a step in formulating a suitable system for topical delivery. The results showed that the antimicrobial activity of lauric acid against *P. acnes* was well maintained and even enhanced at a low concentration of the fatty acid, indicating that a liposome formulation containing lauric acid could be an effective treatment for acne vulgaris and other *P. acnes*-associated diseases [46]. Carpo *et al.* carried out an *in vitro* study of the sensitivity and resistance of organisms in culture isolates from skin infections and the mechanisms of action of monolaurin as compared with six common antibiotics: penicillin, oxacillin, fusidic acid, mupirocin, erythromycin and vancomycin. Sensitivity rates to 20 mg/ml monolaurin were determined for *S. aureus*, *Streptococcus* spp., *E. vulneris*, *Enterobacter* spp. and *Enterococcus* spp. The results showed that monolaurin has statistically significant *in vitro* broad-spectrum activity against Gram-positive and Gram-negative bacterial isolates from superficial skin infections and most of the bacteria did not show resistance to it [47].

A randomized, double blind, placebo-controlled clinical trial was carried out with the objective of investigating the antiviral and wound-healing effects of a hydrogel containing either monocaprin or a combination of monocaprin and a low dose of doxycycline against herpes labialis. The rationale for the combination was that monocaprin efficiently inactivates HSV while tetracyclines are inhibitors of matrix metalloproteinases (MMP) that are part of the inflammatory response and contribute to the breakdown of tissue in ulcers. Subjects were divided into two groups: 1) with prodromal symptoms of cold sore and 2) with a vesicle. Both groups applied the hydrogel five times a day for five days. For the monocaprin and doxycycline groups the mean number of days to healing were 5.5 days (prodromal) and 5.1 days (vesicles/sores) respectively, a significantly shorter time than for the placebo groups with 7.25 and 7.5 days respectively ($p < 0.05$). The results indicate that combining monocaprin with low-dose doxycycline offers an effective treatment for cold sores, significantly reducing the time to healing and pain compared with the placebo and monocaprin alone. The latter had some effect, though not reaching significance [48].

It has been suggested that fatty acids could be used in the prevention of dental caries or for treatment of infections in the oral mucosa [27, 49]. In a recent study, monocaprin solutions were tested *in vitro* against a number of microorganisms commonly found in the oral cavity. *C. albicans* was the most sensitive to monocaprin but *Streptococcus mutans* also showed appreciable sensitivity. This indication, that monocaprin may have potential as a topical agent against *Candida*, was subsequently tested in an open study of denture disinfection in 32 patients attending a geriatric daycare centre. A significant, but short-term, reduction in counts of *Candida* on the fitting surface of full dentures was observed [27]. The principal drawback to this study was that subjects had to rinse their dentures with a monocaprin solution and then rinse the monocaprin away before reinserting the denture. The time for anticandidal activity was thus short and no direct treatment of the oral mucosa itself was possible. It was concluded from this preliminary clinical study that a more suitable vehicle for retaining monocaprin on the denture fitting surface could be helpful; tissue conditioners/soft liners and denture adhesives might be the most suitable vehicles. Current products have no anticandidal activity, which somewhat limits their usefulness. Preliminary studies *in vitro* on a monocaprin tissue conditioner formulation have already shown a slow release of monocaprin from two such conditioners and a cidal effect on strains of *C. albicans*, *C. dubliniensis* and *C. glabrata*. A similar *in vitro* study incorporating monocaprin into a denture adhesive showed comparable results with a cidal effect of monocaprin on cultures of strains of the same three species of yeast in contact with the modified adhesive [50]. The results indicate that monocaprin may have potential as a broad-spectrum antimicrobial agent for topical use and for inhibiting biofilm accumulation on mucosal surfaces and on equipment and devices, such as catheters, that are difficult to disinfect once in use. Reducing the need for topical, or systemic, antifungal drugs to treat denture stomatitis would be a considerable therapeutic advance.

The antibiotic mupirocin is active against certain Gram-negative bacteria and a wide range of Gram-positive bacteria including MRSA. Mupirocin is used as a topical treatment for bacterial skin infections but there are indications that resistance to mupirocin can develop if it is used for extended periods of time, or indiscriminately used [51]. Rouse *et al.* evaluated combinations of monolaurin with lactic, mandelic, malic or benzoic acid as possible alternatives to mupirocin. The *in vitro* activity of 13 monolaurin formulations against 30 methicillin-susceptible *S. aureus* (MSSA) isolates and 30 MRSA isolates was determined. The *in vivo* activity of three monolaurin formulations was compared to that of mupirocin using a murine model of MRSA nasopharyngeal colonization. The results showed that a monolaurin ointment containing 3% monolaurin in petrolatum was more effective in eradication of MRSA than mupirocin ointment, with efficacy rates of 71% and 50% respectively. The authors concluded that monolaurin formulations warrant evaluation for *S. aureus* nasal decolonization in humans and have a potential to prevent and reduce bacterial colonization of the nose and skin [52].

The drugs presently used against RSV and other viruses causing respiratory infection are only recommended for high-risk children or severely ill infants, due to efficacy and safety concerns [53,54]. Therefore new prophylactic or therapeutic compounds for general use against respiratory viruses would be of value. Palmitoleic and oleic acids have been found to be highly active against RSV, even in low concentrations. However, because of the instability of these long-chain unsaturated fatty acids, other microbicidal compounds such as lauric acid, monolaurin and particularly monocaprin are more practical candidates as active ingredients in topical formulations against RSV, parainfluenza and influenza virus infections [17]. Pharmaceutical formulations were designed containing those lipids and/or alcohols which are most active against RSV and parainfluenza virus type 2 *in vitro* according to previous studies. The formulations were composed so that they were suitable for treatment of respiratory infections, for example as solutions for mouth and throat rinse, nose and throat spray or nasal drops. The amount of the active compounds used was in the lowest possible concentration and the activity was increased by lowering the pH of the formulations. The activity of all formulations was tested against RSV as well as parainfluenza virus type 2 and influenza A virus *in vitro*. The most active formulations were tested in a mouse model for RSV infection, first with respect to toxicity and second with respect to their activity in preventing or reducing infection. The results showed that treatment with a pharmaceutical dosage form containing monocaprin and lauric acid as virucidal ingredients significantly reduced the viral load in the nasal mucosa of rats intranasally infected with RSV. This conclusion is supported by the observation that nasal mucosas of rats receiving saline showed redness, indicating infection which was not seen in rats treated with dosage forms. The solutions were well tolerated by the rats and did not cause weight loss or other physical abnormalities. More studies are needed to further explore the use of lipids as topical treatments against RSV infection in experimental animals or in humans [55].

In a study of the susceptibility of *Salmonella* spp., *E. coli* and *H. pylori* to fatty acids and monoglycerides, none of the lipids showed significant antibacterial activity against *Salmonella* spp. and *E. coli* but 8 of 12 lipids tested showed high activity against *H. pylori*, monocaprin and monolaurin being the most active. The high activity of monoglycerides against *H. pylori* suggests that they may be useful as active ingredients in pharmaceutical formulations for treatment of stomach ulcers caused by *H. pylori* [8, 56, 57].

7.3.3 Lipids as Penetration Enhancers in Pharmaceutical Dosage Forms

Fatty acids have been shown to interact with the lipids in stratum corneum and have been used as penetration enhancers in transdermal dosage forms to improve the delivery of drugs [58]. The penetration-enhancing effect of fatty acids is dependent upon their structure, alkyl chain length and degree of unsaturation, and fatty acids of chain length ranging from C10 to C20 have been evaluated as penetration enhancers. Of the saturated fatty acids, palmitic acid has been found to be the most effective, but of the monounsaturated fatty acids oleic acid showed the highest permeation-enhancing effect [58]. The penetration-enhancing effect of oleic acid has been known and utilized for a number of years. Sodium caprate has also been used as a penetration enhancer and recently it has been considered to promote absorption of poorly permeable drugs across the intestinal epithelium. Evidence from preclinical and clinical studies indicate that sodium caprate in solid dosage forms, designed for the initial simultaneous release of high concentrations of penetration enhancer and drug, can effectively increase the oral bioavailability of drugs that are poorly absorbed following oral administration, including peptides, oligonucleotides and polysaccharides, as well as poorly absorbed antibiotics and bisphosphonates. Mode-of-action studies suggest that the actions of high concentrations of the penetration enhancer used *in vivo* are by a transient transcellular perturbation in addition to the paracellular pathway. It seems that damage caused to the intestinal epithelium may not be physiologically relevant as it is transient, mild and rapidly reversible. Finally, although some of the safety concerns with the use of sodium caprate have been, in part, addressed in recent clinical studies and by its presence in a marketed antibiotic suppository, certain safety aspects may require additional consideration [59].

The effect of polyunsaturated fatty acids (PUFAs) on drug delivery through the skin has been investigated. The effects of EPA (C20 : 5), docosahexaenoic acid (DHA, C22 : 6) and several commonly used penetration enhancers on the *in vitro* percutaneous absorption of the antihypertensive atenolol from a series of formulations were evaluated. The best permeation profile for the atenolol formulations was obtained with formulations containing either the conventional penetration-enhancer transcutol or the fatty acids. The results showed that PUFAs increased the apparent diffusion coefficient of the drugs and suggested that atenolol transdermal delivery could be feasible [60]. A number of studies have evaluated the effect of cod-liver oil extract on the penetration of drugs through skin [61]. Loftsson and coworkers evaluated the effect of fatty acid extract from cod-liver oil on HSV-1 and the transdermal delivery of acyclovir. The fatty acid extract from cod-liver oil increased the flux of acyclovir 50- to 70-fold, with maximum enhancement being obtained at 5% concentration of the fatty acid extract [62]. Tsutsumi *et al.* reported that a cod-liver oil extract increased the buccal permeation of ergotamine, a drug used clinically in the treatment of migraines but with a rather poor bioavailability following oral administration. The results indicated that the cod-liver oil extract mainly enhanced the permeation of the nonionized form of ergotamine [63].

7.4   Microbicidal Lipids in Agriculture and Aquaculture

There is a growing awareness that antibiotics should be used more carefully in the food industry. Antibiotics have been used in attempts to control bacterial disease in aquaculture, which can be a direct threat to human health as well as to the environment. Since the 1940s antibiotics have been used as growth promoters in animal feeds, with the rationale that the effect of antibiotics on intestinal microbial populations allows a faster and more efficient growth of the animal, but it has also been proposed that the mechanism could be inhibition of inflammatory response [64]. Concerns regarding the development of antimicrobial resistance and the transference of antibiotic-resistance genes from animal to human microbiota has led to withdrawal of approval for antibiotics as growth promoters in the EU from 2006 [65]. It is important to develop processes to control the transmission of bacteria from food, particularly poultry, to humans [66]. New and more efficient methods to decontaminate or prevent the contamination of food such as raw meat and poultry by pathogens such as *Campylobacter*, *Salmonella*, and *E. coli* would therefore be desirable. *C. jejuni* is the most common cause of foodborne infection. Thormar *et al.* tested the effectiveness of 11 fatty acids and monoglycerides against *C. jejuni*, and found that monocaprin was the most active in killing the bacterium [13]. A procedure was developed to manufacture 200 mM concentrated emulsions of monocaprin that upon dilution with tap water caused a more than 6 to 7 $\log_{10}$ reduction in viable *C. jejuni* count in one minute at room temperature. The monocaprin emulsions killed a variety of *Campylobacter* isolates from humans and poultry and also killed strains of *C. coli* and *C. lari*, indicating a broad anticampylobacter activity. The results showed that lowering the pH to 4 to 5 caused a greater than 6 to 7 $\log_{10}$ reduction in viable bacterial counts of *Salmonella* spp. and *E. coli* in 10 minutes. *C. jejuni* was also more susceptible to monocaprin emulsions at low pH [13]. Subsequently it was studied whether *Campylobacter* infection in chickens would be affected by adding monocaprin emulsions to their drinking water and feed. It was found that addition of monocaprin to water and feed did not prevent the spread of *Campylobacter* from artificially infected to noninfected 24-day-old chickens, but *Campylobacter* counts in cloacal swabs were significantly reduced, particularly during the first two days of treatment. There was also a significant reduction in the *Campylobacter* counts in cloacal swabs of naturally infected 36-day-old broilers that were treated for three days prior to slaughter. Addition of monocaprin to drinking water and feed two to three days before slaughter might therefore be considered as a means of reducing *Campylobacter* infection in broilers, possibly in combination with other antibacterial agents such as short-chain organic acids. The authors showed that addition of 5 and 10 mM monocaprin emulsions to *Campylobacter*-spiked chicken feed significantly reduced bacterial contamination. The results indicated that monocaprin emulsions could potentially be used to control the spread of foodborne bacteria from poultry to humans [67].

As short-chain fatty acids (SCFAs) have shown bacteriostatic activity they have been considered as biocontrol agents in animal production. The bacteriostatic activities of SCFAs have been studied most intensively in enterobacteria such as *Salmonella* spp., *E. coli* and *Shigella flexneri* [68]. The results are affected by the administration mode, the type and concentration of acid and the condition of the animals. Researchers have developed methods to microencapsulate SCFAs that prevent the absorption of these acids in the upper gastrointestinal tract in order to ensure their release further down [69, 70]. Most commercial products that are used in the field contain propionic (C3 : 0) and formic acids (C1 : 0), either in powder form or encapsulated in silica beads. Coated butyric acid (C4 : 0) was tested in chicken feed for its effect on *Salmonella* colonization but although the number of chickens shedding *Salmonella* was significantly lower in the group receiving butyric acid, cecal colonization at slaughter age was equal to that of the control group of chickens [71]. Van Immerseel *et al.* [72] tested the effect of feed supplementation with microencapsulated SCFAs on *Salmonella* invasion. The administration of the butyric acid-impregnated microbeads resulted in significantly decreased colonization by *S. enteritidis* in the cecum but feed supplementation with acetic acid (C2 : 0) and to a lesser extent formic acid led to an increased colonization of the cecum and other internal organs. The SCFAs appear to be not as effective against *Salmonella* as MCFAs (C6 to C12) [13, 73]. Another study evaluated SCFAs as biocontrol agents in animal production, with emphasis on polyhydroxyalkanoates (PHAs), which are polymers of beta-hydroxy SCFAs. These biopolymers can be depolymerized by many different microorganisms that produce extracellular PHA depolymerases. Several studies have provided some evidence that PHAs can also be degraded upon passage through the gastrointestinal tract of animals and consequently adding these compounds to the feed might result in biocontrol effects similar to those described for SCFAs [74]. Based on reports that poly-β-hydroxyalkanoate polymers can be degraded into β-hydroxy-SCFA and on the observation that β-hydroxybutyrate has the same positive effect towards challenged brine shrimp as other SCFAs, it has been suggested that β-hydroxybutyrate could be used as a biocontrol agent to fight luminescent vibriosis, that is *Vibrio harveyi* and closely related bacteria such as *Vibrio campbellii* and *Vibrio parahaemolyticus*, by addition to the culture water or feed [75].

In commercial rabbit farming diarrhoea caused by *E. coli* is the major reason for high morbidity of young rabbits. In an experiment in which rabbits were infected experimentally with *E. coli*, rabbits that were fed a diet supplemented with caprylic acid or triacylglycerols of caprylic and capric acids had significantly reduced numbers of coliform bacteria in the faeces and a lower rate of mortality compared to rabbits that were fed a nonsupplemented diet. These results indicate that including antimicrobial lipids in the diet may improve the resistance of rabbits to enterocolitis [76]. A further study of the effects of different lipids in a supplemented diet on the mortality of early-weaned rabbits showed that the mortality of rabbits fed a diet supplemented with oil containing triacylglycerols of caprylic and capric acids (23%) was significantly lower than that of rabbits fed palm fat (45%), and nonsignificantly lower than that of rabbits fed coconut oil (37%) [77]. A study carried out in piglets showed that a diet supplemented with lipids and lipolytic enzymes had a suppressive effect on the pig proximal gut flora, which is essential to obtaining a growth promotion comparable to that obtained with antibacterials, without the risks of the latter [78].

The use of lipids instead of antibiotics as animal-feed supplements has many advantages. In addition to those previously reported, it has been discovered that fatty acids such as capric acid have antimethanogenic activity and potential to reduce the emission of methane from ruminants to the atmosphere [79].

It has been suggested recently that the cost of adding lipids to animal feeds could be high, but Desbois and Smith have pointed out that single-celled algae could be an inexpensive source of fatty acids [66, 80]. Cell lysates from the marine diatom, *Phaeodactylum tricornutum*, have antibacterial activity and the monounsaturated fatty acid palmitoleic acid and the polyunsaturated fatty acids EPA (C20 : 5) and hexadecatrienoic acid (HTA, C16 : 3) have been identified as the compounds responsible for this activity [6, 80]. Algae are presently being seriously considered as feedstocks for next-generation biofuel production as their high productivity and the associated high lipid yields make them attractive [81].

There has been interest in assessing the antifungal activity of fatty acids against phytopathogenic fungi in agriculture. The effects of the fatty acids linolenic acid (C18 : 3), linoleic acid (C18 : 2), erucic acid (C22 : 1) and oleic acid (C18 : 1) on the growth of the plant pathogenic fungi *Rhizoctonia solani*, *Pythium ultimum*, *Pyrenophora avenae* and *Crinipellis perniciosa* were examined in *in vitro* studies. Linolenic and linoleic acids exhibited activity against all of the fungi; oleic acid had no significant effect on the growth of *R. solani* or *P. avenae*, but gave significant reductions in mycelial growth of *P. ultimum* and reduced the growth of *C. perniciosa* significantly; erucic acid had no effect on fungal growth at any concentration examined [82]. Liu *et al.* [83] evaluated the antifungal activities of nine fatty acids – butyric, caproic, caprylic, capric, lauric, myristic, palmitic, oleic and linoleic acid – against four phytopathogenic fungi – *Alternaria solani*, *Colletotrichum lagenarium*, *Fusarium oxysporum* sp. *cucumerinum* and *Fusarium oxysporum* sp. *Lycopersici* – by measuring mycelial growth and spore germination via Petri-dish assay. Except for oleic acid, the fatty acids tested were observed to inhibit the mycelial growth of one or more tested fungi. In addition to the suppression of mycelial growth, butyric, caproic, caprylic, capric, lauric and palmitic acid showed an inhibitory effect against spore germination; the extent of inhibition varied with both the type of fatty acid and the fungus. In particular, capric acid displayed a strong inhibitory effect on mycelial growth and spore germination of *C. lagenarium*. The saturated fatty acids showed stronger antifungal activity than the unsaturated fatty acids. Mixing palmitic and oleic acids in the soil where poor plant growth had been observed enhanced the growth of tomato and cucumber seedlings [83]. The antifungal activity demonstrated by the fatty acids in these studies suggests that fatty acids might be applicable as alternative approaches to integrated control of phytopathogens.

7.5 Lipids in Therapy

7.5.1 Effect of Lipids on Infectious and Inflammatory Diseases

A large number of researchers have investigated the effect of fatty acids on the progress or prevention of diseases. There are reports from the early part of the twentieth century about the effect of lipids on infectious diseases, such as leprosy and tuberculosis, which are caused by mycobacterial organisms. Among the lipids that were found to be effective were salts of the unsaturated fatty acids from chaulmoogra oil, which is derived from the seeds of Taraktogenos kurzii king tree, and from cod-liver oil [84, 85]. However after the discovery in the 1940s of isoniazide and the sulfones (derivatives of 4,4′-diaminodiphenylsulfone), which were found to be effective against mycobacterial organisms, the interest in the lipids faded away. Interest in the activity of lipids against infectious diseases may be rekindled as various marine organisms such as algae, echinoderms, sponges, tropical rays and many other invertebrates that are a source of fatty acids have recently been investigated. The most intriguing marine invertebrates seem to be sponges that commonly contain very long-chain fatty acids. Little is known about the biomedical potential of these unusual sponge fatty acids, in particular how strong their bioactivity is compared to that reported for the more common fatty acids. The main interest has been focussed on the possible antimalarial, antimycobacterial and antifungal properties of these fatty acids [86–89].

There has for a long time been interest in the beneficial effects of fish oil on human health, mainly cod-liver oil, but recently the activity of fatty acids from other marine sources such as seal oil has been investigated [90]. The omega-6 fatty acid arachidonic acid (C20 : 4) gives rise to eicosanoid mediators that have well-known roles in inflammation, and arachidonic acid metabolism is a long-recognized target for commonly used antiinflammatory therapies. A number of studies have demonstrated the benefit of fish oil in arthritis. A metaanalysis published in 1995 by Fortin *et al.* [91], where the results of 10 published studies were evaluated, showed that dietary fish oil supplementation for three months resulted in a modest, statistically significant improvement in reduced tender joint count and morning stiffness as compared with heterogeneous dietary control oils. Fish oil has a high content of EPA and DHA, but EPA and DHA have been found to be strong inhibitors of the inflammatory consequences of eicosanoids derived from arachidonate. They also give rise to mediators that are less inflammatory than those produced from arachidonic acid or are antiinflammatory. In addition to modifying the lipid mediator profile, omega-3 PUFAs exert effects on other aspects of inflammation such as leukocyte chemotaxis, expression of adhesion molecules and production of inflammatory cytokines. The effects of EPA and DHA are not differentiated in most studies. It has been shown that EPA and DHA have different effects on leukocyte functions such as phagocytosis, chemotactic response and cytokine production. DHA and EPA modulate differently expression of genes in lymphocytes. Activation of intracellular signaling pathways involved with lymphocyte proliferation is also differently affected by these two fatty acids [92]. Recently, EPA- and DHA-derived lipid mediators called resolvins, E- and D-series, have been described. They have potent antiinflammatory properties in model systems. These mediators might explain many of the antiinflammatory actions of omega-3 fatty acids that have been described. In addition to modifying the profile of lipid-derived mediators, fatty acids can also influence peptide mediator; that is, cytokine production. To a certain extent this action may be due to the altered profile of regulatory eicosanoids, but it seems likely that eicosanoid-independent actions are a more important mechanism [93].

Dose-dependent gastrointestinal and cardiovascular side effects can limit the use of nonsteroidal antiinflammatory drugs (NSAIDs) in the management of rheumatoid arthritis (RA). A dual-centre, double-blind placebo-controlled randomized study of nine months' duration was carried out to determine whether cod-liver oil supplementation helps reduce daily NSAID requirement of patients with RA. Ninety-seven patients with RA were randomized to take either 10 grams of cod-liver oil containing 2.2 grams of omega-3 essential fatty acids (EFAs) or air-filled identical placebo capsules. Fifty-eight patients (60%) completed the study. Thirty nine per cent of the patients in the cod-liver oil group and 10% in the placebo group were able to reduce their daily NSAID requirement by more than 30% ($P = 0.002$, chi-squared test). No differences between the groups were observed in the clinical parameters of RA disease activity or in the side effects observed. The study suggests that cod-liver oil supplements containing omega-3 fatty acids can be used as NSAID-sparing agents in RA patients [94].

Dietary supplementation with fish oil is not an efficient way to obtain sufficient blood levels of the PUFA and a number of studies have focussed on the formulation and evaluation of a drug delivery system for topical use, containing antiarthritis drugs, ibuprofen or ketoprofen, in additon to DHA and EPA. The efficacy of topical preparations can be limited because of poor rates of transcutaneous delivery, chiefly attributable to the barrier function of the skin. Heard *et al.* [95] investigated a topical antiarthritis system, as the permeation of ibuprofen or ketoprofen plus EPA and DHA was determined from a fish-oil vehicle across pig-ear skin *in vitro*. At 12 hours, the ketoprofen : ibuprofen ratio of the moles permeated was 0.27, the ratio of EPA permeated simultaneously with ketoprofen and ibuprofen was 0.22 and the ratio of DHA permeated simultaneously with ketoprofen and ibuprofen was 0.24. This showed that simultaneous permeation of NSAIDs and EFAs, EPA and DHA, from a formulation containing fish oil is feasible. In addition, for both NSAIDs, the relative rates of permeation of EPA and DHA were in proportion to their levels in the fish oil and the permeation rate of each individual fatty acid was higher when the permeation rate of the solute was greater. This suggests that the greater the rate of permeation of the NSAID, the greater the rate of permeation of the vehicle [95].

Evidence of the clinical efficacy of EPA and DHA is reasonably strong in some settings, for example in rheumatoid arthritis, but is weak in others, for example in inflammatory bowel diseases and asthma. Large, well-designed trials are required to assess the therapeutic potential of long-chain omega-3 PUFAs in inflammatory diseases [96].

7.5.2 Effect of Lipids on Psoriasis

Zulfakir *et al.* [97] have reviewed the possible use of topically delivered EPA in the treatment of psoriasis. The basis for using EPA and/or its metabolites in psoriasis is given by reports suggesting that it has antiinflammatory properties. EPA use in psoriasis has been demonstrated in trials using oral, intravenous and topical preparations, with generally positive outcomes. Depth profile analysis revealed that EPA and its metabolite, 15-hydroxyeicosapentaenoic acid (15-HEPE), are deposited in the epidermis, particularly in the metabolically active basal layer, which is considered advantageous in psoriasis therapy. Currently there are many unknowns about psoriasis aetiology and it is not known what effect blocking of different cytokines has on the disease progression. More information is needed about the effects of EPA on cellular immunity other than via prostaglandin and leukotriene synthesis to fully understand the mode of action of EPA. EPA could have a potential role in the treatment of psoriasis, in particular in topical treatments either as an active antiinflammatory agent by itself or as a dual-action permeation enhancer for other antipsoriatic treatments. The challenges include optimizing the delivery of EPA to the skin and determining the derivatives of EPA which would give maximal effects, and overcoming pharmacokinetic and formulation problems to optimally deliver EPA to its intended target cells [98].

7.5.3 Effect of Lipids on Cardiovascular Diseases

There have been numerous published studies on the effect of lipids on hyperlipidemia and cardiovascular diseases [99–102]. A study published in 2009 reported a metaanalysis to quantitatively evaluate all the randomized trials of fish oils in hyperlipidaemic subjects. The final analysis comprised 47 studies in otherwise untreated subjects taking fish oils with weighted average daily intake of 3.25 grams of EPA and/or DHA. The conclusion was that fish-oil supplementation produced a clinically significant dose-dependent reduction of fasting blood triglycerides but not total HDL or LDL cholesterol in hyperlipidaemic subjects [103].

The first and only EU- and FDA-approved omega-3-derived prescription drug is Omacor/Lovaza. A capsule of Omacor (omega-3 acid ethyl esters) contains at least 900 mg of the ethyl esters of omega-3 fatty acids. These are predominantly a combination of ethyl esters of EPA, approximately 465 mg, and DHA, approximately 375 mg. Omacor is prescribed as an adjunct to diet for the treatment of elevated levels of triglycerides. The synthesis of triglycerides is inhibited through reduced production of triglycerides in the liver, as EPA and DHA are poor substrates for the enzymes responsible for triglyceride synthesis. EPA and DHA also inhibit esterification of other fatty acids. Omacor increases peroxisomal beta-oxidation of fatty acids in the liver. Omacor is also approved in European and certain Asian markets for the secondary prevention of post-myocardial infarction, the period following the initial survival of a hearth attack [104].

There is an association between nonalcoholic fatty liver disease (NAFLD) and the polycystic ovary syndrome (PCOS). A randomized, crossover study was carried out on 25 women with PCOS with a mean age of 32.7 years and mean body mass index of 34.8 kg/m^2, comparing 4 grams per day of omega-3 fatty acids with placebo over eight weeks. Omega-3 fatty acids significantly decreased liver fat content compared with placebo. There was also a reduction in triglycerides, systolic blood pressure and diastolic blood pressure in subjects taking omega-3 fatty acids compared with placebo. Omega-3 fatty acids particularly decreased hepatic fat in women with hepatic steatosis, defined as liver fat percentage greater than 5% [105]. It was concluded that omega-3 fatty acid supplementation has a beneficial effect on liver fat content and other cardiovascular risk factors in women with PCOS, including those with hepatic steatosis. Whether this translates into a reduction in cardiometabolic events warrants further study.

7.5.4   Effect of Lipids on Cognitive Function

Research has established that DHA plays a fundamental role in brain structure and function. Neurological disorders, including Alzheimer's disease, Parkinson's disease and major depression, display a neuroinflammatory component. Hence it has been suggested that omega-3 PUFAs are promising candidates in the prevention and treatment of neurological disorders [106]. Cunnane *et al.* [107] reviewed the possible link between oily fish intake, DHA and declining cognitive function associated with ageing, with emphasis on three types of human study: epidemiological studies of fish or DHA intake, measurement of blood or brain DHA levels, and clinical trials of supplements containing DHA in ageing-related cognitive decline, Alzheimer's disease or other forms of dementia [107]. The authors concluded that in the absence of easily accessible and specific biomarkers of Alzheimer's disease, further trials are needed to assess the impact of various doses and proportions of EPA and DHA on intermediate biomarkers, such as oxidative stress, inflammation and amyloid beta concentration in older persons.

Chih-Chiang Chiu *et al.* carried out a preliminary randomized, double-blind, placebo-controlled study on the effects of omega-3 fatty acids monotherapy in Alzheimer's disease and mild cognitive impairment [108]. Twenty-three participants with mild or moderate Alzheimer's disease and 23 with mild cognitive impairment were randomized to receive omega-3 PUFAs, 1.8 grams/day, or placebo (olive oil). There was no significant difference in the change in the cognitive portion of the Alzheimer's Disease Assessment Scale (ADAS-cog) during follow-up in these two groups. However, the omega-3 fatty acids group showed a significant improvement in ADAS-cog compared to the placebo group in participants with mild cognitive impairment ($p = 0.03$), which was not observed in those with Alzheimer's disease. Higher proportions of EPA on red-blood-cell membranes were also associated with better cognitive outcome ($p = 0.003$). The authors stated that more detailed studies should be considered with a greater homogeneity of participants, especially those with mild Alzheimer's disease and mild cognitive impairment.

In the past 10 years there have been more than a dozen intervention studies conducted using various preparations of long-chain omega-3 PUFA in unipolar and bipolar depression. The majority of these studies administered long-chain omega-3 PUFA as an adjunct therapy.

The results of these studies have been conflicting as some have reported positive results and others no significant effect. The mechanisms that have been invoked to account for the benefits of long-chain omega-3 PUFA in depression include reductions in prostaglandins derived from arachidonic acid, which lead to decreased brain-derived neurotrophic factor levels and/or alterations in blood flow to the brain [109].

Lin and Su carried out a metaanalysis of 10 double-blind, placebo-controlled trials of antidepressant efficacy of omega-3 fatty acids in patients with mood disorders receiving omega-3 PUFAs, with the treatment period lasting four weeks or longer. The authors concluded that although the results showed significant antidepressant efficacy of omega-3 PUFAs, it is still premature to validate this finding due to publication bias and heterogeneity, and that more large-scale, well-controlled trials are needed to find out the favorable target subjects, therapeutic dose of EPA and the composition of omega-3 PUFAs in treating depression [110]. Hamazaki has published a summary of the effects of fish oils on aggressive behavior and hostility sense. Fourteen intervention studies have reported the effects of fish oils on aggressive behavior but the research methods were too heterogeneous to apply metaanalysis [111].

7.6 Lipids in Cosmetics

Lipids have been used in cosmetics for a long time and among the earliest examples of lipid use in cosmetics was the use of eggs in preparations for skin and hair, first whole eggs and later separated parts: egg yolks and egg-yolk oil [112]. Lipids have been used as excipients/carriers in a variety of cosmetic preparations such as creams, ointments, lipsticks and sunscreen products. The role of lipids in cosmetics is varied and ranges from emollient, moisturizer, antioxidant, penetration enhancer and occlusive skin conditioner to antibacterial agent. Cosmetics are used to improve the appearance of the user, but more recently also to benefit their target, for example the hair, skin and mucous membranes. The fatty acids most often used in cosmetics are straight-chain acids ranging from C10 to C18. Linoleic acid and linolenic acid are used in cosmetic products as they are considered to influence the metabolic processes of the skin and to promote the activity of vitamins A and E and recovery barrier properties of the stratum corneum. Liposomes and solid lipid nanoparticles represent promising carrier systems for cosmetic active ingredients due to their numerous advantages over existing conventional formulations. Glyceryl monoesters of fatty acids are frequently used in cosmetics, mostly as skin-conditioning agents, emollients and/or emulsifying agents. The glyceryl monoesters used in cosmetics are usually not pure monoesters, but mostly mixtures with mono-, di- and triesters. The fatty acids linked to the glycerol molecule can be either saturated or unsaturated, with a chain length ranging from C8 to C22.

7.6.1 Sources of Lipids Used in Cosmetics

The sources of lipids used in cosmetics are very diverse and lipids can be isolated from plants, fruits, animals and bacteria. Examples of vegetable oils are almond oil, apricot oil, avocado oil, borage oil, castor oil, coffee oil, corn oil, macadamia-nut oil, olive oil, safflower oil, sesame oil, soybean oil, walnut oil, wheat-germ oil, tomato-seed oil, peanut oil and cottonseed oil [113]. These oils are principally made up of triglycerides, but some also contain other lipophilic substances such as fatty acids, fatty alcohols, vitamins and phytosterols. Vegetable fats have also been used extensively in cosmetics. They are mixtures of heterogeneous compositions with a high content of saturated triglycerides. Examples of vegetable fats are cocoa oil, coconut oil, oil from the fruits of the African oil palm, elaeis guineensis and shea butter from the seeds of the African shea tree. Ricinoleic acid, the primary constituent in castor oil, derived from the seed of the Ricinus communis plant, is along with certain of its salts and esters used as a skin-conditioning agent, emulsion stabilizer and surfactant in cosmetics. The fungicidal undecylenic acid is a fatty acid derived from Ricinus communis oil. The highest reported use concentration (81%) for castor oil is associated with lipstick.

Animal oils used in cosmetics for various applications include fish oils. Waxes are esters formed by a combination of fatty acids and high-molecular-weight monohydroxy alcohols. Waxes used in cosmetics include, for example, jojoba oil extracted from the seeds of the jojoba shrub, bees wax, spermaceti extracted from sperm oil present in the head cavities of the sperm whale, and lanolin, which is a greasy yellow substance coming mostly from the wool of domestic sheep and secreted by their sebaceous glands.

Bacteria also produce lipids that have potential for use in cosmetics and pharmaceuticals, with *E. coli* the most studied bacterium regarding fatty acid synthesis [114]. Fatty acids can be synthesized by two related but distinct fatty acid synthase (FAS) pathways. Human cells rely on a type I FAS, whereas plants, bacteria and other microorganisms contain type II FAS pathways. This difference exposes the type II FAS enzymes as possible targets for antimicrobial drugs that have little or no side effects in the human host. Several inhibitors of type II FAS enzymes have been discovered, many of which have antibacterial activity. Extensive biochemical and structural studies have shed light on how these compounds inhibit their target enzymes, thereby laying the foundation for the design of inhibitors with increased potency [115].

7.6.2   Antimicrobial Activity of Lipids in Cosmetics

Chemical analysis of the composition of the lipids in the skin shows that free fatty acids account for 35 to 45%, of which about 20% are saturated, with palmitic acid being the most abundant. About 20% of the fatty acids are unsaturated, of which oleic acid seems most abundant. When testing *in vitro* sensitivity of several bacterial species it was noticed that *S. pyogenes* was the most sensitive of the bacteria tested and *S. aureus* somewhat less sensitive. Gram-negative bacteria, such as *Bacterium coli*, *Pseudomonas pyocyanea* and micrococci, were resistant. The activity against *S. pyogenes* and *S. aureus* was attributed to the unsaturated fatty acid fraction, apparently in oleic acid.

Cosmetics are at risk of bacterial contamination during use as they are often stored in large multidose-containers. The use of preservatives in the formulation of cosmetics is therefore essential. The advantage of using microbicidal lipids as preservatives in cosmetics is that they are nontoxic and cause little irritation to the skin and mucous membranes. The primary reason for their use as preservatives in cosmetics is their low toxicity rather than high antimicrobial activity. There are several examples of the use of lipids, mainly the monoglyceride monolaurin, as part of a preservative system in cosmetics, for example in shampoos, conditioners and moisturizers. Monolaurin needs to be used with another antimicrobial compound to achieve a wider spectrum of activity. However, since cosmetics are complex formulations with many ingredients it is important to be aware that the inter-action of lipids with other compounds in the formulation can lead to a decrease or loss of activity [2, 4]. There are several factors that need to be considered, for example solubility of the preservative, pH, interaction with other ingredients and antimicrobial spectrum of the preservative.

7.6.3 Other Activities of Lipids in Cosmetics

In the past, lipids have been used as penetration enhancers in cosmetics and recently they have been used as nanoparticles such as solid lipid nanospheres, liposomes, nanosomes and nanostructured lipid carriers. The first two cosmetic products containing lipid nanoparticles were marketed in 2005 and within three years about 30 cosmetic products containing lipid nanoparticles were available [116].

Probably the most frequent use of lipids in cosmetics is to hydrate the skin. The skin is our first line of defence against external pathogens and a weak skin leads to greater possibilities of infection. The stratum corneum is approximately 0.1 to 1.0 mm thick, depending on the body site. The stratum corneum is made up of 15 to 25 layers of dead keratinized cells that are about 75% protein, 15 to 20% lipids and fats and 5 to 10% water. A reduction in the lipid concentration will result in a drier skin with weaker defence properties. The lipids in the skin also serve the purpose of keeping the normal skin microflora in balance [117].

The skin lipids have been considered to impose their effect on the skin by forming an inert protective layer on its surface. However, as topically applied lipids penetrate the skin [118], structural lipids that are normal to the stratum corneum have been suggested to be more efficient than other types of lipid in correcting hydration and scaling disorders in the skin [119, 120]. The lipids commonly used in moisturizers are mono-, di- and triglycerides, waxes, long-chain esters, fatty acids, lanolin and mineral oils. There are a large variety of fatty acids among the glycerides used in moisturizers, with stearic acid (C18 : 0), oleic acid (C18 : 1) and linoleic acid (C18 : 2) being the most common. Topically applied polyunsaturated lipids have also been suggested to influence cutaneous inflammation because of a possible antiinflammatory action [121].

7.7 Lipids in Health Food

Traditionally, lipids have been used in foods for their nutritional value, but also as emulsifiers and to influence the consistency. Seafood and marine oils are an excellent source of fatty acids and are used extensively in health foods. The beneficial health effects of marine lipids are now well established and are mostly attributed to the omega-3 long-chain PUFAs, especially EPA and DHA [99, 100, 122].

The antimicrobial effect of lipids combined with their lack of toxicity makes them suitable candidates for preservatives in food [123]. It has been suggested that antimicrobial lipids could be added to formula milk to reduce the incidence of gastrointestinal infections in infants. Lipids previously shown to have antiviral and antibacterial activity in buffers were added to human milk, bovine milk and infant formulas to determine whether increased protection from infection could be provided to infants as part of their diet. Fatty acids and monoglycerides with chain lengths varying from 8 to 12 carbons were found to be more strongly antiviral and antibacterial when added to milk and formula than long-chain monoglycerides. Lipids added to milk and formula inactivated a number of pathogens, including RSV, HSV-1, *Haemophilus influenzae* and Group B streptococcus [17, 124]. Another study evaluated the combined effect of caprylic acid and mild heat on *Cronobacter* spp. (*Enterobacter sakazakii*) in infant formula reconstituted in sterile water and showed a synergistic action between the lipid and increased temperature. The bactericidal activity of caprylic acid increased with the temperature as the viable counts of *Cronobacter* spp. were reduced by approximately 7.8 $\log_{10}$ in samples treated with 30 mM caprylic acid for 60 minutes at 45 °C, 20 minutes at 50 °C, or 10 minutes at 55 °C. The solubility of caprylic acid increased with increasing temperatures, which could affect its enhanced ability to disrupt cell membranes [125]. The results presented in these studies suggest that the addition of medium-chain fatty acids or monoglycerides to milk and infant formulas increases protection from infection, thereby reducing the risk of infant gasteroenteritis, and that fatty acids and monoglycerides could have a role as natural antimicrobial agents for use in the dairy industry.

Research indicates that monoglycerides could be used as preservatives in certain classes of minimally processed refrigerated foods when intrinsic antimicrobial activity is inadequate [126–128]. As fatty acids and monoglycerides are effective against Gram-positive bacteria and fungi but have less activity against Gram-negative bacteria they are commonly used in combination with other preservatives. Of the lipids studied as potential preservatives in food, monolaurin has been most extensively studied. It has been found to be active against commonly found microbial contamination such as *Listeria monocytogenes*, *E. coli*, *S. aureus* and *Bacillus subtilis*. Monolaurin has in several studies been combined with other lipids such as monocaprin, lauric acid and capric acid, or known preservatives such as nisin, sodium dehydroacetate, ethylenediaminetetraacetic acid (EDTA), sorbic acid and lactic acid, and the results have shown a synergistic effect against test microorganisms [129–131]. It should be noted that the antimicrobial effectiveness of monolaurin is influenced by other food components as the antibacterial activity is reduced by fat or starch but remains unchanged in the presence of protein [132, 133].

It has been reported that enhanced antimicrobial activities against *S. aureus* are obtained by using microemulsions containing monolaurin [134]. Oleic acid has been found to be fungistatic and has been used in preserving foodstuffs including bakery products with sweet soft cream as a topping, which are prone to quick spoilage by yeasts under conditions of nonrefrigeration [25]. One drawback to using lipids as preservatives in food is that they can affect the taste of the product, but for monolaurin for example the taste is undetected if the concentration is lower than 500 µg/ml [127].

The long-chain omega-3 polyunsaturated fatty acid EPA was evaluated for its antimicrobial action against a range of foodborne and food-spoilage pathogens. EPA exhibited antimicrobial activity against *B. subtilis*, *L. monocytogenes*, *S. aureus* and *Pseudomonas aeruginosa*, with significant bactericidal and bacteriostatic effects against both *P. aeruginosa* and *S. aureus* [135]. The antimicrobial activity of bioconversion extracts of EPA and DHA showed antibacterial activities against four Gram-positive bacteria, *B. subtilis*, *L. monocytogenes*, *S. aureus* (ATCC 6538) and *S. aureus* (KCTC 1916), and seven Gram-negative bacteria, *Enterobacter aerogenes*, *E. coli* O157 : H7, *E. coli* O157 : H7 (human), *P. aeruginosa*, *Salmonella enteritidis* and *S. typhimurium*. The growth inhibition by both bioconverted EPA and DHA was similar against Gram-positive bacteria, while the bioconverted extract of DHA was more effective than EPA against Gram-negative bacteria as determined by MIC [136].

7.8 Conclusions

Although few pharmaceutical formulations containing lipids as the active substance are currently on the market, the large amount of research being carried out on the beneficial effects of lipids, and the promising results already published, are likely to lead to commercial use. Large, well-designed clinical trials are needed to assess the therapeutic effect of fatty acids, for example against pathogenic microorganisms, many of which have developed resistance to common antibiotics. It can therefore be expected that further work on the beneficial properties of lipids will in the near future lead to the marketing of pharmaceutical formulations containing lipids for the prevention or treatment of diseases.

Lipids have for a long time been an essential part of cosmetic formulations, with function ranging from moisturizing to preservative. Contemporary cosmetic formulations are becoming more complex and lipids are being used as carriers in liposomes and micro- and nanoparticles. Lipids have traditionally been used in food as excipients and for their nutritional value, but they are suitable candidates for preservatives because of their antimicrobial properties and low toxicity. The beneficial health effects of marine lipids, which are mostly attributed to the omega-3 fatty acids, have been confirmed.

References

1. Sands, J.A. (1977) Inactivation and inhibition of replication of enveloped bacteriophage phi6 by fatty-acids. *Antimicrob. Agents Chemother.*, **12**, 523–528.
2. Kabara, J.J., Swieczkowski, D.M., Conley, A.J. and Truant, J.P. (1972) Fatty-acids and derivatives as antimicrobial agents. *Antimicrob. Agents Chemother.*, **2**, 23–28.
3. Kabara, J.J., Vrable, R. and Liekenjie, M.S.F. (1977) Antimicrobial lipids: natural and synthetic fatty-acids and monoglycerides. *Lipids*, **12**, 753–759.
4. Kabara, J.J. (1984) Antimicrobial agents derived from fatty acids. *J. Am. Oil Chem. Soc.*, **61**, 397–403.
5. Conley, A.J. and Kabara, J.J. (1973) Antimicrobial action of esters of polyhydric alcohols. *Antimicrob. Agents Chemother.*, **4**, 501–506.
6. Desbois, A.P., Mearns-Spragg, A. and Smith, V.J. (2009) A fatty acid from the diatom *Phaeodactylum tricornutum* is antibacterial against diverse bacteria including multi-resistant *Staphylococcus aureus* (MRSA). *Mar. Biotechnol.*, **11**, 45–52.
7. Bergsson, G., Arnfinnsson, J., Karlsson, S.M. *et al.* (1998) In vitro inactivation of *Chlamydia trachomatis* by fatty acids and monoglycerides. *Antimicrob. Agents Chemother.*, **42**, 2290–2294.
8. Bergsson, G., Steingrimsson, O. and Thormar, H. (2002) Bactericidal effects of fatty acids and monoglycerides on *Helicobacter pylori*. *Int. J. Antimicrob. Agents*, **20**, 258–262.
9. Bergsson, G., Steingrimsson, O. and Thormar, H. (1999) In vitro susceptibilities of *Neisseria gonorrhoeae* to fatty acids and monoglycerides. *Antimicrob. Agents Chemother.*, **43**, 2790–2792.
10. Altieri, C., Bevilacqua, A., Cardillo, D. and Sinigaglia, M. (2009) Effectiveness of fatty acids and their monoglycerides against Gram-negative pathogens. *Int. J. Food Sci. Technol.*, **44**, 359–366.
11. Wojtczak, L. and Wieckowski, M.R. (1999) The mechanisms of fatty acid-induced proton permeability of the inner mitochondrial membrane. *J. Bioenerg. Biomembr.*, **31**, 447–455.

12. Schonfeld, P., Wieckowski, M.R. and Wojtczak, L. (2000) Long-chain fatty acid-promoted swelling of mitochondria: further evidence for the protonophoric effect of fatty acids in the inner mitochondrial membrane. *FEBS Lett.*, **471**, 108–112.

13. Thormar, H., Hilmarsson, H. and Bergsson, G. (2006) Stable concentrated emulsions of the 1-monoglyceride of capric acid (monocaprin) with microbicidal activities against the food-borne bacteria *Campylobacter jejuni, Salmonella* spp. and *Escherichia coli. Appl. Environ. Microbiol.*, **72**, 522–526.

14. Shibasaki, I. and Kato, N. (1978) Combined effects on antibacterial activity of fatty acids and their esters against Gram-negative bacteria, in *Symposium on the Pharmacological Effect of Lipids* (ed. J.J. Kabara), The American Oil Chemists' Society, Champaign, IL, USA, pp. 15–24.

15. Thormar, H., Isaacs, C.E., Brown, H.R. *et al.* (1987) Inactivation of enveloped viruses and killing of cells by fatty-acids and monoglycerides. *Antimicrob. Agents Chemother.*, **31**, 27–31.

16. Hilmarsson, H., Larusson, L.V. and Thormar, H. (2006) Virucidal effect of lipids on visna virus, a lentivirus related to HIV. *Arch. Virol.*, **151**, 1217–1224.

17. Hilmarsson, H., Traustason, B.S., Kristmundsdottir, T. and Thormar, H. (2007) Virucidal activities of medium- and long-chain fatty alcohols and lipids against respiratory syncytial virus and parainfluenza virus type 2: comparison at different pH levels. *Arch. Virol.*, **152**, 2225–2236.

18. Isaacs, C.E., Kim, K.S. and Thormar, H. (1994) Inactivation of enveloped viruses in human bodily fluids by purified lipids. *Annals NY Acad. Sci.*, **724**, 457–464.

19. Thormar, H., Isaacs, C.E., Kim, K.S. and Brown, H.R. (1994) Inactivation of visna virus and other enveloped viruses by free fatty-acids and monoglycerides. *Annals NY Acad. Sci.*, **724**, 465–471.

20. Sands, J.A., Landin, P., Auperin, D. and Reinhardt, A. (1979) Enveloped virus inactivation by fatty acid derivatives. *Antimicrob. Agents Chemother.*, **15**, 134–136.

21. Bergsson, G., Arnfinnsson, J., Steingrimsson, O. and Thormar, H. (2001) Killing of Gram-positive cocci by fatty acids and monoglycerides. *APMIS*, **109**, 670–678.

22. Hilmarsson, H., Kristmundsdottir, T. and Thormar, H. (2005) Virucidal activities of medium- and long-chain fatty alcohols, fatty acids and monoglycerides against herpes simplex virus types 1 and 2: comparison at different pH levels. *APMIS*, **113**, 58–65.

23. Bartolotta, S., Garcia, C.C., Candurra, N.A. and Damonte, E.B. (2001) Effect of fatty acids on arenavirus replication: inhibition of virus production by lauric acid. *Arch. Virol.*, **146**, 777–790.

24. McLain, N., Ascanio, R., Baker, C. *et al.* (2000) Undecylenic acid inhibits morphogenesis of *Candida albicans*. *Antimicrob. Agents Chemother.*, **44**, 2873–2875.

25. Davidson, W.S., Saxena, R.K. and Gupta, R. (1999) The fungistatic action of oleic acid. *Curr. Sci.*, **76**, 1137–1140.

26. Bergsson, G., Arnfinnsson, J., Steingrimsson, O. and Thormar, H. (2001) In vitro killing of *Candida albicans* by fatty acids and monoglycerides. *Antimicrob. Agents Chemother.*, **45**, 3209–3212.

27. Thorgeirsdottir, T.O., Kristmundsdottir, T., Thormar, H. *et al.* (2006) Antimicrobial activity of monocaprin: a monoglyceride with potential use as a denture disinfectant. *Acta Odontol. Scand.*, **64**, 21–26.

28. Puri, A., Loomis, K., Smith, B. *et al.* (2009) Lipid-based nanoparticles as pharmaceutical drug carriers: from concepts to clinic. *Crit. Rev. Ther. Drug Carr. Syst.*, **26**, 523–580.

29. Schafer-Korting, M., Mehnert, W.G. and Korting, H.C. (2007) Lipid nanoparticles for improved topical application of drugs for skin diseases. *Adv. Drug Deliv. Rev.*, **59**, 427–443.

30. Thormar, H. and Hilmarsson, H. (2007) The role of microbicidal lipids in host defense against pathogens and their potential as therapeutic agents. *Chem. Phys. Lipids*, **150**, 1–11.

31. Clarke, S.R., Mohamed, R., Bian, L. *et al.* (2007) The *Staphylococcus aureus* surface protein IsdA mediates resistance to innate defenses of human skin. *Cell Host Microbe*, **1**, 199–212.

32. Lukowski, G., Lindequist, U., Mundt, S. *et al.* (2008) Inhibition of dermal MRSA colonization by microalgal micro- and nanoparticles. *Skin Pharmacol. Physiol.*, **21**, 98–105.

33. Bourne, N., Ireland, J., Stanberry, L.R. and Bernstein, D.I. (1999) Effect of undecylenic acid as a topical microbicide against genital herpes infection in mice and guinea pigs. *Antiviral Res.*, **40**, 139–144.

34. Bousquet, E., Spadaro, A., Santagati, N.A. *et al.* (2002) Determination of undecylenic and sorbic acids in cosmetic preparations by high performance liquid chromatography with electrochemical detection. *J. Pharm. Biomed. Anal.*, **30**, 947–954.
35. Shafran, S.D., Sacks, S.L., Aoki, F.Y. *et al.* (1997) Topical undecylenic acid for herpes simplex labialis: a multicenter, placebo-controlled trial. *J. Infect. Dis.*, **176**, 78–83.
36. Kristmundsdottir, T., Arnadottir, S.G., Bergsson, G. and Thormar, H. (1999) Development and evaluation of microbicidal hydrogels containing monoglyceride as the active ingredient. *J. Pharm. Sci.*, **88**, 1011–1015.
37. Thorgeirsdottir, T.O., Thormar, H. and Kristmundsdottir, T. (2003) Effects of polysorbates on antiviral and antibacterial activity of monoglyceride in pharmaceutical formulations. *Pharmazie*, **58**, 286–287.
38. Thorgeirsdottir, T.O., Thormar, H. and Kristmundsdottir, T. (2005) The influence of formulation variables on stability and microbicidal activity of monoglyceride monocaprin. *J. Drug Deliv. Sci. Technol.*, **15**, 233–236.
39. Thormar, H., Bergsson, G., Gunnarsson, E. *et al.* (1999) Hydrogels containing monocaprin have potent microbicidal activities against sexually transmitted viruses and bacteria in vitro. *Sex. Transm. Infect.*, **75**, 181–185.
40. Neyts, J., Kristmundsdottir, T., De Clercq, E. and Thormar, H. (2000) Hydrogels containing monocaprin prevent intravaginal and intracutaneous infections with HSV-2 in mice: Impact on the search for vaginal microbicides. *J. Med. Virol.*, **61**, 107–110.
41. Arnfinnsdottir, A.V., Geirsson, R.T., Kristmundsdottir, T. *et al.* (2004) Effects of a monocaprin hydrogel on the vaginal mucosa and bacterial colonisation. 34th Congress of Nordic Federation of Societies of Obstetrics and Gynecology, Helsinki, Finland, p. 91.
42. Li, Q.S., Estes, J.D., Schlievert, P.M. *et al.* (2009) Glycerol monolaurate prevents mucosal SIV transmission. *Nature*, **458**, 1034–1038.
43. Strandberg, K.L., Peterson, M.L., Lin, Y.C. *et al.* (2010) Glycerol monolaurate inhibits *Candida* and *Gardnerella vaginalis* in vitro and in vivo but not *Lactobacillus*. *Antimicrob. Agents Chemother.*, **54**, 597–601.
44. Kitahara, T., Aoyama, Y., Hirakata, Y. *et al.* (2006) In vitro activity of lauric acid or myristylamine in combination with six antimicrobial agents against methicillin-resistant *Staphylococcus aureus* (MRSA). *Int. J. Antimicrob. Agents*, **27**, 51–57.
45. Nakatsuji, T., Kao, M.C., Fang, J.Y. *et al.* (2009) Antimicrobial property of lauric acid against *Propionibacterium acnes*: its therapeutic potential for inflammatory acne vulgaris. *J. Invest. Dermatol.*, **129**, 2480–2488.
46. Yang, D.R., Pornpattananangkul, D., Nakatsuji, T. *et al.* (2009) The antimicrobial activity of liposomal lauric acids against *Propionibacterium acnes*. *Biomaterials*, **30**, 6035–6040.
47. Carpo, B.G., Verallo-Rowell, V.M. and Kabara, J.J. (2007) Novel antibacterial activity of monolaurin compared with conventional antibiotics against organisms from skin infections: an in vitro study. *J Drugs Dermatol.*, **10**, 991–998.
48. Skulason, S., Kristmundsdóttir, T., Holbrook, W.P. and Thormar, H. (2008) Novel combination of monocaprin and doxycycline in treatment of cold sores, Abstr. 379. Proc. 35th Annual Meeting of the Controlled Release Society.
49. Kurihara, H., Goto, Y., Aida, H. *et al.* (1999) Antibacterial activity against cariogenic bacteria and inhibition of insoluble glucan production by free fatty acids obtained from dried *Gloiopeltis furcata*. *Fish. Sci.*, **65**, 129–132.
50. Holbrook, W.P., Skulason, S., Soto, M. and Kristmundsdottir, T. (2007) Further studies of monocaprin as a potential topical agent for treating oral candidosis, Abstr.695. IADR/AADR/CADR 85th General Session, New Orleans, LA, USA.
51. Simor, A.E., Stuart, T.L., Louie, L. *et al.* (2007) Mupirocin-resistant, methicillin-resistant *Staphylococcus aureus* strains in Canadian hospitals. *Antimicrob. Agents Chemother.*, **51**, 3880–3886.
52. Rouse, M.S., Rotger, M., Piper, K.E. *et al.* (2005) In vitro and in vivo evaluations of the activities of lauric acid monoester formulations against *Staphylococcus aureus*. *Antimicrob. Agents Chemother.*, **49**, 3187–3191.

53. Simoes, E.A. (2002) Immunoprophylaxis of respiratory syncytial virus: global experience. *Respir. Res.*, **3** (Suppl 1), S26–S33.
54. Wyde, P.R., Chetty, S.N., Timmerman, P. *et al.* (2003) Short duration aerosols of JNJ 2408068 (R170591) administered prophylactically or therapeutically protect cotton rats from experimental respiratory syncytial virus infection. *Antiviral Res.*, **60**, 221–231.
55. H. Hilmarsson (2008) Microbicidal activity of lipids, their effect on mucosal infections in animals and their potential as disinfecting agents, PhD Thesis, Faculty of Life and Environmental Sciences, University of Iceland, Reykjavik.
56. Petschow, B.W., Batema, R.P. and Ford, L.L. (1996) Susceptibility of *Helicobacter pylori* to bactericidal properties of medium-chain monoglycerides and free fatty acids. *Antimicrob. Agents Chemother.*, **40**, 302–306.
57. Thompson, L., Cockayne, A. and Spiller, R.C. (1994) Inhibitory effect of polyunsaturated fatty-acids on the growth of *Helicobacter-pylori*: a possible explanation of the effect of diet on peptic-ulceration. *Gut*, **35**, 1557–1561.
58. Ibrahim, S. and Li, S. (2010) Efficiency of fatty acids as chemical penetration enhancers: mechanisms and structure enhancement relationship. *Pharm. Res.*, **27**, 115–125.
59. Maher, S., Leonard, T.W., Jacobsen, J. and Brayden, D.J. (2009) Safety and efficacy of sodium caprate in promoting oral drug absorption: from in vitro to the clinic. *Adv. Drug Deliv. Rev.*, **61**, 1427–1449.
60. Puglia, C. and Bonina, F. (2008) Effect of polyunsaturated fatty acids and some conventional penetration enhancers on transdermal delivery of atenolol. *Drug Deliv.*, **15**, 107–112.
61. Tsutsumi, K., Obata, Y., Takayama, K. *et al.* (1998) Effect of the cod-liver oil extract on the buccal permeation of ionized and nonionized forms of ergotamine using the keratinized epithelial-free membrane of hamster cheek pouch mucosa. *Int. J. Pharm.*, **174**, 151–156.
62. Loftsson, T., Thormar, H., Olafsson, J.H. *et al.* (1998) Fatty acid extract from cod-liver oil:activity against herpes simplex virus and enhancement of transdermal delivery of acyclovir in-vitro. *Pharm. Pharmacol. Commun.*, **4**, 287–291.
63. Tsutsumi, K., Obata, Y., Takayama, K. *et al.* (1998) Effect of cod-liver oil extract on the buccal permeation of ergotamine tartrate. *Drug Dev. Ind. Pharm.*, **24**, 757–762.
64. Niewold, T.A. (2007) The nonantibiotic anti-inflammatory effect of antimicrobial growth promoters, the real mode of action? A hypothesis. *Poult. Sci.*, **86**, 605–609.
65. Castanon, J.I.R. (2007) History of the use of antibiotic as growth promoters in European poultry feeds. *Poult. Sci.*, **86**, 2466–2471.
66. Desbois, A. and Smith, V. (2010) Antibacterial free fatty acids: activities, mechanisms of action and biotechnological potential. *Appl. Microbiol. Biotechnol.*, **85**, 1629–1642.
67. Hilmarsson, H., Thormar, H., Thrainsson, J.H. and Gunnarsson, E. (2006) Effect of glycerol monocaprate (monocaprin) on broiler chickens: an attempt at reducing intestinal *Campylobacter* infection. *Poult. Sci.*, **85**, 588–592.
68. Van Immerseel, F., De Buck, J., Pasmans, F. *et al.* (2003) Invasion of *Salmonella enteritidis* in avian intestinal epithelial cells in vitro is influenced by short-chain fatty acids. *Int. J. Food Microbiol.*, **85**, 237–248.
69. Boyen, F., Haesebrouck, F., Vanparys, A. *et al.* (2008) Coated fatty acids alter virulence properties of *Salmonella Typhimurium* and decrease intestinal colonization of pigs. *Vet. Microbiol.*, **132**, 319–327.
70. Timbermont, L., Lanckriet, A., Dewulf, J. *et al.* (2010) Control of *Clostridium perfringens*-induced necrotic enteritis in broilers by target-released butyric acid, fatty acids and essential oils. *Avian Pathol.*, **39**, 117–121.
71. Van Immerseel, F., Boyen, F., Gantois, I. *et al.* (2005) Supplementation of coated butyric acid in the feed reduces colonization and shedding of *Salmonella* in poultry. *Poult. Sci.*, **84**, 1851–1856.
72. Van Immerseel, F., Fievez, V., De Buck, J. *et al.* (2004) Microencapsulated short-chain fatty acids in feed modify colonization and invasion early after infection with *Salmonella enteritidis* in young chickens. *Poult. Sci.*, **83**, 69–74.
73. Vandeplas, S., Dauphin, R.D., Beckers, Y. *et al.* (2010) *Salmonella* in chicken: current and developing strategies to reduce contamination at farm level. *J. Food Prot.*, **73**, 774–785.

74. Defoirdt, T., Boon, N., Sorgeloos, P. *et al.* (2009) Short-chain fatty acids and poly-beta-hydroxyalkanoates: (New) Biocontrol agents for a sustainable animal production. *Biotechnol. Adv.*, **27**, 680–685.
75. Defoirdt, T., Boon, N., Sorgeloos, P. *et al.* (2007) Alternatives to antibiotics to control bacterial infections: luminescent vibriosis in aquaculture as an example. *Trends Biotechnol.*, **25**, 472–479.
76. Skrivanova, E., Molatova, Z. and Marounek, M. (2008) Effects of caprylic acid and triacylglycerols of both caprylic and capric acid in rabbits experimentally infected with enteropathogenic *Escherichia coli* O103. *Vet. Microbiol.*, **126**, 372–376.
77. Skrivanova, E., Skrivanova, V., Volek, Z. and Marounek, M. (2009) Effect of triacylglycerols of medium-chain fatty acids on growth rate and mortality of rabbits weaned at 25 and 35 days of age. *Vet. Med.*, **54**, 19–24.
78. Dierick, N.A., Decuypere, J.A., Molly, K. *et al.* (2002) The combined use of triacylglycerols (TAGs) containing medium chain fatty acids (MCFAs) and exogenous lipolytic enzymes as an alternative to nutritional antibiotics in piglet nutrition-II. In vivo release of MCFAs in gastric cannulated and slaughtered piglets by endogenous and exogenous lipases; effects on the luminal gut flora and growth performance. *Livest. Prod. Sci.*, **76**, 1–16.
79. Goel, G., Arvidsson, K., Vlaeminck, B. *et al.* (2009) Effects of capric acid on rumen methano-genesis and biohydrogenation of linoleic and alpha-linolenic acid. *Animal*, **3**, 810–816.
80. Desbois, A.P., Lebl, T., Yan, L.M. and Smith, V.J. (2008) Isolation and structural characterisation of two antibacterial free fatty acids from the marine diatom. *Phaeodactylum tricornutum*, *Appl. Microbiol. Biotechnol.*, **81**, 755–764.
81. Williams, P.J.le B. and Laurens, L.M.L. (2010) Microalgae as biodiesel & biomass feedstocks: Review & analysis of the biochemistry, energetics & economics. *Energy Environ. Sci.*, **3**, 554–590.
82. Walters, D., Raynor, L., Mitchell, A. *et al.* (2004) Antifungal activities of four fatty acids against plant pathogenic fungi. *Mycopathol.*, **157**, 87–90.
83. Liu, S.Y., Ruan, W.B., Li, J. *et al.* (2008) Biological control of phytopathogenic fungi by fatty acids. *Mycopathol.*, **166**, 93–102.
84. Walker, E.L. and Sweeney, M.A. (1920) The chemotherapeutics of the chaulmoogric acid series and other fatty acids in leprosy and tuberculosis. I. Bactericidal action, active principle, specificity. *J. Infect. Dis.*, **26**, 238–264.
85. Hanel, F. und S. Piller (1950) Ungesättigte Fettsäuren in der Therapie der Lungentuberkulose. *Beitr. Klin. Tuberk.*, **103**, 239–245.
86. Shipar, M.A.H. (2007) Physical and chemical characteristics, major fatty acids, antimicrobial activity and toxicity analysis of red shrimp (*Metapenaeus brevicornis*) brain lipid. *Food Chem.*, **102**, 649–655.
87. Carballeira, N.M. (2008) New advances in fatty acids as antimalarial, antimycobacterial and antifungal agents. *Prog. Lipid Res.*, **47**, 50–61.
88. Barnathan, G. (2009) Non-methylene-interrupted fatty acids from marine invertebrates: occurrence, characterization and biological properties. *Biochimie*, **91**, 671–678.
89. Maki, K.C., Reeves, M.S., Farmer, M. *et al.* (2009) Krill oil supplementation increases plasma concentrations of eicosapentaenoic and docosahexaenoic acids in overweight and obese men and women. *Nutr. Res.*, **29**, 609–615.
90. Zhu, F.S., Liu, S., Chen, X.M. *et al.* (2008) Effects of n-3 polyunsaturated fatty acids from seal oils on nonalcoholic fatty liver disease associated with hyperlipidemia. *World J. Gastroenterol.*, **14**, 6395–6400.
91. Fortin, P.R., Lew, R.A., Liang, M.H. *et al.* (1995) Validation of a meta-analysis: The effects of fish oil in rheumatoid arthritis. *J. Clin. Epidemiol.*, **48**, 1379–1390.
92. Gorjao, R., Azevedo-Martins, A.K., Rodrigues, H.G. *et al.* (2009) Comparative effects of DHA and EPA on cell function. *Pharmacol. Ther.*, **122**, 56–64.
93. Calder, P.C. (2009) Polyunsaturated fatty acids and inflammatory processes: new twists in an old tale. *Biochimie*, **91**, 791–795.
94. Galarraga, B., Ho, M., Youssef, H.M. *et al.* (2008) Cod liver oil (n-3 fatty acids) as an non-steroidal anti-inflammatory drug sparing agent in rheumatoid arthritis. *Rheumatology*, **47**, 665–669.

95. Heard, C.M., Gallagher, S.J., Harwood, J. and Maguire, P.B. (2003) The in vitro delivery of NSAIDs across skin was in proportion to the delivery of essential fatty acids in the vehicle: evidence that solutes permeate skin associated with their solvation cages? *Int. J. Pharmaceut.*, **261**, 165–169.
96. Calder, P.C. (2008) Polyunsaturated fatty acids, inflammatory processes and inflammatory bowel diseases. *Mol. Nutr. Food Res.*, **52**, 885–897.
97. Zulfakar, M.H., Edwards, M. and Heard, C.M. (2007) Is there a role for topically delivered eicosapentaenoic acid in the treatment of psoriasis? *Eur. J. Dermatol.*, **17**, 284–291.
98. Zulfakar, M.H., Abdelouahab, N. and Heard, C.M. (2010) Enhanced topical delivery and ex vivo anti-inflammatory activity from a betamethasone dipropionate formulation containing fish oil. *Inflamm. Res.*, **59**, 23–30.
99. Erkkilä, A., deMello, V.D.F., Risérus, U. and Laaksonen, D.E. (2008) Dietary fatty acids and cardiovascular disease: an epidemiological approach. *Prog. Lipid Res.*, **47**, 172–187.
100. Schmidt, E.B., Arnesen, H., Christensen, J.H. *et al.* (2005) Marine n-3 polyunsaturated fatty acids and coronary heart disease. Part II. Clinical trials and recommendations. *Thromb. Res.*, **115**, 257–262.
101. Lopez-Huertas, E. (2010) Health effects of oleic acid and long chain omega-3 fatty acids (EPA and DHA) enriched milks. A review of intervention studies. *Pharmacol. Res.*, **61**, 200–207.
102. Richard, D., Kefi, K., Barbe, U. *et al.* (2008) Polyunsaturated fatty acids as antioxidants. *Pharmacol. Res.*, **57**, 451–455.
103. Eslick, G.D., Howe, P.R.C., Smith, C. *et al.* (2009) Benefits of fish oil supplementation in hyperlipidemia: a systematic review and meta-analysis. *Int. J. Cardiol.*, **136**, 4–16.
104. Hoy, S.M. and Keating, G.M. (2009) Omega-3 ethylester concentrate A review of its use in secondary prevention post-myocardial infarction and the treatment of hypertriglyceridaemia. *Drugs*, **69**, 1077–1105.
105. Cussons, A.J., Watts, G.F., Mori, T.A. and Stuckey, B.G.A. (2009) Omega-3 fatty acid supplementation decreases liver fat content in polycystic ovary syndrome: a randomized controlled trial employing proton magnetic resonance spectroscopy. *J. Clin. Endocrinol. Metab.*, **94**, 3842–3848.
106. Orr, S.K. and Bazinet, R.P. (2008) The emerging role of docosahexaenoic acid in neuroinflammation. *Curr. Opin. Investig. Drugs*, **9**, 735–743.
107. Cunnane, S.C., Plourde, M., Pifferi, F. *et al.* (2009) Fish, docosahexaenoic acid and Alzheimer's disease. *Prog. Lipid Res.*, **48**, 239–256.
108. Chiu, C.-C., Su, K.-P., Cheng, T.-C. *et al.* (2008) The effects of omega-3 fatty acids monotherapy in Alzheimer's disease and mild cognitive impairment: a preliminary randomized double-blind placebo-controlled study. *Prog. Neuro-Psychopharmacol. Biol. Psychiatry*, **32**, 1538–1544.
109. Stahl, L.A., Begg, D.P., Weisinger, R.S. and Sinclair, A.J. (2008) The role of omega-3 fatty acids in mood disorders. *Curr. Opin. Investig. Drugs*, **9**, 57–64.
110. Lin, P.Y. and Su, K.P. (2007) A meta-analytic review of double-blind, placebo-controlled trials of antidepressant efficacy of omega-3 fatty acids. *J. Clin. Psychiatry*, **68**, 1056–1061.
111. Hamazaki, T. and Hamazaki, K. (2008) Fish oils and aggression or hostility. *Prog. Lipid Res.*, **47**, 221–232.
112. Szuhaj, B.F. (ed.) (1989) *Lecithins: Sources, Manufacture and Uses*, American Oil Chemists' Society, Champaign, IL, USA.
113. Alvarez, A.M.R. and Rodriguez, M.L.G. (2000) Lipids in pharmaceutical and cosmetic preparations. *Grasas Aceites*, **51**, 74–96.
114. Rock, C.O. and Cronan, J.E. (1996) *Escherichia coli* as a model for the regulation of dissociable (type II) fatty acid biosynthesis. *Biochim. Biophys. Acta*, **1302**, 1–16.
115. Lu, J.Z.Q., Lee, P.J., Waters, N.C. and Prigge, S.T. (2005) Fatty acid synthesis as a target for antimalarial drug discovery. *Comb. Chem. High Throughput Screen*, **8**, 15–26.
116. Pardeike, J., Hommoss, A. and Muller, R.H. (2009) Lipid nanoparticles (SLN, NLC) in cosmetic and pharmaceutical dermal products. *Int. J. Pharm.*, **366**, 170–184.
117. Burtenshaw, J.M.L. (1942) The mechanisms of self disinfection of the human skin and its appendages. *J. Hyg.*, **42**, 184–209.

118. Escobar, S.O., Achenbach, R., Iannantuono, R. and Torem, V. (1992) Topical fish oil in psoriasis: a controlled and blind study. *Clin. Exp. Dermatol.*, **17**, 159–162.
119. Imokawa, G., Akasaki, S., Hattori, M. and Yoshizuka, N. (1986) Selective recovery of deranged water-holding properties by stratum-corneum lipids. *J. Invest. Dermatol.*, **87**, 758–761.
120. Man, M.Q., Feingold, K.R. and Elias, P.M. (1993) Exogenous lipids influence permeability barrier recovery in acetone-treated murine skin. *Arch. Dermatol.*, **129**, 728–738.
121. Loden, M. (2005) The clinical benefit of moisturizers. *J. Eur. Acad. Dermatol. Venereol.*, **19**, 672–688.
122. Barcelo-Coblijn, G., Murphy, E.J., Othman, R. *et al.* (2008) Flaxseed oil and fish-oil capsule consumption alters human red blood cell n-3 fatty acid composition: a multiple-dosing trial comparing 2 sources of n-3 fatty acid. *Am. J. Clin. Nutr.*, **88**, 801–809.
123. Freese, E., Sheu, C.W. and Galliers, E. (1973) Function of lipophilic acids as antimicrobial food additives. *Nature*, **241**, 321–325.
124. Isaacs, C.E., Litov, R.E. and Thormar, H. (1995) Antimicrobial activity of lipids added to human milk, infant formula, and bovine-milk. *J. Nutr. Biochem.*, **6**, 362–366.
125. Jang, H.I. and Rhee, M.S. (2009) Inhibitory effect of caprylic acid and mild heat on *Cronobacter* spp. (*Enterobacter sakazakii*) in reconstituted infant formula and determination of injury by flow cytometry. *Int. J. Food Microbiol.*, **133**, 113–120.
126. Wang, L.L. and Johnson, E.A. (1997) Control of *Listeria monocytogenes* by monoglycerides in foods. *J. Food Protect.*, **60**, 131–138.
127. Zhang, H., Wei, H.W., Cui, Y.N. *et al.* (2009) Antibacterial interactions of monolaurin with commonly used antimicrobials and food components. *J. Food Sci.*, **74**, M418–M421.
128. Mbandi, E., Brywig, M. and Shelef, L.A. (2004) Antilisterial effects of free fatty acids and monolaurin in beef emulsions and hot dogs. *Food Microbiol.*, **21**, 815–818.
129. Mansour, M. and Milliere, J.B. (2001) An inhibitory synergistic effect of a nisin-monolaurin combination on *Bacillus* sp vegetative cells in milk. *Food Microbiol.*, **18**, 87–94.
130. Anang, D.M., Rusul, G., Bakar, J. and Ling, F.H. (2007) Effects of lactic acid and lauricidin on the survival of *Listeria monocytogenes*, *Salmonella enteritidis* and *Escherichia coli* O157 : H7 in chicken breast stored at 4 degrees C. *Food Cont.*, **18**, 961–969.
131. McLay, J.C., Kennedy, M.J., O'Rourke, A.L. *et al.* (2002) Inhibition of bacterial foodborne pathogens by the lactoperoxidase system in combination with monolaurin. *Int. J. Food Microbiol.*, **73**, 1–9.
132. Kato, N. and Shibasaki, I. (1975) Comparison of antimicrobial activities of fatty-acids and their esters. *J. Ferment. Technol.*, **53**, 793–801.
133. Ababouch, L.H., Bouqartacha, F. and Busta, F.F. (1994) Inhibition of *Bacillus cereus* spores and vegetative cells by fatty acids and glyceryl monododecanoate. *Food Microbiol.*, **11**, 327–336.
134. Zhang, H., Feng, F.Q., Fu, X.W. *et al.* (2007) Antimicrobial effect of food-grade GML microemulsions against *Staphylococcus aureus*. *Eur. Food Res. Technol.*, **226**, 281–286.
135. Shin, S.Y., Bajpai, V.K., Kim, H.R. and Kang, S.C. (2007) Antibacterial activity of eicosapentaenoic acid (EPA) against foodborne and food spoilage microorganisms. *LWT-Food Sci. Technol.*, **40**, 1515–1519.
136. Shin, S.Y., Bajpai, V.K., Kim, H.R. and Kang, S.C. (2007) Antibacterial activity of bioconverted eicosapentaenoic (EPA) and docosahexaenoic acid (DHA) against foodborne pathogenic bacteria. *Int. J. Food Microbiol.*, **113**, 233–236.

8

Antimicrobial Lipids as Disinfectants, Antiseptics and Sanitizers

Halldor Thormar and Hilmar Hilmarsson

Faculty of Life and Environmental Sciences, University of Iceland, Reykjavik, Iceland

8.1 Introduction

Disinfectants are antimicrobial agents used to destroy microorganisms on nonliving objects or surfaces such as countertops, dishes and so on. The process known as disinfection kills most or all pathogenic organisms in a given time [1], literally making them noninfectious

Lipids and Essential Oils as Antimicrobial Agents   Halldor Thormar
© 2011 John Wiley & Sons, Ltd

[2]. Disinfectants are distinguished from antibiotics, which inhibit microbial growth within the body, and antiseptics, which destroy microorganisms on the surface of living tissue such as skin and thereby reduce the probability of infection. Sanitizers are defined as substances which reduce the number of microorganisms to a level that is safe by public health standards, rather than causing a near total kill of pathogenic organisms as in the case of disinfectants. By one definition, sanitizers cause at least a 5 $\log_{10}$ reduction in the number of microorganisms within 30 seconds [1]. Sterilization, on the other hand, refers to a complete destruction of all microbial life, including spores, by a chemical or physical process [3]. A large number of different chemical compounds are used as disinfectants and antiseptics with different modes of action. Some have a wide spectrum of antimicrobial activity, while others have a more restricted spectrum, killing a smaller range of pathogens. Although known to some extent for centuries, a widespread use of antiseptics followed the work of Joseph Lister in 1843 on the use of phenol (carbolic acid) as an antiseptic in surgical practice [4]. For comparison of the relative effectiveness of disinfectants, the phenol coefficient test has been used since 1903 [5] as a screening test in which the potency of a disinfectant is compared with that of phenol against a standard microbe, usually *Salmonella typhi* or *Staphylococcus aureus* [6]. The higher the phenol coefficient value, the more effective the disinfectant is considered to be [1, 7].

In 1881, Robert Koch published a comprehensive report on the disinfecting action of a variety of chemical compounds against the spores of anthrax bacilli, with a particular emphasis on carbolic acid [8]. As pointed out by Geppert in 1889 [9], a faulty technique led Koch to a wrong conclusion concerning the sporicidal action of disinfectants on anthrax spores. This was due to carry-over of disinfectant into the assay medium, in which the sporicidal action would continue. To prevent this, Geppert advised the use of a neutralizing agent to stop the action of the disinfectant. However, in his study Koch also tested the effects of a number of compounds on the growth of anthrax bacilli. He found that a 1 : 1000 dilution of potassium soap completely inhibited the growth of the bacilli in nutrient broth.

This was the first indication that soaps might have an antimicrobial effect and that natural soaps, which are potassium or sodium salts of fatty acids, could be used as disinfectants as well as for cleaning, a function that had been known for thousands of years.

Soapmaking has been practiced since antiquity, the earliest recorded evidence being from about 2800 BC in ancient Bablylon [10, 11]. During the long history of soapmaking, the process has not fundamentally changed; it consists in heating fats with a strong alkaline solution, either sodium or potassium hydroxide, leading to a chemical reaction called saponification. In this reaction, the fats – triglycerides of fatty acids – are hydrolysed, yielding free glycerol and soap – potassium or sodium salts of the fatty acids. Originally, lye made from wood ashes, so-called potash, was used as a source of alkali but in the nineteenth century, chemical methods were invented to manufacture sodium and potassium hydroxide on a large scale [11]. A large variety of natural soaps are known, depending on the type of fat or oil, on the method used in the soapmaking process and on whether sodium hydroxide or potassium hydroxide is used. Reacting fats with potassium hydroxide makes soft or liquid soaps, whereas sodium hydroxide produces hard soaps. Many types of fat and oil have been used, both animal fats such as tallow from cattle or sheep and a wide choice of vegetable oils such as palm, coconut, olive, linseed, cottonseed and castor oil. Combinations of oils, such as a mix of coconut, palm and olive oils, have also been used. Since the fatty acid contents of the various fats and oils vary greatly, the characteristics of

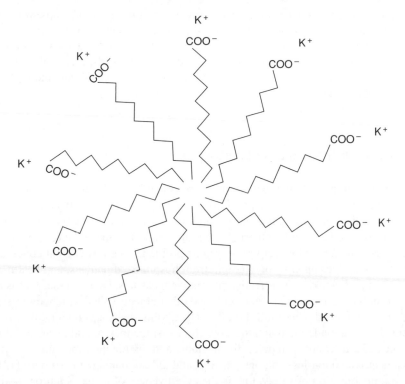

Figure 8.1 *A micelle of soap in water. The hydrophobic fatty acid chains do not interact with water molecules but are attracted to each other, or to particles of grease and dirt, to form a cluster away from the water. The hydrophilic carboxyl groups interact with water and keep the micelles in solution.*

the soaps likewise vary. Soaps made from vegetable oils, such as palm oil, are typically softer than soaps made from animal fat, but Castile soap, which was originally made in Spain from pure olive oil, is an example of a white hard soap.

The cleaning action of soaps depends on their amphiphilic nature. The soap molecule has a lipophilic, hydrophobic end, made up of a hydrocarbon chain which forms a complex with nonpolar dirt and grease, as well as a hydrophilic anionic end, which makes the complex soluble in water in the form of micelles (Figure 8.1). The micellar suspension can then be rinsed off with clean water, thus removing the dirt.

The chemistry of soap manufacturing stayed essentially the same until 1916, when the first synthetic detergent was developed in Germany due to a shortage of fats during the First World War. Detergent production began in the United States in the early 1930s and increased greatly after the Second World War, mostly due to lack of fats and oils for making soap and to the fact that synthetic detergents were in many ways superior to soaps as cleaning agents, particularly in hard water, due to its content of calcium ions. In the 1950s the production of detergents had surpassed that of soap and at present they have all but replaced pure natural soap, even in products used for personal hygiene [12].

Although sodium and potassium salts of fatty acids in the form of soaps were the first lipid-derived compounds to show antibacterial action, the activity was soon found to be associated with the fatty acid moiety of the soap molecule. Pure fatty acids were shown to be potent antimicrobial agents, as were monoglycerides of fatty acids, despite their low water solubility. This chapter will review the utilization of soaps, fatty acids and monoglycerides as disinfectants and sanitizers.

8.2   Soaps as Disinfectants and Antiseptics

The report by Koch in 1881 that potassium soap had antibacterial activity against anthrax bacilli elicited a great interest in the possibility that soaps might not only be used for cleaning but also as disinfectants against pathogenic bacteria. At this time infectious diseases like cholera, typhoid fever and diptheritis were rampant in Europe and infection of wounds by streptococci and staphylococci was a serious problem. There was great concern about the spread of infectious agents, for example from hospitalized patients, and health professionals were eagerly looking for effective preventive measures. One necessary measure was disinfection of bedlinens, patients' garments and other objects soiled with body waste, such as faeces, pus, blood and other fluids [13, 14]. The finding that soaps might kill some of the pathogenic bacteria made them an attractive choice, since they would clean and disinfect at the same time. In contrast to other disinfectants and disinfecting measures, soaps were harmless, nontoxic, odourless and inexpensive, and did not damage the laundry [13].

However, the question of safety, that is the effectiveness of a soap solution in killing a specific pathogenic bacterium under a variety of conditions, was the most important factor in determining the choice of a soap as a disinfectant. Behring in 1890 [15] emphasized the much greater resistance of bacterial spores, such as antrax spores, than of nonsporeforming bacteria. He listed a number of bacteria which fell into the latter category, such as the cholera, typhoid and diphtheria bacilli, and staphylococci and streptococci. As described in detail in Chapter 2, numerous studies were carried out in the last decade of the nineteenth century and the first decade of the twentieth century to determine the susceptibility of these important pathogenic bacteria to various types of soap under various well-defined conditions [16–21]. The reliability of the results was considered most important, so that a given soap solution could be used with confidence as a dependable disinfectant against a specific bacterium [21]. It was found that not only were the susceptibilities of different bacteria highly variable but so too were the antibacterial activities of various types of soap. Cholera bacteria were consistently found to be susceptible to soap solutions, particularly of potassium soaps, at low concentrations and in a short time [16]. Similarly, a potassium-soap solution was found to kill typhoid bacilli, but at a higher soap concentration and in a longer time [17]. Since a piece of linen contaminated with typhoid bacilli was found to be sterilized by a 6% soap solution in 15 minutes and by a 1% soap solution in one hour, it was concluded that soaps were suitable for disinfection of soiled wash from patients infected by these two most common pathogenic bacteria [17]. After a cholera outbreak in Germany in the summer of 1892, the government issued a regulation recommending a 6% solution of potassium soft soap for disinfection, not only of soiled clothing but also of train compartments used by patients [13].

At this time there was great concern regarding the safety of the public water supply during epidemics. Thus, during an outbreak of cholera in the Netherlands in the early 1890s, there was widespread fear of using water from the River Vecht for bathing in the home. In order to make the water safer for bathing and to reassure consumers, the feasibility of adding disinfectants to the water, which would kill cholera bacteria fast and safely, was considered. The disinfectant had to be harmless, inexpensive, easy to use and convenient. Experiments were carried out adding regular soap to a bathtub of water containing the number of cholera bacilli that might be expected in contaminated tap water. Potassium soft soap and medicinal soap (Sapo Medicatus) were found to lower the bacterial count, but only when used in quantities more than 10-fold those normally used for bathing. Only by using sublimate soap, which contained 1% mercury chloride, was it possible to kill large numbers of cholera bacilli in bathwater under normal conditions. The seriousness and dedication with which these experiments were carried out is demonstrated in a report by Albert Hendrik Nijland, a military doctor in Amsterdam, who carried out this investigation in the early 1890s [22]. To simulate the real situation in which a human body might influence the results, Nijland filled his bathtub with water from the Vecht. After adding cholera bacilli from a 20-hour agar culture, he stepped into the water and mixed the bacteria thoroughly with the water by moving around in the bathtub. After that, samples were taken for bacterial counts. He then washed himself with sublimate soap, not using more soap than he normally would. During his bathing, in which the soap was mixed with the water, samples were taken every five minutes for bacterial counts. The counts showed that the number of cholera bacilli was reduced from about 700/ml water at the beginning of the bath to undetectable after 15 minutes. It was concluded that by using sublimate soap, all cholera bacteria were killed in the water within a normal bathing time. Because sublimate is a toxic compound, the final concentration of sublimate in the bathwater was calculated and found to be 0.19 mg/litre, or 200-fold less than was considered safe by the Dutch Pharmacopeia.

In 1896, Reithoffer [18] tested the disinfection of careful handwash using various soap concentrations. A 5% solution of the most effective soap sufficed to kill almost instantly all the cholera bacteria on the hands. Handwashing with a more concentrated soap solution was needed for disinfection of typhoid bacteria: 10% for one minute or 5% for three minutes. For practical purposes, all the soaps were useless as antiseptics against coli bacteria and staphylococci.

Despite promising reports on the disinfecting potential of soaps against cholera and to some extent against typhoid bacteria at the end of the nineteenth century, they do not seem to have reached a widespread use as disinfectants at that time. Possibly the reason for this was their low activity against many other common pathogens.

In a study of the disinfecting action of soaps in dishwater, a 0.5% solution of brown soap, such as is used for washing dishes in army messes, was found to reduce the number of streptococci from innumerable to 15 in two minutes, and the number of *Bacterium influenzae* and pneumococci to undectable within half a minute. The number of staphylococci and typhoid bacilli was not reduced in 10 minutes, and colon bacilli, which are most likely to be spread by dirty dishwater, were not killed by the soap solution. It was concluded from this study that soap solutions kill respiratory bacteria in dishwater, which, however, is not a likely vehicle for their spread in nature. On the other hand, soap did not kill colon bacilli in dirty dishwater, though it is a likely vehicle for spreading intestinal bacteria. It was concluded that disinfection by boiling was needed to prevent the spread of these bacteria through contaminated dishwater [23].

Walker in 1931 [24] discussed the apparent lack of interest in soaps as antiseptics and disinfectants, in spite of their documented germicidal activity against a number of pathogenic bacteria such as streptococci, pneumococci, meningococci, gonococci, diptheria bacilli, *B. influenzae* and *Spirochaeta pallida*, the causative agent of syphilis. In the paper, Walker presented the results of experiments which showed that soaps killed these bacteria in a couple of minutes, in 10- to 20-fold higher dilutions than those to be expected in normal handwashing. This indicated that thorough washing with soap sufficed to destroy the above bacteria on the hands. Therefore, the germicidal action of soaps compared favorably with that of many newly synthesized chemical disinfectants, and, when properly used for cleaning of the hands or for dishwashing, soaps would play an important role in preventing the spread of disease. Still, soaps were hardly mentioned as germicides in standard textbooks on surgery and bacteriology, perhaps because of their low activity against common bacteria such as staphylococci.

As mentioned earlier, synthetic detergents with added antibacterial chemicals almost completely replaced pure natural soaps as disinfectants and antiseptics during and after the Second World War. The marked selectivity of soaps in their effect on bacteria, particularly their lack of action against staphylococci, and certain other undesirable characteristics such as precipitation by calcium in hard water, limited their usefulness as cleaners and disinfectants. By the 1930s many new synthetic detergents were already shown to be antibacterial [25]. However, in order to achieve a more powerful disinfection of inanimate objects and a better antiseptic action in the cleaning of hands, a number of different antibacterial compounds have been added to detergents and soaps, such as hexachlorophane, chlorhexidine digluconate, triclosan and many other chemical disinfectants [26–30]. A further discussion of these antibacterial soaps and detergents is outside the scope of this review.

8.3   Use of Bactericidal Lipids to Reduce Microbial Contamination of Food Products

The importance of microbial contamination of food products is twofold. First, foodborne infections are an important public health problem, caused mainly by contamination of processed meat, such as poultry, with *Campylobacter jejuni* and *Samonella* spp. Other bacteria, spread by contaminated food and water and causing infections in humans, include *Escherichia coli* strain 0157, *Listeria monocytogenes*, *Shigella dysenteriae*, *Clostridium* spp. and, rarely, *Bacillus cereus*. In the first decade of the twenty-first century, contaminated fresh produce has increasingly been implicated in outbreaks of foodborne infections in the United States, as well as in other parts of the world [31,32]. Second, microorganisms attach to processed carcasses, such as the skin of poultry, and cause spoilage of the product upon storage. As is well known, spoilage occurs even under refrigeration, since psychrotrophic bacteria, such as *Pseudomonas* spp., and yeasts are common food contaminants.

A variety of chemical and physical treatments have been assessed for reduction of surface contamination by pathogenic and spoilage organisms on poultry carcasses and skin

[33–49]. Trisodium phosphate (TSP), which in 1992 was approved by the United States Department of Agriculture for use in poultry processing, was found to reduce the number of *Campylobacter* on chicken carcasses by 1.2 to 1.5 $\log_{10}$ [35]. A similar reduction of viable *Campylobacter* counts has been observed for most chemical treatments.

Hinton and Ingram [39] studied the effect of oleic acid on the bacterial flora of poultry skin, both foodborne pathogens and spoilage bacteria. Skin samples were removed from commercial broiler carcasses and washed in solutions of oleic acid prepared by dissolving potassium oleate paste in water. The number of aerobic bacteria in rinsates from the skin pieces after washing for one minute with 2, 4, 6, 8, or 10% oleic acid was determined and found to be significantly reduced compared to rinsates from control skin washed in peptone water. However, the reduction was less than 2 $\log_{10}$, even after washing with a 10% solution of oleic acid. The reduction in the number of aerobic bacteria remaining in the skin was also less than 2 $\log_{10}$, compared with the control. The washing was somewhat more efficient in removing *Campylobacter* and enterococci from the skin. The findings indicated that washing with oleic acid/potassium oleate significantly reduced the number of spoilage and pathogenic bacteria on poultry skin, although some bacteria remained attached to the skin even after two washes in 10% oleic acid/oleate.

In a similar study, Hinton and Ingram [40] showed that washing with mixtures of tripotassium phosphate (K_3PO_4) and potassium oleate was more effective in reducing microbial contamination of broiler skin than washing in either K_3PO_4 or potassium oleate alone. Washing the skin by stirring in K_3PO_4–oleate mixture for 30 minutes significantly reduced the number of bacteria that remained attached to the skin, as measured after suspension of the skin samples in a Waring blender. However, 2.7 $\log_{10}$ CFU of aerobic bacteria and 1.5 $\log_{10}$ CFU of *Enterobacteriacae*, both per gram of skin, remained attached to the skin after this thorough washing. This demonstrated that some bacteria were firmly attached to the poultry skin, as was also shown by Lillard [50], who recovered bacteria from poultry carcasses after 40 rinses in peptone water.

In a later study, Hinton and Ingram [41] determined the ability of mixtures of tripotassium phosphate and lauric and myristic acids to reduce bacterial flora on pieces of broiler skin by washing in a stomacher for two minutes at high speed. Washing of skin samples in the mixtures significantly reduced the numbers of total aerobic bacteria, *Campylobacter*, *E. coli* and *Pseudomonas* recovered from skin rinsates after washing. Mixtures with lauric acid appeared to be significantly more active than mixtures with myristic acid. In a similar study [42], Hinton and Ingram determined the effect of mixtures of potassium hydroxide and lauric acid on bacteria attached to pieces of broiler skin. They found that washing with mixtures of 1% potassium hydroxide and 0.5, 1.0, 1.5 or 2.0% lauric acid was more effective than washing in 1% potassium hydroxide alone, and that washing with the mixtures significantly reduced the number of total aerobic bacteria and enterococci on the skin. From these studies they concluded that mixtures of either tripotassium phosphate or potassium hydroxide with fatty acids might be useful for reducing microbial contamination on poultry carcasses. However, as pointed out by the authors, testing of these compounds in commercial poultry processing facilities would be necessary, since the vigorous washing in the stomacher might help reduce the bacterial number on the skin.

Solutions intended for sanitizing of carcasses in the slaughterhouse, particularly poultry, were patented by Guthry in 1994 [51]. The solutions preferably contained a mixture of caprylic and capric acids as the bactericidal ingredients, citric acid as a chelating agent and hydrochloric acid to maintain the pH of the solution at pH 2 to 4. A test solution was found to sanitize chicken drumettes, artificially contaminated with *Salmonella*, after treatment for 30 minutes at room temperature, with 19 out of 20 drumettes testing negative.

In an international patent application from 1995 [52], a disinfectant composition was claimed which contained fatty acid monoesters and was intended to reduce microbial contamination of processed meat, particularly poultry carcasses. In example 1 of this PCT application it was shown that glycerol monolaureate, monocaprate and monocaprylate, when used alone, did not have antimicrobial activity against *E. coli* and *S. aureus*. In contrast, compositions containing glycerol monolaureate, an organic acid, an anionic surfactant and a nonionic surfactant were highly antimicrobial against these bacteria. Compositions containing monolaureate were also effective in reducing the number of spoilage bacteria on poultry carcasses treated at 0 °C for about one hour. Compositions containing mixtures of glycerol monolaureate, propylene glycol monocaprate and propylene glycol monocaprylate reduced the number of *Salmonella typhimurium* on artificially infected breast skin from poultry carcasses, but in all cases by less than 1.5 $\log_{10}$. All the compositions had to contain an organic acid and a surfactant in addition to the fatty acid monoesters in order to show antimicrobial activity.

In a European patent from 2009 [53], antimicrobial compositions were formulated for use as disinfectants of plants or plant parts. The compositions contained a fatty acid monoester, an enhancer – for example an organic acid – and two or more anionic surfactants. The fatty acid monoesters included in compositions tested on seeds, fruits and vegetables were glycerol monolaureate and propylene glycol monocaprylate. Diluted formulations containing propylene glycol monocaprylate were found to reduce the bacterial counts of *E. coli* and *S. typhimurium* on artificially contaminated alfalfa seeds by 4 to 6 $\log_{10}$ after treatment for 15 minutes at 50 °C. Similarly, diluted formulations containing either propylene glycol monocaprylate or glycerol monolaurate reduced total aerobic counts and coliform counts on naturally contaminated vegetables by 2 to 4 $\log_{10}$ after chopped vegetables were washed by shaking in the formulation for 10 minutes at room temperature.

8.4   Killing of Foodborne Bacteria by Glycerol Monocaprate (Monocaprin) Emulsions

In a study of the bactericidal action of lipids on foodborne bacteria [54], fatty acids and monoglycerides with chain lengths C8 to C18 were tested *in vitro* against *Campylobacter jejuni*. Capric acid (C10 : 0) and its monoglyceride, monocaprin, were found to reduce the viable bacterial counts by almost 7 $\log_{10}$ in 10 minutes at 37 °C, whereas the other fatty acids and monoglycerides tested had no or very little activity. A solution of monocaprin in ethanol was added to warm water under constant forceful stirring in order to make a 200 mM (5%) monocaprin-in-water emulsion. In some cases, a solution of Tween 20 or Tween 40 was added to the emulsion. The concentrated emulsions became clear at room temperature and remained stable upon storage for at least one year.

Monocaprin-in-water emulsions were extremely active in killing *Campylobacter jeuni* isolates from humans, chickens, ducks and turkeys, as well as strains of *C. coli* and *C. lari*. Thus, monocaprin in concentrations as low as 1.25 mM (0.03%) reduced the number of viable *C. jejuni* by more than 6.5 $\log_{10}$ after treatment in water for one minute at room temperature. The 200 mM monocaprin emulsions, without or with Tween, could therefore be diluted 160-fold in water without losing activity against *Campylobacter*, and remained stable and active for weeks at room temperature.

Salmonella spp. and *E. coli* are not killed by fatty acids and monoglycerides in water at neutral pH, even after treatment at high concentrations for 10 minutes [55]. Monocaprin emulsions diluted in water to a concentration of 5 mM did not reduce the viable counts of these bacteria in 30 minutes at 37 °C. However, upon dilution in 60 mM citrate–lactate buffer adjusted to pH 4.1 to 5.0, 1.25 mM monocaprin emulsions effectively killed clinical isolates of *Salmonella enterica* serovar Typhimurium and serovar Enteritidis and the virulent *E. coli* O157, reducing the viable counts by more than 6.7 $\log_{10}$ in 10 minutes at room temperature. Also, low concentrations of monocaprin emulsions were more active against *C. jejuni* in citrate–lactate buffer at low pH than in water at neutral pH [54].

Monocaprin is a natural compound found in plant oils, such as coconut oil, and is harmless to the human body in concentrations which kill bacteria. Monoglycerides, including monocaprin, are classified as GRAS (generally recognized as safe) by the United States Food and Drug Administration (FDA) [56] and were approved as safe by the United States Environmental Protection Agency (EPA) in 2004 [57]. They are approved as food additives by the European Union (EU) (E471), as are the other ingredients of monocaprin emulsions: Tween 20 (E432), Tween 40 (E434), citrate (E331) and lactic acid (E270) [58]. The demonstrated activity against foodborne bacteria of monocaprin emulsions in dilute citrate–lactate buffer at low pH therefore suggested that acidified monocaprin emulsions could be used as sanitizers in the food industry, particulary in the processing of poultry.

8.4.1 Reduction of Viable Campylobacter in Drinking Water and Feed of Chickens by Addition of Monocaprin Emulsions

One way to reduce the frequency of *Campylobacter* infections in humans caused by contaminated poultry would be to prevent or reduce intestinal colonization of the bacteria in broiler chickens. Although the routes by which the bacterium spreads into broiler houses are largely unknown and may vary from one location to another, spread by contaminated water and/or feed is a possibility [59]. Because of the strong anticampylobacter action of monocaprin in water, experiments were carried out to test whether or not the drinking water of chickens could be made safe by addition of diluted emulsions of monocaprin. A faecal isolate of *C. jejuni* was suspended in tapwater at a concentration of 6 to 7 $\log_{10}$ CFU/ml and the suspension divided into two aliquots. Monocaprin emulsion was added to one aliquot to a concentration of 2.5 mM (0.06%) and the mixture was kept for 10 minutes before being tested in decimal dilutions on blood agar plates for viable bacteria. Five 15-day-old chickens received 0.1 ml of the mixture by mouth. Bacterial suspensions without monocaprin were similarly tested on blood agar and given by mouth to five chickens. The chickens were given regular drinking water and feed for six days and were then euthanized and samples were taken from the ceca for *Campylobacter* counts. No viable *Campylobacter* were detectable in monocaprin-treated water samples, either on blood agar or by inoculation into chickens. In contrast, chickens in the control group which received 0.1 ml of untreated water were all heavily colonized with *Campylobacter*, the ceca containing 6.5 $\log_{10}$ CFU/g [60]. This experiment showed that drinking water of chickens contaminated with *Campylobacter* could be disinfected by addition of 0.06% monocaprin emulsion.

Further experiments were done to test whether addition of monocaprin emulsions to chicken feed contaminated with *Campylobacter* would reduce the number of viable bacteria in the feed [54]. Pellets of commercial broiler feed were mixed with an equal weight of 10 mM (0.24%) or 5 mM (0.12%) monocaprin emulsion, diluted either in water or in 0.06 M citrate–lactate buffer at pH 4.1. An equal volume of a *Campylobacter* suspension was then added to the mixture, to a final monocaprin concentration of 0.12 and 0.06%, respectively, and incubated at 37 °C for 30 minutes. Mixtures of pellets with water or of buffer without monocaprin were used as controls. After centrifugation of the mixtures, decimal dilutions of the supernatant were tested for viable *Campylobacter*. Monocaprin in a final concentration of 0.12% at pH 7 reduced the *Campylobacter* counts by 4.2 $\log_{10}$ compared with the control and was significantly more effective than 0.06% monocaprin, which reduced the counts by 2.4 $\log_{10}$. However, the 0.06% monocaprin emulsion significantly reduced the *Campylobacter* counts compared with the control ($P < 0.01$). There was not a significant difference between the reduction of *Campylobacter* counts by monocaprin in water or in buffer. This may be due to the fact that the pH of the buffered solution increased from 4.1 to 5.5 by mixture with the pellets. The microbicidal activity of monocaprin mixed with the *Campylobacter*-spiked chicken feed was much less than in water, most likely due to adsorption of monocaprin molecules to proteins and/or lipids in the feed, which might block its activity. However, the 2 to 4 $\log_{10}$ reduction in bacterial counts in monocaprin-treated feeds observed in this study suggests a possible use of monocaprin emulsions for disinfection of animal feeds.

The growth rate of 12-day-old chickens given drinking water and feed containing 0.12% monocaprin for 12 consecutive days was equal to the growth rate of chickens not receiving monocaprin in their water and feed (Figure 8.2). A macroscopic and microscopic examination of the intestinal tract and the liver of each group of chickens showed no abnormalities. It was concluded that monocaprin in a concentration at least two times higher than that needed to kill *Campylobacter* in drinking water and feed had no adverse effects on 12-day-old chickens [60].

8.4.2 Reduction of Viable Campylobacter on Poultry Carcasses

Another way to reduce the spread of *Campylobacter* from contaminated poultry to humans might be to reduce the number of viable *Campylobacter* on carcasses during processing in the slaughterhouse. A risk-assessment study has estimated that a 2 $\log_{10}$ reduction in the number of viable *Campylobacter* on chicken meat will lead to a 25- to 30-fold reduction in the number of human cases of campylobacteriosis [61, 62]. As pointed out previously, a number of different chemicals have been tested with the aim of reducing surface contamination by pathogenic bacteria, including *Campylobacter*, on poultry carcasses and skin. Among the compounds tested were salts of fatty acids, either alone or in combination with potassium phosphate or potassium hydroxide [39–42].

Because of the strong bactericidal activtity of monocaprin against *Campylobacter*, experiments were done to test the reduction of viable *Campylobacter* on poultry carcasses by dipping into monocaprin emulsions for varying lengths of time and at various temperatures and pHs [63]. All the experiments were done on carcasses naturally contaminated with

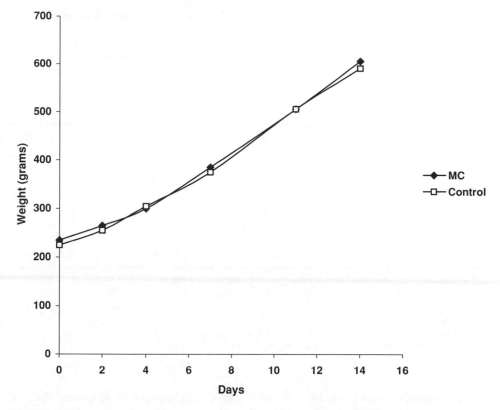

Figure 8.2 *Mean weight in grams of a group of five chickens treated with an emulsion of 5 mM (0.12%) monocaprin in drinking water and feed for 12 days compared with a group of five untreated controls. The chickens were received at 10 days of age and the monocaprin treatment was started on day 2.*

Campylobacter, either in the laboratory using legs – that is, thigh with attached drumstick – and pieces of neck skin, or in a slaughterhouse using whole carcasses. In the laboratory, chicken legs were obtained immediately after the carcasses were taken off the processing line and kept on ice until they arrived at the laboratory within 30 minutes. The legs were immersed into a container with 1 litre of monocaprin emulsion, drained for one minute and lightly blotted with paper towel. They were then transferred into plastic bags with buffered peptone water (BPW) and handmassaged for one minute. Samples from the BPW rinsates were tested for viable *Campylobacter* by inoculation of decimal dilutions on to Campy-Cefex agar plates [64]. Untreated legs or legs immersed in water were used as controls. The results are summarized in Table 8.1, which shows that immersion in acidified 20 mM (0.5%) monocaprin emulsion for one minute at 20 °C or for five minutes at 5 °C caused a greater than 2 $\log_{10}$ reduction in the number of *Campylobacter* on chicken legs. Similar results were obtained for neck skins immersed in 20 mM acidified monocaprin emulsions at 20 °C for one minute.

Table 8.1 Campylobacter *counts on naturally contaminated chicken legs after immersion in an emulsion of monocaprin (MC). Comparison with controls immersed in water or not treated. The table shows one, two or three trials for each treatment. (Data from [63].)*

MC emulsion	Temp.	Time	Number of viable *Campylobacter* ($\log_{10}$ CFU)/100gram[a]		
			MC emulsion	Water	Untreated
10 mM in water at pH 7	20 °C	10 min	2.16 ± 0.31^b	3.03 ± 0.23	3.35 ± 1.02
	20 °C	10 min	3.50 ± 0.41^b	4.98 ± 0.58	4.80 ± 0.13
10 mM in buffer at pH 4.1[c]	20 °C	1 min	4.13 ± 0.83	ND[d]	4.87 ± 0.82
	20 °C	1 min	4.08 ± 0.30	ND	4.34 ± 0.32
20 mM in buffer at pH 4.1[e]	20 °C	1 min	1.87 ± 0.01^f	ND	4.47 ± 0.05
	20 °C	1 min	2.16 ± 0.34^f	ND	4.87 ± 0.82
	20°C	1 min	2.37 ± 1.12	ND	4.34 ± 0.32
20 mM in buffer at pH 4.1[e]	5 °C	5 min	1.54 ± 1.11^f	ND	5.42 ± 0.46
10 mM in buffer at pH 4.1[c]	5 °C	5 min	4.75 ± 0.25^g	ND	5.97 ± 0.25
	5 °C	5 min	2.63 ± 0.67^g	ND	5.44 ± 1.02

[a]$\log_{10}$ colony forming units (CFU). Means for three legs ± standard deviation (SD).
[b]Significant reduction in bacterial counts compared with water and untreated control ($P < 0.05$).
[c]3.6 mM citrate–lactate buffer
[d]ND: not done
[e]7.2 mM citrate–lactate buffer
[f]Significant reduction in bacterial counts compared with untreated control ($P < 0.01$).
[g]Significant reduction in bacterial counts compared with untreated control ($P < 0.05$).

In experiments carried out in the slaughterhouse, *Campylobacter*-contaminated chicken carcasses were taken off the processing line after evisceration and before spraying with cold water. The carcasses were dipped into a bucket containing 20 mM emulsion of monocaprin at pH 4.1 and 20 °C. After manual agitation for one minute the carcasses were put back on the processing line, where they were rinsed for two hours by spraying with water at 0 to 4 °C. After spraying, the legs were removed from each carcass and brought to the laboratory for *Campylobacter* testing. Legs from untreated carcasses were used as controls. The number of viable *Campylobacter* was significantly reduced by this treatment, although by less than 2 $\log_{10}$.

The relatively low efficacy of bactericidal compounds on chicken skin, compared with their activity in water, is most likely due to the firm attachment of bacteria to the skin [40]. The bacteria may be entrapped in folds, crevices or pores, making them less accessible to bactericidal compounds in the surrounding fluid [65–67]. Short-term surface freezing of the skin of *Campylobacter*-contaminated carcasses and exposure to either steam or dry heat have been found to be most effective in reducing bacterial counts [68, 69]. This indicates the importance of a direct effect on the skin in killing or releasing the bacteria. Another factor which may play a role in reducing the bactericidal action of compounds, such as monocaprin, on the skin of poultry is their tendency to form complexes with proteins and lipids. Such molecules in the skin may act as competitors to the lipoproteins of the bacterial cell wall and thus counteract the bactericidal effect [70].

Although dipping of chicken carcasses into 20 mM monocaprin emulsion at pH 4.1 and at 20 °C significantly reduced the *Campylobacter* counts compared with untreated control carcasses, the mean reduction varied considerably between replicate trials. There was also a considerable variation in the *Campylobacter* counts among carcasses, both treated and controls. This variation may be caused by a difference in the access of monocaprin emulsions to the bacteria, attached to or entrapped in the skin of individual birds. The conditions in broiler houses and the duration and magnitude of contamination could possibly play a role. It would therefore be of interest to study monocaprin treatment of carcasses from different broiler houses. Since poultry chilling systems vary in commercial processing plants and ice-water-immersion chilling and air chilling at the end of the processing line are more commonly used than spraying with cold water, it would be of interest to try treatment with monocaprin emulsions in slaughterhouses which use either of these chilling methods. Spray-washing of carcasses with monocaprin emulsions at 5 °C for five minutes or at 20 °C for one minute might be carried out by inserting a washer into the processing line before the final chilling. Based on the experiments carried out in the laboratory (Table 8.1), either treatment might result in a significant reduction in viable *Campylobacter* on the meat.

8.4.3 Reduction of Psychrotrophic Spoilage Bacteria on Chicken Carcasses

For almost 50 years, various treatments of poultry carcasses have been tried in order to reduce the load of spoilage bacteria and extend the shelf-life of fresh poultry products [38, 39, 41, 43, 71–80]. These include treatments with organic acids, such as acetic, lactic and citric acids; fatty acids, such as oleic, lauric and myristic acids; tripotassium phosphate; acidified sodium chlorite; sodium acid pyrophosphate; sodium tripolyphosphate; electrolysed oxidizing water; hydrogen peroxide; and essential oils. A review of the literature shows a great variability in the effect of these chemical compounds on the bacterial load of poultry carcasses, as well as in their effect on meat quality, for example on colour, texture, smell and taste, and thus on consumer acceptability [80]. The reduction in bacterial counts caused by each compound was in most studies 1 to 2 log_{10} or less but in some studies was 2 to 3 log_{10} or even greater. Similarly, the effects of the different chemical compounds on the quality of the meat varied among the studies. The results from studies of a particular compound did not only vary between studies but also within each study. However, in most studies treatment with chemical compounds significantly reduced the microbial load. The FDA has approved the use of trisodium phosphate, acidified sodium chlorite and citric acid for treatment of poultry carcasses in processing plants, but in the past the EU did not permit treatment of poultry carcasses with bactericidal chemicals. However, EU legislation effective from 1 January 2006 permits use of chemical compounds to remove bacterial surface contamination from poultry meat during processing [47, 80]. Such compounds must be harmless to the consumer when used in an effective concentration and they should be without adverse effects on the sensory quality of the products, either visible or detectable by smell and taste. Several compounds are being studied with regard to their ability to meet these criteria [80].

Post-chill dipping of chicken legs into 5 mM monocaprin emulsions, acidified to pH 4.1 by 0.06 M citrate–lactate buffer, significantly reduced the number of psychrotrophic bacteria on the legs during storage at 3 °C (Figure 8.3), but did not affect meat quality [63]. Similar results were obtained from prechill treatments, in which carcasses were immersed in 20 mM monocaprin emulsion at pH 4.1 for one minute before being sprayed with cold water for two hours (Figure 8.4). However, as with the various compounds mentioned above, the

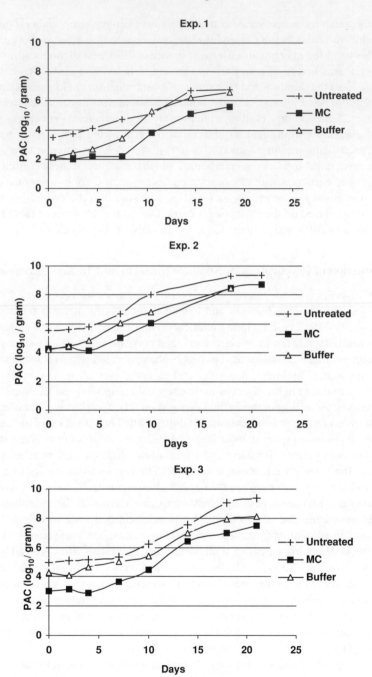

Figure 8.3 *Psychrotrophic aerobic colony counts (PACs) on chicken legs packed after immersion for five minutes at room temperature in an emulsion of 5 mM (0.12%) monocaprin (MC) and 0.02% Tween 40 in 60 mM citrate–lactate buffer at pH 4.1 (■), after immersion in buffer alone (Δ), and without immersion (+). The legs were stored at 3 °C and PACs were counted at intervals up to 21 days of storage. Each point represents PAC/gram for each package (three legs). The results of three separate experiments are shown. (Data from [63].)*

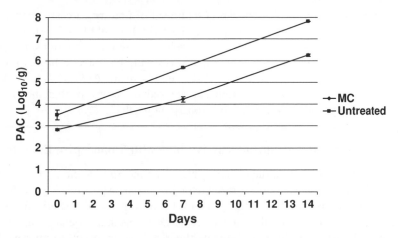

Figure 8.4 *Psychrotrophic aerobic colony counts (PACs) on chicken carcasses immersed for one minute at 20 °C in an emulsion of 20 mM (0.5%) monocaprin (MC) and 0.08% Tween 40 in 7.2 mM citrate-lactate buffer at pH 4.1. After immersion the carcasses were sprayed with cold water for two hours and the legs were packaged and stored at 3 °C. PACs were counted on days 0, 7 and 14. Each point represents the means for four legs showing the standard deviation as vertical lines. Carcasses not immersed in MC (untreated) were tested as controls. PACs were significantly lower (P < 0.01) for legs from MC-treated compared with untreated carcasses on all days. (Data from [63].)*

results of replicate trials of monocaprin emulsions carried out over a period of two years varied greatly, and only in 8 of 12 trials was there a significant reduction of psychrotrophic aerobic colony counts (PACs) on the chicken carcasses [63]. The cause of this variability in efficacy of identical treatments is not clear, but it may be the same as for treatments against *Campylobacter*, as outlined above. Bacteria are firmly attached to the surface of poultry carcasses when they arrive in the slaughterhouse [50] and there may be differences in both attachment and entrapment in the skin depending on the conditions in different broiler houses at different times. There may also be differences in the composition of the microbial flora on carcasses from different broiler houses. It can therefore not be assumed that a microbicidal treatment will have identical effects on the microflora of carcasses from different sources [81].

Since immersion in monocaprin emulsions at low pH reduces not only *Campylobacter* contamination on chicken carcasses but also in most cases the number of naturally occurring psychrotrophic bacteria causing spoilage, it might be feasible to use this treatment to simultaneously reduce *Campylobacter* contamination and the number of spoilage microorganisms on poultry carcasses. Compared with other means that have been used to reduce bacterial contamination on poultry carcasses, treatment with acidified monocaprin emulsions may be an attractive choice because monocaprin is derived from medium-chain triglycerides common in various fats, from both animals and plants. Thus, as previously pointed out, monoglycerides such as monocaprin are natural compounds which are harmless to the body in concentrations that kill bacteria, are classified as GRAS by the FDA and are approved as food additives by the EU. However, more experiments are needed to further establish the efficacy of monocaprin treatment of poultry carcasses contaminated with pathogenic and/or psychrotrophic spoilage bacteria. It would be of particular interest to test the effect on carcasses of broilers that have been reared under various housing conditions. Also, treatment with monocaprin emulsions should be tested in broiler processing plants using different chilling systems, as discussed above for *Campylobacter*. A longer prechill immersion at 20 °C might result in a greater reduction of PAC than the one-minute dip used in this study, and thus render more consistent results.

8.4.4   Sanitation of Kitchen Surfaces by Washing with Monocaprin Emulsions

Cross-contamination of bacteria such as *Campylobacter* and *Salmonella* during food preparation in the kitchen is considered a risk factor in human exposure to these foodborne pathogens, an important route of contamination being transfer of bacteria from poultry carcasses via cutting boards or other unwashed surfaces to salad vegetables and ready-to-eat foods [61, 82–88]. To reduce or eliminate this risk it is important to improve kitchen hygiene, for example by promptly washing contaminated surfaces with solutions which quickly and effectively kill the foodborne pathogens. Several studies have been done on cleaning procedures for plastic and wooden cutting boards and other surfaces in the kitchen that come into contact with chicken carcasses and other potentially contaminated foodstuffs [89–94]. Cleaning with disinfectants such as sodium hypochlorite (household bleach) and quaternary ammonium has been found to significantly reduce the number of viable bacteria on contaminated kitchen surfaces and dishcloths, whereas cleaning with common kitchen detergents and hot water is much less effective. Monolaurin and acetic acid were found to have a synergistic effect in killing planktonic and one-day adherent cells of *Listeria monocytogenes* on stainless steel chips. Exposure to 0.01% monolaurin reduced the number of one-day adherent cells by less than 1 $\log_{10}$ in 25 minutes at 25 °C, whereas addition of 1% acetic acid increased the reduction to more than 5 $\log_{10}$. A seven-day biofilm of *L. monocytogenes* cells was much more resistant to the combined effect of monolaurin and acetic acid [95].

Monocaprin emulsions at low pH rapidly kill foodborne pathogens such as *Campylobacter*, *Salmonella* and *E. coli* in water [54] and it has been shown that they also efficiently kill these bacteria on a variety of surfaces commonly found in kitchens, although not to the same degree as in water [96, 97]. Monocaprin emulsions can be mixed with liquid soaps without losing their bactericidal activity and thus they make the soaps more effective as cleaners/sanitizers or disinfectants in the home and in food-preparing and food-processing facilities. Thus, 15 commercial washing-up liquids (WULs) of various brands were mixed with emulsions of 5% monocaprin in water in the ratio of one part WUL to four parts monocaprin emulsion (unpublished). The WULs varied in their anionic and nonionic contents and six of them were labelled 'antibacterial'. About half of the 15 WULs tested were clear after 1 : 4 mixture with monocaprin emulsion and remained clear after storage for several weeks at room temperature. Other WULs became cloudy after mixture with monocaprin emulsions and separated into two phases upon storage but could be remixed by shaking. The reason for the difference in solubility of monocaprin in the different brands of WUL is not known and did not seem to depend on their anionic/nonionic contents. All of the WUL–monocaprin mixtures, diluted in water to 1.25% WUL and 0.25% monocaprin, reduced viable *Campylobacter* counts below the detection limit in one minute, or more than

$6.7 \log_{10}$. Water with 1.25% WUL did not show noticeable activity against *Campylobacter* in two minutes, as compared with controls, whereas 5% WUL showed a reduction in bacterial counts of about $1 \log_{10}$. One WUL labelled 'antibacterial with real Eucalyptus Extracts' showed a considerably greater activity against *Campylobacter* upon contact for two minutes; that is, a $3.2 \log_{10}$ reduction in viable counts. Other WULs labelled 'antibacterial' did not show more activity against *Campylobacter* than WULs that were not claimed to be antibacterial.

To mimic handling of *Campylobacter*-contaminated chicken on cutting boards in the kitchen, contaminated meat juice was spread on plastic and wooden board surfaces, which were then washed with 10 mM (0.25%) monocaprin emulsions in water at neutral pH or in citrate–lactate buffer at pH 4.1 [96]. Both caused a significant reduction in viable *Campylobacter* counts and in most cases bacteria were not detectable in swabs from surfaces after washing with monocaprin emulsions. However, the emulsions were generally less active against *Campylobacter* on wooden than on plastic surfaces. Notably, mixtures of a commercial WUL and monocaprin emulsions were as active against *Campylobacter* as monocaprin emulsions alone. On the other hand, 5% WUL in water was not more microbicidal on plastic boards than water alone. It was more difficult to kill *Salmonella* than *Campylobacter* in chicken-meat juice on plastic cutting boards, but washing with 20 mM monocaprin emulsions containing 90 mM citrate–lactate buffer at pH 4.1 reduced the viable bacteria counts in *Salmonella*-spiked chicken-meat juice by 4 to 5 $\log_{10}$, compared with washing the plastic boards in water or in 1% WUL solution [97].

To simulate routine cleaning procedures in a kitchen, 20 mM monocaprin emulsions at pH 4.1 were applied for two minutes to laminated plastic kitchen counters which had been soiled with about 8 $\log_{10}$ CFU/ml of *S. enteritidis* or *E. coli* in diluted nutrient broth, either fresh or dried on the surface and thus leaving visible dirt spots. As a control, the surfaces were washed with 1% WUL in water: a soap solution routinely used in the kitchen. Several notable results emerged from these tests [97]. First, cleaning with monocaprin emulsions, with or without WUL, reduced the number of viable *E. coli* and *Salmonella* below the detectable level (less than 1.7 $\log_{10}$ CFU/ml), that is by more than 6 $\log_{10}$, in dry spots on a laminated plastic surface. Second, the number of viable *Salmonella* was reduced below the detectable level on wet laminated plastic and viable *E. coli* was reduced by 4 to 5 $\log_{10}$. Third, cleaning with 1 to 2% WUL in water reduced the number of viable bacteria by only 1 to 2 $\log_{10}$.

Since *E. coli* was more resistant to monocaprin emulsions than *Salmonella* on a wet kitchen counter, killing of *E. coli* by monocaprin emulsions was tested on other surfaces commonly found in kitchens or in comparable facilities used for food processing and food preparation that were considered at risk of being contaminated with foodborne bacteria [97]. There were significant differences in the reduction of viable *E. coli* counts on the wet surfaces, but on all of them the reduction was 4 to 5 $\log_{10}$ after washing with 0.5% monocaprin emulsion for two minutes, as compared with surfaces washed with 1% WUL in water (Table 8.2). The monocaprin emulsions were most effective against *E. coli* on glass and stainless steel and least effective on a plastic cutting board. The differences in microbicidal activities against *E. coli* on various kitchen surfaces were confirmed when mixtures of monocaprin emulsions and commercial WULs were used. Notably, the mixtures were found to be more active against *E. coli* in dry spots than on wet surfaces. It has been reported that drying causes a significant reduction in viable *Salmonella* and *E. coli* counts on

Table 8.2 *Killing of* E. coli *on various surfaces soiled with 0.5 ml of a 22-hour culture of* E. coli *in nutrient broth diluted 10-fold in water. The wet surfaces were washed with 1 ml of a 20 mM (0.5%) monocaprin emulsion in 7.2 mM citrate–lactate buffer at pH 4.1. As a control, contaminated surfaces were washed in the same way with 1% WUL in water. After washing for two minutes, the fluid was wiped off with paper towel and the surface was swabbed for 30 seconds with cotton swabs moistened in buffered peptone water (BPW). The swabs were soaked in 1 ml of BPW and samples (100 μl) were streaked in duplicate on to blood agar plates. (Data from [97].)*

Surface	Number of viable E. coli in swabs $\log_{10}$ CFU[a]/ml[b]
Plastic board	2.67 ± 0.42
Laminated plastic	1.80 ± 0.17[c]
Tile	2.33 ± 0.15
Stainless steel	<1.70[d] $\pm$ NA[e]
Glass	$<1.70 \pm$ NA
1% WUL, all surfaces	6.42 ± 0.28
Control[f]	8.10 ± 0.10

[a]Colony-forming units.
[b]Means for three experiments ± standard deviation.
[c]Significantly less than the number of viable E. coli in swab from plastic board ($P < 0.05$) and in swab from tile ($P < 0.005$).
[d]<1.7: The number of viable E. coli was below the detection limit.
[e]NA: not available.
[f]The number of viable E. coli ($\log_{10}$) in 0.5 ml of nutrient broth added to the surface.

laminated surfaces soiled with diluted broth cultures. However, about 90% of the bacteria were still viable after one hour and live bacteria were detectable on the surface for at least 24 hours [98]. In our study [97], drying of *E. coli* in nutrient broth on a laminated plastic surface for one hour reduced the number of viable bacteria by 1 $\log_{10}$, but their survival on the dry surface was not followed over a longer period of time. In contrast, the number of viable *E. coli* was reduced below the detection limit by rinsing with MC emulsions for two minutes.

It was shown by these studies that emulsions of monocaprin in citrate–lactate buffer at pH 4.1 are highly efficient in killing the foodborne bacteria *Campylobacter*, *Salmonella* and *E. coli* in various environments. Since monocaprin is a surface-active molecule it has a considerable cleaning effect in water, but the cleaning effect seemed to be enhanced when monocaprin emulsions were mixed with a liquid soap like WUL. Mixing with WUL did not reduce the microbicidal activity of monocaprin against *Campylobacter* and *Salmonella* or against *E. coli* in dry spots [96, 97]. Considering the combined cleaning and sanitizing effect of WUL and monocaprin, mixtures of these compounds may be highly effective as sanitizing cleaners in kitchens and other food-preparing and food-processing facilities. They may also be useful as general cleaners in the home, and even in schools and child-care centres [99]. Compared with other compounds that have been used as sanitizers or disinfectants in the home, acidified monocaprin emulsions may be an attractive choice because monocaprin is derived from natural sources and is both environmentally safe and harmless to the body in concentrations which kill bacteria.

8.5 Conclusions

After the discovery in the late 1800s that soaps kill certain bacteria, there was a considerable interest in their utilization as disinfectants. However, it was soon realized that bacteria differ widely in their susceptibility to the germicidal effect of soaps, which could therefore not be used as general disinfectants. Their use was limited to the bacteria which showed the highest susceptibility. As a consequence, soaps never reached a widespread utilization as disinfectants. With the advent of synthetic detergents in the 1930s and their increasing popularity in the 1940s and 1950s, the role of natural soaps as disinfectants diminished, particularly since the detergents could be made germicidal by addition of chemicals such as hexachlorophane, chlorhexidine digluconate and triclosan.

In recent years there has been a renewed interest in using antimicrobial fatty acids and monoglycerides to reduce bacterial contamination of foodstuffs and as sanitizers or disinfectants in home kitchens and in food-preparing and food-processing facilities. Glycerol monocaprate (monocaprin) has been found to be highly microbicidal, even after mixture with liquid soaps. It has been suggested that such mixtures could be useful as sanitizing cleaners in the home and in food-preparing and food-processing facilities, as well as in various settings such as schools and child-care centres. Compared with other compounds that have been used as sanitizers or disinfectants, monocaprin, either in emulsions in water or in liquid soaps, may be an attractive choice because it is derived from natural sources and is both environmentally safe and harmless to the body in concentrations which kill bacteria.

References

1. http://en.wikipedia.org/wiki/Disinfectant (Accessed July 2009).
2. Kanz, E. und Kanz, C. (1969) Zur Prüfung Beurteilung 'desinfizierender Seifen'. *Gesundheitswesen und Desinfektion*, **61**, 145–158.
3. McDonnell, G. and Russell, A.D. (1999) Antiseptics and disinfectants: activity, action, and resistance. *Clin. Microbiol. Rev.*, **12**, 147–179.
4. http://en.wikipedia.org/wiki/Antisepsis (Accessed July 2009).
5. Riedel, S. and Walker, J.T.A. (1903) Standardization of disinfectants. *J. Royal Sanitary Inst., London*, **24**, 424–441.
6. Wright, E.S. and Mundy, R.A. (1960) Defined medium for phenol coefficient tests with *Salmonella typhosa* and *Staphylococcus aureus*. *J. Bacteriol*, **80**, 279–280.
7. Varley, J.C. and Reddish, G.F. (1936) Phenol coefficient as measure of practical value of disinfectants. *J. Bacteriol.*, **32**, 215–225.
8. Koch, R. (1881) Ueber Desinfection. *Mittheil. des kaiserl. Gesundheitsamtes*, **1**, 234–282.
9. Geppert, J. (1889) Zur Lehre von den Antiseptics, Eine Experimentaluntersuchung. *Berl. klin. Wochenschr.*, **26**, 789–794, 819–821.
10. http://en.wikipedia.org/wiki/Soap (Accessed July, 2009).
11. Hunt, J.A. (1999) A short history of soap. *Pharmaceut. J.*, **263**, 985–989.
12. Soaps and detergents: history, http://www.sdahq.org/sdalatest/html/soaphistory2.htm (Accessed July, 2009).
13. Beyer, T. (1896) Ueber Wäschedesinfection mit dreiprocentiger Schmierseifenlösungen und mit Kalkwasser. *Zeitschr. f. Hyg.*, **22**, 228–262.
14. Förster, D. (1900) Versuche über Wäschedesinfektion. *Hyg. Rundschau*, **10**, 513–529.
15. Behring, E.A. (1890) Ueber Desinfection, Desinfectionsmittel und Desinfectionsmethoden. *Zeitschr. f. Hyg.*, **9**, 395–478.

16. Jolles, M. (1893) Ueber die Desinfectionsfähigkeit von Seifenlösungen gegen Cholerakeime. *Zeitschr. f. Hyg.*, **15**, 460–473.
17. Jolles, M. (1895) Weitere Untersuchungen über die Desinfectionsfähigkeit von Seifenlösungen. *Zeitschr. f. Hyg.*, **19**, 130–138.
18. Reithoffer, R. (1896) Ueber die Seifen als Desinfectionsmittel. *Arch. f. Hyg.*, **27**, 350–364.
19. Serafini, A. (1898) Beitrag zum experimentellen Studium der Desinfectionsfähigkeit gewöhnlicher Waschseifen. *Arch. f. Hyg.*, **33**, 369–398.
20. Konrádi, D. (1902) Über die baktericide Wirkung der Seifen. *Arch. f. Hyg.*, **44**, 101–112.
21. Reichenbach, H. (1908) Die desinfizierenden Bestandteile der Seifen. *Zeitschr. f. Hyg.*, **59**, 296–316.
22. Nijland, A.H. (1893) Ueber das Abtödten von Cholerabacillen in Wasser. *Arch. f. Hyg.*, **18**, 335–372.
23. Nichols, H.J. (1919–1920) Bacteriologic data on the epidemiology of respiratory diseases in the army. *J. Lab. Clin. Med.*, **5**, 502–511.
24. Walker, J.A. (1931) The germicidal and therapeutic applications of soaps. *J. Am. Med. Assoc.*, **97**, 19–20.
25. Baker, Z., Harrison, R.W. and Miller, B.F. (1941) Action of synthetic detergents on the metabolism of bacteria. *J. Exp. Med.*, **73**, 249–271.
26. Quinn, H., Voss, J.G. and Whitehouse, H.S. (1954) A method for the *in vivo* evaluation of skin sanitizing soaps. *Appl. Microbiol*, **2**, 202–204.
27. Lowbury, E.J.L., Lily, H.A. and Bull, J.P. (1964) Disinfection of hands: removal of transient organisms. *Brit. Med. J.*, **2**, 230–233.
28. Lily, H.A. and Lowbury, E.J.L. (1971) Disinfection of the skin: an assessment of some new preparations. *Brit. Med. J.*, **3**, 674–676.
29. Lily, H.A. and Lowbury, E.J.L. (1978) Transient skin flora. Their removal by cleansing or disinfection in relation to their mode of deposition. *J. Clin. Pathol.*, **31**, 919–922.
30. Tanner, J., Swarbrook, S. and Stuart, J. (2008) Surgical hand antisepsis to reduce surgical site infection. *Cochrane Database Syst. Rev.* **1** (Art. No. CD004288).
31. Doyle, M.P. and Erickson, M.C. (2008) Summer meeting 2007. The problem with fresh produce: an overview. *J. Appl. Microbiol.*, **105**, 317–330.
32. Lynch, M.F., Tauxe, R.V. and Hedberg, C.W. (2009) The growing burden of foodborne outbreaks due to contaminated fresh produce: risk and opportunities. *Epidemiol. Infect.*, **137**, 307–315.
33. Izat, A.L., Colberg, M., Thomas, R.A. *et al.* (1990) Effects of lactic acid in processing waters on the incidence of salmonellae on broilers. *J. Food Quality*, **13**, 295–306.
34. Kim, J-W., Slavik, M.F., Pharr, M.D. *et al.* (1994) Reduction of *Salmonella* on post-chill chicken carcasses by trisodium phosphate (NA_3PO_4) treatment. *J. Food Safety*, **14**, 9–17.
35. Slavik, M.F., Kim, J-W., Pharr, M.D. *et al.* (1994) Effect of trisodium phosphate on *Campylobacter* attached to post-chill chicken carcasses. *J. Food Protect.*, **57**, 324–326.
36. Kim, J-W. and Slavik, M.F. (1996) Cetylpyridium chloride (CPC) treatment on poultry skin to reduce attached *Salmonella*. *J. Food Protect.*, **59**, 322–326.
37. White, P.L., Baker, A.R. and James, W.O. (1997) Strategies to control *Salmonella* and *Campylobacter* in raw poultry products. *Rev. Scient. Techn.*, **16**, 525–541.
38. Dickens, J.A., Berrang, M.E. and Cox, N.A. (2000) Efficacy of an herbal extract on the microbiological quality of broiler carcasses during a simulated chill. *Poult. Sci.*, **79**, 1200–1203.
39. Hinton, Jr, A. and Ingram, K.D. (2000) Use of oleic acid to reduce the population of the bacterial flora of poultry skin. *J. Food Protect.*, **63**, 1282–1286.
40. Hinton, Jr, A. and Ingram, K.D. (2003) Bactericidal activity of tripotassium phosphate and potassium oleate on native flora of poultry skin. *Food Microbiol.*, **20**, 405–410.
41. Hinton, Jr, A. and Ingram, K.D. (2005) Microbicidal activity of tripotassium phosphate and fatty acids toward spoilage and pathogenic bacteria associated with poultry. *J. Food Protect.*, **68**, 1462–1466.
42. Hinton, Jr, A. and Ingram, K.D. (2006) Antimicrobial activity of potassium hydroxide and lauric acid against microorganisms associated with poultry processing. *J. Food Protect.*, **69**, 1611–1615.

43. Hinton, Jr, A., Northcutt, J.K., Smith, D.P. *et al.* (2007) Spoilage bacteria of broiler carcasses washed with electrolyzed oxidizing or chlorinated water using an inside-outside bird washer. *Poult. Sci.*, **86**, 123–127.
44. Oyarzabal, O., Hawk, C., Bilgill, S. *et al.* (2004) Effects of postchill application of acidified sodium chlorite to control *Campylobacter* spp. and *Escherichia coli* on commercial broiler carcasses. *J. Food Protect.*, **67**, 2288–2291.
45. Zhao, T. and Doyle, M.P. (2006) Reduction of *Campylobacter jejuni* on chicken wings by chemical treatments. *J. Food Protect.*, **69**, 762–767.
46. Sexton, A., Raven, G., Holds, G. *et al.* (2007) Effect of acidified sodium chlorite treatment on chicken carcases processed in South Australia. *Int. J. Food Microbiol.*, **115**, 252–255.
47. Tandrup Nielsen, C., Brondsted, L.O., Rosenquist, H. and Christensen, B.B. (2007) Chemical decontamination of *Campylobacter jejuni* on chicken skin and meat. *Zoonoses and Public Health*, **54** (Suppl. 1), 121.
48. Northcutt, J., Smith, D., Ingram, K.D. *et al.* (2007) Recovery of bacteria from broiler carcasses after spray washing with acidified electrolyzed water or sodium hypochlorite solutions. *Poult. Sci.*, **86**, 2239–2244.
49. Gellynck, X., Messens, W., Halet, D. *et al.* (2008) Economics of reducing *Campylobacter* at different levels within the Belgian poultry meat chain. *J. Food Protect.*, **71**, 479–485.
50. Lillard, H.S. (1989) Incidence and recovery of of salmonellae and other bacteria from commercially processed poultry carcasses at selected pre- and post-evisceration steps. *J. Food Protect.*, **52**, 88–91.
51. Guthry, E. Disinfecting product, United States Patent 5364650, 11/15/1994.
52. J. F. Andrews and J. F. Munson (1995) Disinfectant composition, WO 95/07616, 23 March 1995.
53. K. Tautvydas and J. F. Andrews, Fruit, vegetable and seed disinfectants, European Patent EP 1 231 838 B1, Bulletin 2009/49, 02/12/2009.
54. Thormar, H., Hilmarsson, H. and Bergsson, G. (2006) Stable concentrated emulsions of 1-monoglyceride of capric acid (monocaprin) with microbicidal activities against *Campylobacter, Salmonella* and *Escherichia coli*. *Appl. Environ. Microbiol*, **72**, 522–526.
55. Bergsson, G., Steingrímsson, Ó. and Thormar, H. (2002) Bactericidal effects of fatty acids and monoglycerides on *Helicobacter* pylori. *Int. J. Antimicrob. Agents*, **20**, 258–262.
56. Food and Drug Administration (1999) Code of Federal Regulations, Title 21, Vol. 3, Part 184, Sec. 184.1505, page 505, US Government Printing Office, Washington DC.
57. US Environmental Protection Agency (2004) Pesticides: Regulating pesticides, Fatty acid monoesters with glycerol or propanediol fact sheet (011288). *Federal Register*, **69** (72), 19844–19845. Registration approval, issued 10/20/04. Accessed in August 2009.
58. European Union, Richtlinie 95/2 /EG des Europäishen Parlaments und des Rates über andere Lebensmittelzusatzstoffe als Farbstoffe und Süssungsmittel, Seite 49 und 53, 20 Februar 1995. Accessed in August 2009.
59. CHRO (2007) 14th International Workshop on *Campylobacter, Helicobacter* and Related Organisms, Rotterdam, The Netherlands, 2–5 September 2007. *Zoonoses and Public Health*, **54** (Suppl. 1), 378, 397 (http://chro2007.nl/).
60. Hilmarsson, H., Thormar, H., Thráinsson, J.H. *et al.* (2006) Effect of glycerol monocaprate (monocaprin) on broiler chickens: an attempt at reducing intestinal *Campylobacter* infection. *Poult. Sci*, **85**, 588–592.
61. Rosenquist, H., Nielsen, N.L., Sommer, H.M. *et al.* (2003) Quantitative risk assessment of human campylobacteriosis associated with thermophilic *Campylobacter* species in chickens. *Int. J. Food Microbiol.*, **83**, 87–103.
62. Boysen, L. and Rosenquist, H. (2007) Reducing numbers of thermotolerant *Campylobacter* by physical decontamination in broiler processing and optimized hygienic practices. *Zoonoses and Public Health*, **54** (Suppl. 1), 52.
63. Thormar, H., Hilmarsson, H., Thráinsson, J.H. *et al.* (2011) Treatment of fresh poultry carcasses with emulsions of glycerol monocaprate (monocaprin) to reduce contamination with *Campylobacter* and psychrotrophic bacteria. *Brit. Poultry Sci.,* in press.

64. Stern, N.J., Wojtona, B. and Kwiatek, K. (1992) A differential-selective medium and dry ice-generated atmosphere for recovery of *Campylobacter jejuni*. *J. Food Protect.*, **55**, 514–517.
65. Mandrell, R.E. and Wachtel, M.R. (1999) Novel detection techniques for human pathogens that contaminate poultry. *Curr. Op. Biotechnol.*, **10**, 273–278.
66. Chantarapanont, W., Berrang, M. and Frank, F.J. (2003) Direct microscopic observation and viability determination of *Campylobacter jejuni* on chicken skin. *J. Food Protect.*, **66**, 2222–2230.
67. Chantarapanont, W., Berrang, M. and Frank, J.F. (2004) Direct microscopic observation of viability of *Campylobacter jejuni* on chicken skin treated with selected chemical sanitizing agents. *J. Food Protect*, **67**, 1146–1152.
68. Corry, J.E., James, C., O'Neill, D. *et al.* (2003) Physical methods, readily adapted to existing commercial processing plants, for reducing numbers of campylobacters on raw poultry. *Int. J. Med. Microbiol.*, **293** (Suppl. 35), 32.
69. James, C., James, S.J., Hannay, N. *et al.* (2007) Decontamination of poultry carcasses using steam or hot water in combination with rapid cooling, chilling or freezing of carcass surfaces. *Int. J. Food Microbiol.*, **114**, 195–203.
70. Kodicek, E. (1949) The effect of unsaturated fatty acids on Gram-positive bacteria. *Soc. Exp. Biol. Symp.*, **3**, 217–232.
71. Thatcher, F.S. and Loit, A. (1961) Comparative microflora of chlortetracycline-treated and non-treated poultry with special reference to public health aspects. *Appl. Microbiol.*, **9**, 39–45.
72. Lillard, H.S. and Thomson, J.E. (1983) Efficacy of hydrogen peroxide as a bactericide in poultry chiller water. *J. Food Sci.*, **48**, 125–126.
73. VanDer Marel, G.M., Van Logtestijn, J.G. and Mossel, D.A.A. (1988) Bacteriological quality of broiler carcasses as affected by in-plant lactic acid decontamination. *Int. J. Food Microbiol.*, **6**, 31–42.
74. Zeitoun, A.A.M. and Debevere, J.M. (1990) The effect of treatment with buffered lactic acid on microbial decontamination and on shelf life of poultry. *Int. J. Food Microbiol.*, **11**, 305–312.
75. Hwang, C-A. and Beuchat, L. (1995) Efficacy of a lactic acid/sodium benzoate wash solution in reducing bacterial contamination of raw chicken. *Int. J. Food Microbiol.*, **27**, 91–98.
76. Rathgeber, B.M. and Waldroup, A.L. (1995) Antibacterial activity of a sodium acid pyrophosphate product in chiller water against selected bacteria on broiler carcasses. *J. Food Protect.*, **58**, 530–534.
77. Vareltzis, K., Soultos, N., Koidis, P. *et al.* (1997) Antimicrobial effects of sodium tripolyphosphate against bacteria attached to the surface of chicken carcasses. *Technology*, **30**, 665–669.
78. Kemp, K.G., Aldrich, M.L. and Waldroup, A.L. (2000) Acidified sodium chlorite antimicrobial treatment of broiler carcasses. *J. Food Protect.*, **63**, 1087–1092.
79. Chouliara, E., Karatapanis, A., Savvaidis, I.N. and Kontominas, M.G. (2007) Combining effect of oregano essential oil and modified atmosphere packaging on shelf-life extension of fresh chicken breast meat, stored at 4 °C. *Food Microbiol.*, **24**, 607–617.
80. Del Rio, E., Panizo-Morán, M., Prieto, M. *et al.* (2007) Effect of various chemical decontamination treatments on natural microflora and sensory characteristics of poultry. *Int. J. Food Microbiol.*, **115**, 268–280.
81. Gill, C.O. and Badoni, M. (2004) Effects of peroxyacetic acid, acidified sodium chlorite or lactic acid solutions on the microflora of chilled beef carcasses. *Int. J. Food Microbiol.*, **91**, 43–50.
82. De Boer, E. and Hahné, M. (1990) Cross-contamination with *Campylobacter jejuni* and *Salmonella* spp. from raw chicken products during food preparation. *J. Food Protect.*, **53**, 1067–1068.
83. Zhao, P., Zhao, T., Doyle, M.P. *et al.* (1998) Development of a model for evaluation of microbial cross-contamination in the kitchen. *J. Food Protect.*, **61**, 960–963.
84. Kusumaningrum, H.D., Riboldi, G., Hazeleger, W.C. and Beumer, R.R. (2003) Survival of foodborne pathogens on stainless steel surfaces and cross-contamination to foods. *Int. J. Food Microbiol.*, **85**, 227–236.
85. Mattick, K., Durham, K., Domingue, G. *et al.* (2003) The survival of foodborne pathogens during domestic washing-up and subsequent transfer onto washing-up sponges, kitchen surfaces and food. *Int. J. Food Microbiol.*, **85**, 213–226.

86. Kusumaningrum, H.D., van Asselt, E.D., Beume, R.R. and Zwietering, M.H. (2004) A quantitative analysis of cross-contamination of *Salmonella* and *Campylobacter* spp. via domestic kitchen surfaces. *J. Food Protect.*, **67**, 1892–1903.

87. Cliver, D.O. (2006) Cutting boards in *Salmonella* cross-contamination. *J. AOAC Int.*, **89**, 538–542.

88. Luber, P., Brynestad, S., Topsch, D. *et al.* (2006) Quantification of *Campylobacter* species cross-contamination during handling of contaminated fresh chicken parts in kitchens. *Appl. Environ. Microbiol.*, **72**, 66–70.

89. Scott, E., Bloomfield, S.F. and Barlow, C.G. (1984) Evaluation of disinfectants in the domestic environment under 'in use' conditions. *J. Hyg. Camb.*, **92**, 193–203.

90. Ak, N.O., Cliver, D.O. and Kaspar, C.W. (1994) Decontamination of plastic and wooden cutting boards for kitchen use. *J. Food Protect.*, **57**, 23–40.

91. Cogan, T.A., Bloomfield, S.F. and Humphrey, T.J. (1999) The effectiveness of hygiene procedures for prevention of cross-contamination from chicken carcasses in the domestic kitchen. *Lett. Appl. Microbiol.*, **29**, 354–358.

92. Cogan, T.A., Slader, J., Bloomfield, S.F. and Humphrey, T.J. (2002) Achieving hygiene in the domestic kitchen: the effectiveness of commonly used cleaning procedures. *J. Appl. Microbiol.*, **92**, 885–892.

93. Barker, J., Naeeni, M. and Bloomfield, S.F. (2003) The effects of cleaning and disinfection in reducing *Salmonella* contamination in a laboratory model kitchen. *J. Appl. Microbiol.*, **95**, 1351–1360.

94. Kusumaningrum, H.D., Paltinaite, R., Koomen, A.J. *et al.* (2003) Tolerance of *Salmonella Enteritidis and Staphylococcus aureus* to surface cleaning and household bleach. *J. Food Protect.*, **66**, 2289–2295.

95. Oh, D.-H. and Marshall, D.L. (1996) Monolaurin and acetic acid inactivation of *Listeria monocytogenes* attached to stainless steel. *J. Food Protect.*, **59**, 249–252.

96. Thormar, H. and Hilmarsson, H. (2010) Killing of *Campylobacter* on contaminated plastic and wooden cutting boards by glycerol monocaprate (monocaprin). *Lett. Appl. Microbiol.*, **51**, 319–324.

97. Thormar, H. and Hilmarsson, H. (2010) Killing of *Salmonella and Escherichia coli* on contaminated surfaces by glycerol monocaprate (monocaprin). Submitted for publication.

98. Scott, E. and Bloomfield, S.F. (1990) The survival and transfer of microbial contamination via cloths, hands and utensils. *J. Appl. Bacteriol.*, **68**, 271–278.

99. Cosby, C.M., Costello, C.A., Morris, W.C. *et al.* (2008) Microbiological analysis of food contact surfaces in child care centers. *Appl. Environ. Microbiol.*, **74**, 6918–6922.

9

Chemistry and Bioactivity of Essential Oils

Christine F. Carson and Katherine A. Hammer

Discipline of Microbiology and Immunology, School of Biomedical, Biomolecular and Chemical Sciences, The University of Western Australia, Crawley, WA, Australia

Lipids and Essential Oils as Antimicrobial Agents   Halldor Thormar
© 2011 John Wiley & Sons, Ltd

9.1   Introduction

Essential oils are natural, volatile, complex plant compounds, oily or lipid-like in nature and frequently characterized by a strong fragrance [1, 2]. They have a low solubility in water but are soluble in fats, alcohol, organic solvents and other hydrophobic substances and are generally liquid at room temperature. They are stored in specialized plant cells, usually oil cells or ducts, resin ducts, glands or trichomes (glandular hairs) [3,4] and may be extracted from the leaves, flowers, buds, seeds, fruits, roots, wood or bark of plants by a variety of methods, including solvent and supercritical fluid extraction, expression under pressure, fermentation or enfleurage, but either low- or high-pressure [1] steam or hydro-distillation are used predominantly for commercial production [2, 5]. Essential oils make up only a small proportion of the wet weight of plant material, usually approximately 1% or less [3, 6]. The presence, yield and composition of essential oils may be influenced by many factors, including climate, plant nutrition and stress [7]. In commercial production settings, selection and breeding programmes are often instigated to improve yields and foster desired compositions [8].

Essential oils are also called ethereal oils, volatile oils, plant oils or aetheroleum but the term 'essential oil' will be used throughout this chapter. The 'essential' part of the term 'essential oil' is thought to be derived from a phrase attributed to Phillippus Aureolus Theophrastus Bombastus von Hohenheim (1491–1541), or Paracelsus as he became known, a Swiss physician who named the active component of a drug preparation 'quinta essentia' [2, 9, 10]. The term 'essential oil' groups together a wide range of chemical compounds on the basis of their historic use and method of extraction, usually steam distillation, and belies the variety and complexity of compounds found within them [11].

Some plant families are particularly well known for their oil-bearing species. These include Apiaceae (also known as Umbelliferae), Asteraceae (also known as Compositae), Cupressaceae, Hypericaceae (sometimes included as a subfamily of the Guttiferae/Clusiaceae), Lamiaceae (previously known as Labiatae), Lauraceae, Fabaceae (also known as Leguminosae), Liliaceae, Myrtaceae, Pinaceae, Piperaceae, Rosaceae, Rutaceae, Santalaceae, Zingiberaceae and Zygophyllaceae [4, 8].

Essential oils are often described as secondary plant metabolites. Traditionally, secondary plant metabolites have been all those compounds synthesized by the plant which do not appear to be essential for plant growth and development and/or those compounds without an obvious function [12]. They are also not universally synthesized in all plants. In contrast, primary metabolites are produced by all plants, usually have an obvious function and are part of the essential metabolic processes of respiration and photosynthesis [13]. This artificial and rather simplistic division is also naïve because the natural functions of many secondary plant metabolites are unknown simply because they have never been investigated; this lack of evidence or knowledge is then interpreted as a lack of function [14]. Greater interest in and investigation of secondary metabolites in recent years has led to the discovery that they have roles in defence, signalling and as intermediates in secondary metabolism [15–17].

This chapter will focus on the chemistry of the compounds found in essential oils and the biological activity that they possess. To better illustrate the advances in our comprehension of this biological activity and the mechanisms by which they are exerted, the biological activity will focus on the antimicrobial and anticancer activities of essential oils and components, two areas in which research efforts have been especially focussed. A brief discussion of the applications of essential oils in pharmaceuticals and foods follows.

9.2   Chemistry of Essential Oils

Essential oils are not simple compounds or even simple mixtures of several individual compounds. They may contain up to approximately 100 components, although many contain about 20 to 60 [3, 6, 18, 19]. The compounds found in essential oils are from a variety of chemical classes, predominantly terpenes, but phenylpropanoids and other compounds also occur although at lesser frequency and often, but not always, in smaller proportions [20]. They are all hydrocarbons and their oxygenated derivatives, and they may also contain nitrogen or sulfur. They are generally low-molecular-weight compounds with limited solubility in water [21, 22].

The classification and nomenclature of essential oil compounds is complicated by the fact that many were isolated and studied before the instigation of systematic chemical nomenclature. Consequently, many are known by nonsystematic or trivial or common names [11]. These are sometimes but not always based on their source, such as eucalyptol, limonene, pinene and thymol, names which hint at historical botanical origins of these compounds. In terms of shedding light on their chemistry, the long history and widespread use of these nonsystematic names further obfuscates the chemical nature and characteristics of essential oils and their components.

9.2.1   Terpenes

Terpenes, also known as isoprenes, or terpenoids or isoprenoids when they contain oxygen, are the largest group of natural compounds, with over 30 000 known structures [4, 13, 23, 24]. The name 'terpene' comes from the fact that the first described members of this class were isolated from turpentine, the monoterpene-rich liquid obtained from the resin of various *Pinus* spp. [23].

Traditionally, terpenes have been regarded as polymers of isoprene (C_5H_8) joined together in a repetitive head-to-tail manner. This is largely a legacy of work by the German chemist Otto Wallach, who was the first to recognize that many terpene compounds could be hypothetically constructed in this fashion [12]. This concept, known as the isoprene rule, was the first step in rationalizing the enormous variety of terpenes, since it accounted for the structure of many, but not all, terpenes. However, head-to-head combinations also occur, as do tail-to-tail and head-to-middle combinations [12, 25, 26]. This variation in initial arrangement of the isoprene units coupled with the numerous rearrangements and substitutions that can occur afterwards mean that the isoprene origins of the final compound are often obscured, or at the very least not obvious to the nonchemist. Furthermore, although terpenes may be viewed as polymers of isoprene, the biosynthesis of terpenes does not occur by the successive addition of single isoprene units. Leopold Ruzicka, recipient of the 1939 Nobel Prize in Chemistry, addressed many of the limitations of Wallach's isoprene rule when he proposed the biogenetic isoprene rule [4, 7, 27]. His concept was revolutionary since it emphasized the single biochemical origin of terpenes rather than the ultimate structure [12, 28] and this approach has proved the most practical.

Terpenes are classified by the number of isoprene units from which they were biogenetically derived [12], even though loss or addition of carbon atoms may have subsequently occurred [29]. Therefore, hemi-, mono-, sesqui- and diterpenes contain 1, 2, 3 and 4 isoprene units, respectively. Triterpenes and tetraterpenes contain 6 and 8 isoprene units, respectively. Monoterpenes are the most common terpenes found in essential oils, followed by sesquiterpenes. Many essential oils are composed mainly of monoterpenes and sesquiterpenes and their oxygenated derivatives [18, 19, 30–33].

9.2.1.1   Biosynthesis of Terpenes

The synthesis of terpenes in plants occurs via two distinct, mostly compartmentally separated, biological pathways [24]. The mevalonic-acid pathway was the first described and takes place mainly in the cytoplasm, endoplasmic reticulum and mitochondria, producing sesquiterpenes, sterols and ubiquinones predominantly [12, 24, 34]. Remarkably, the existence and role of the second pathway was confirmed and characterized relatively recently. Known as the nonmevalonic-acid or methyl–erythritol phosphate pathway, this pathway takes place in the plastids of plant cells and is largely responsible for the synthesis of hemi-, mono- and diterpenes, in addition to other higher terpenes not found in essential oils such as carotenoids and the phytols of chlorophyll [24, 34]. Separation of the two pathways for the purposes of terpene synthesis is not absolute; metabolites formed in one pathway and plant-cell compartment may cross over to the other pathway and compartment [24]. Both pathways produce isopentenyl diphosphate (IPP) and its isomer dimethyl-allyl diphosphate (DMAPP) [12, 35], the basic building blocks of terpenes. When combined in different ratios these precursors yield geranyl diphosphate (DMAPP + IPP), farnesyl diphosphate (DMAPP + 2 IPP) and geranylgeranyl diphosphate (DMAPP + 3 IPP), the main precursors of mono-, sesqui-/tri- and di-/tetraterpenes, respectively [12, 26, 34, 35]. Exceptions to this occur in the Apiaceae, Asteraceae and Lamiaceae families, where irregular monoterpenes, sesquiterpenes and diterpenes are derived from the head-to-head coupling of 2 DMAPPs, of DMAPP and GPP, and of 2 GPPs, respectively [26].

9.2.1.2   Monoterpenes

Monoterpenes are formed when two C_5 isoprene units are joined, yielding a skeleton with the molecular formula $C_{10}H_{16}$ [3]. Despite this initial simplicity, subsequent substitutions, cyclizations and/or isomerizations result in a remarkable number of monoterpenoid structures. Approximately 1500 monoterpenoids have been described [23], although not all occur in essential oils. Monoterpenes may be cyclic (that is, ring-forming) or acyclic (also known as linear), regular or irregular, and their derivatives include alcohols, esters, phenols, ketones, lactones, aldehydes and oxides [35].

Cyclic monoterpenes include the monocyclic, bicyclic and even tricyclic compounds. The rings are produced in a multistep process called cyclization by enzymes called monoterpene cyclases via the universal intermediate, α-terpinyl cation [4, 36, 37]. Cyclic monoterpenes that contain a benzene ring such as *p*-cymene are known as aromatic monoterpenes and are common components of many essential oils. In this context, the term 'aromatic' refers to the benzene ring, consisting of a ring of delocalized electrons. In many instances the benzene ring makes a significant contribution to the biological activity of the component and to the whole essential oil, especially when a hydroxyl group is attached to the ring, forming a phenol [38]. Use of the term 'aromatic' in this fashion should not be confused with the terms 'aromatic plants' or 'aromatic oils', which are often also used in discussions about medicinal plants and essential oils and refer to the aroma or fragrance of plants and oils.

Acyclic monterpenes found in essential oils may be regular, linear structures in which the head-to-tail arrangement of isoprene units is readily observed, such as the hydrocarbons β-myrcene or the *(E)* and *(Z)* isomers of β-ocimene. Note that the *(E)* and *(Z)* notation for stereoisomers supercedes the *cis–trans* notation for stereoisomers. The obsolete term 'geometric isomer' is strongly discouraged by the International Union of Pure and Applied Chemistry [39]. Other examples of acyclic monoterpenes commonly found in essential oils include geraniol, linalool and citronellol.

Monocyclic monoterpenes include the largest group of naturally occurring monoterpenes [4], those that arise from the *p*-menthane skeleton by cyclization of a regular acyclic monoterpenoid. Important monoterpenes in this group include limonene, α-terpinene, β-terpinene, γ-terpinene and terpinolene, as well as the aromatic hydrocarbon *p*-cymene and its hydroxylated derivatives thymol and carvacrol, both noted for their antimicrobial activity. Other notable compounds in this group are the carbonyls piperitone and pulegone.

The biogenesis of bicyclic monoterpenes occurs by the further cyclization of monocyclic monoterpenes. They may be further categorized on the basis of the skeleton from which they are derived, including bornane, carane, camphane, fenchane, pinane and thujane [7]. α-pinene and β-pinene are common important constituents of essential oils, particularly pine oils, and are bicyclic monoterpenes formed by intramolecular rearrangement of the universal intermediate α-terpinyl cation, producing the bicyclic structure. Alternative cyclizations of the terpinyl cation yield the bicyclic skeletons for the bornane-, camphane- and fenchane-type monoterpenes. Thujane monoterpenes come from either the terpinen-4-yl cation or from the sabinyl cation and include α-thujene, sabinene and α- and β-thujone. δ-3-carene, a carane-type bicyclic monoterpene, is a common component of various essential oils including those from *Pistacia lentiscus* L. and *Juniperus* spp. [40, 41]. Other important members of this group include the cyclic ethers 1,8-cineole, 1,4-cineole and ascaridol.

Tricyclic monoterpenes occur infrequently in essential oils compared to monocyclic and bicyclic monoterpenes. However, pinene oxide and tricyclene are two important examples found in essential oils.

Irregular monoterpenes also occur and fall into two categories. The first is the troponoids or substituted cycloheptane monoterpenes. These are thought to be formed by ring expansion of the *p*-menthane skeleton (forming a seven-membered ring structure) and oxygenation of the side chain(s) [42]. Many are found in the heartwood of trees in the Cupressaceae family of evergreen shrubs and trees [42, 43]. Examples include the thujaplicins (α-, β-and γ-isoforms) and nezukone. β -thujaplicin is also known as hinokitiol.

The second category of irregular monoterpenes is formed by joining isoprene units in the less common non-head-to-tail arrangements [12]. Compounds in this category include artemisia ketone, chrysanthemol and lavandulol. Many irregular monoterpenes are found in the genus *Artemisia* (Asteraceae) [44–46].

9.2.1.3   Sesquiterpenes

In terms of their frequency in essential oils, sesquiterpenes are the second most common, after the dominant monoterpenes. They are formed from the combination of three isoprene units, giving them the molecular formula $C_{15}H_{24}$. They are a structurally diverse group, all deriving from farnesyl pyrophosphate by various cyclization processes often followed by skeletal rearrangement [12]. Of the terpenoids found in essential oils, they are the most structurally diverse, with over 120 different skeletal types [4, 11]. Sesquiterpenes may be linear, branched or cyclic.

Acyclic sesquiterpenes feature in many essential oils and include the isomers nerolidol and farnesol and the α- and β- structural isomers of farnesene. *(E)* isomers occur more commonly in nature than *(Z)* isomers and *(E)*-nerolidol is also found in many commercially important essential oils, such as neroli oil from the flowers of *Citrus aurantium* [47]. Essential oils containing more than 90% *(E)*-nerolidol have been identified [48]. Farnesol is an important component of the commercially important rose flower essential oil [4] and of Australian sandalwood oil, *Santalum spicatum* [49]. Irregular acyclic sesquiterpenes have been identified in *Santolina* spp. (Asteraceae) [50, 51].

Cyclic sesquiterpenes may be mono-, bi- or tricyclic. Monocyclic sesquiterpenes include abscisic acid, α-bisabolene and its oxygenated derivatives, α- and β-bisabolol, both present at high levels in chamomile (*Matricaria chamomilla*) oils [52]. Bicyclic sesquiterpenes include eudesmol, widdrol, guaiol and the group known as azulenes. Azulenes are responsible for the blue colour of some essential oils, such as chamazulene in chamomile oil [52] and *Artemisia aborescens* oil [53]. The bicyclic caryophyllene is present in many essential oils, β-caryophyllene being the most common form, which may also occur as a major component [54, 55]. Cedrene and santalol are examples of tricyclic sesquiterpenes. Cedrene occurs in many essential oils, including various cedarwood oils derived from *Juniperus* spp., *Cupressus* spp. and *Cedrus* spp. [56, 57], while santalols are important constituents of sandalwood (*Santalum album*) oil [58].

9.2.1.4   Diterpenes

Most diterpenes in essential oils are formed by the head-to-tail combinations of four isoprene units followed by rearrangement and/or substitutions. They are very common and important components of plant resins [59] but are also found in small quantities in many essential oils. They have the general molecular formula $C_{20}H_{32}$ and so are much heavier than their mono- and sesquiterpenoid counterparts. Their heavier molecular mass relative to the mono- and sesquiterpenes means they require a greater amount of energy to be liberated from plant parts by steam distillation. Their recovery and the concentration obtained from essential oils increases with increasing steam-distillation times [4] and can be influenced by the extraction method. For example, supercritical CO_2 extraction of essential oils has been shown to increase the concentration of diterpenes recovered from essential oils [60]. As with monoterpenes and sesquiterpenes, they may be acyclic or cyclic.

Acyclic diterpenes include phytol. Phytol forms the hydrophobic side chain of chlorophyll and so is found in the leaves of all green plants [11,12]. It occurs in many essential oils [61–64]. Another important linear diterpene is plaunotol, the main component of the Thai medicinal plant *Croton stellatopilosus* (Euphorbiaceae; formerly known as *C. sublyratus*) [23,65].

A notable cyclic diterpene in essential oils is the monocyclic camphorene (also known as dimyrcene), a component of camphor oil from the tree *Cinnamonum camphora* (Lauraceae), more commonly known as the camphor laurel. Several isomers of camphorene are found in the essential oils distilled from the leaves and twigs of *P. lentiscus* [66] and from mastic gum derived from the same plant [67]. Bicyclic and tricyclic diterpenes also occur in essential oils. Bicyclic diterpenes found in essential oils fall largely into two structural groups, the labdanes and the clerodanes. Labdane representatives include manool and manoyl oxide while sclareol typifies the clerodane class of bicyclic diterpenes. These components can be found in essential oils such as those from *Salvia* spp. including *Salvia sclarea* or clarysage [68–70]. Tricyclic diterpenes that occur in essential oils include phyllocladene and 16-kaurene [4]. Phyllocladene constitutes a significant portion of essential oils from *Araucaria* spp. (up to 61%) from the Araucariaceae family and 16-kaurene constitutes 60% of the essential oil from the ancient Wollemi pine, *Wollemia nobilis*, from the same family [71].

Tetracyclic and pentacyclic diterpenes also occur in essential oils, although they are minor components [72,73].

9.2.1.5 Norterpenes

Carotenoids (C_{40}) are a class of higher terpenes based on eight isoprene units and are important in plants for several reasons, including their role in photosynthesis [74]. They do not occur in essential oils. However, they are relevant to oils because when their carbon backbone is cleaved, usually oxidatively, they yield a range of smaller compounds known as apocarotenoids [75,76]. The most common and widespread group of apocarotenoids occur when carotenoids are cleaved at the 9–10 position, yielding C_{13} products known as norterpenoids or norisoprenoids. These are important minor components of some essential oils, contributing particularly to aroma and flavour [75,77]. Examples include β-ionone, the violet-like aroma found naturally in *Boronia megastigma* [78,79], and β-damascone from *Rosa damascena* [80].

9.2.2 Phenylpropanoids

9.2.2.1 Biosynthesis of Phenylpropanoids

Continuing the biogenetic theme seen with the terpenoids, phenylpropanoids are grouped together on the basis of their common biosynthetic origin from the shikimic acid pathway. This pathway occurs only in microorganisms and plants, never in animals. This adds weight to the possibility that compounds affecting this pathway will have the selective toxicity and safety profile that is advantageous for their use in humans and other animals. The shikimic acid pathway is responsible for the synthesis of many of the phenolic compounds in plants and, beginning with glucose in plants, produces the aromatic amino acids phenylalanine, tyrosine and tryptophan [81]. Shikimic acid is one of the pathway intermediates and lends its name to the whole pathway. Phenylpropanoids arise from the aromatic amino acids phenylalanine and to a minor extent tyrosine [12, 58, 82, 83]. They have a C_6C_3 skeleton composed of a six carbon aromatic ring with a three-carbon side chain. The aromatic ring is also known as a benzene ring. When the three-carbon side chain attached to the phenyl ring is shortened by two carbons, benzenoids are formed [84]. The term is often used to include phenylpropanoids [67, 85, 86].

9.2.2.2   *Phenylpropanoids in Essential Oils*

Only approximately 50 phenylpropanoids have been described. Phenylpropanoids occur in essential oils less frequently and usually less abundantly than terpenoids [1, 20, 58, 87]. However, some of the oils in which phenylpropanoids do occur contain significant proportions of them, such as the eugenol in clove oil, present at 70 to 90% of the oil [88], or the methyleugenol-rich chemotype of the root essential oil of *Anemopsis californica*, or yerba mansa, containing 59% methyleugenol [89]. Plant families in which phenylpropanoids occur more frequently include Apiaceae (Umbelliferae), Lamiaceae, Myrtaceae [90], Piperaceae [91] and Rutaceae [79].

Important phenylpropanoids include the hydroxycinnamic acids, anethole, chavicol, eugenol, and their methylated derivatives, estragol (methyl chavicol) and methyl eugenol, as well as the widely distributed cinnamaldehyde. Myristicin and dillapiole are two other phenylpropanoids that occur commonly in essential oils when phenylpropanoids are present [87, 91–94].

As seen with the terpenoids, the extraction method used to produce essential oils may influence their phenylpropanoid content [95, 96].

9.2.3   Sulfur and Nitrogen Compounds of Essential Oils

More rarely, a few compounds found in essential oils contain one or more sulfur or nitrogen molecules. The presence of sulfur in particular confers an often strong, characteristic odour [3, 97, 98]. Sulfur- and nitrogen-containing compounds occur mainly as aglycones or glucosinolates, or their breakdown products, which include isothiocyanates. Aglycones are the nonsugar portion of a glycoside, a compound made up of a sugar group, termed the glycone, joined to another group. Glucosinolates, historically known as mustard oil glucosides, are sulfur- or nitrogen-containing compounds formed from glucose and one of eight amino acids [99]. When endogenous plant enzymes called myrosinases act on glucosinolates to cleave the glucose group they leave an unstable aglycone which then rearranges to form various breakdown products, including isothiocyanates, thiocyanates and nitriles [99, 100]. It is these breakdown products which may be major constituents of essential oils, such as phenylacetonitrile which makes up 85.9% of the oil from *Lepidium meyenii* (Walp.) [101] and various isothiocyanates, which are the major constituents of mustard (*Brassica rapa*) seed essential oil [102–104]. The Brassicaceae family, with over 350 genera and 3000 species, is an important source of glucosinolates and isothiocyanates [100]. This includes the cruciferous vegetables, such as broccoli, cauliflower, brussel sprouts and various cabbages.

In addition to the isothiocyanates, cyanates and nitrile compounds, other nitrogen-containing compounds occasionally occur in essential oils. The seed oil of *Azadirachta indica* (neem) contains the nitrogen compounds 5,6-dihydro-2,4,6-triethyl-(4H)-1,3,5-dithiazine, 2,6-diethylpyridine, 1H-pyrazole, 1H-benzotriazole and dodecanamide. These compounds were also detected in *A. excelsa* oil, although at lower levels. Notably, the first compound made up 11.7% of the *A. indica* oil [105], illustrating that although nitrogen compounds are usually present at very low concentrations, they may occasionally be major components of oils.

Methyl anthranilate is found in a variety of citrus oils, including lemon and mandarin oils [106], jasmine oil [107], and in the essential oil derived from the flowers of *Murraya exotica* L. (Rutaceae) [108]. Methyl *N*-methyl anthranilate is found in mandarin oil [106, 109] as well as oils from ylang-ylang [110] and the seeds of *Nigella damascena* and *N. sativa* [111]. It is the main component of mandarin petitgrain oil [112, 113]. The nitrogen-containing pyridines and pyrazines have been detected in oils including vetiver and black pepper [114, 115].

Oils from plants in the Alliaceae family are also particularly well known for sulfur-containing compounds; these include plants such as *Allium cepa* L. (onion), *Allium porrum* L. (leek) and *Allium sativum* L. (garlic), in which the sulfur compounds are responsible for the characteristic aroma and taste [116, 117]. Cysteine sulfoxides including alliin predominate in mature, intact *Allium* spp., along with γ-glutamyl cysteines [118]. Upon rupture, such as when chopped or pressed, the action of a class of enzymes known as alliinases catalyses the conversion of cysteine sulfoxides into the volatile thiosulfinates [101, 116] including allicin. Typically, allicin makes up 70 to 80% of the thiosulfinates. Allicin and other thiosulfinates quickly decompose to other compounds, including diallyl sulfide, diallyl disulfide, diallyl trisulfide, dithiins and ajoene, while the γ-glutamyl cysteines are converted to S-allylcysteine through a nonalliin/allicin pathway [118]. It is these end products rather than their precursors that may be found in essential oils.

Other sulfur-containing compounds have been detected, frequently at trace levels, in many essential oils. Mint sulfide has been identified in many oils important in the perfume industry, including peppermint, spearmint, pepper, ylang-ylang, narcissus, geranium, chamomile, davana [98] and rose oil [119], as well as cumin oil [120].

9.3 Biological Activity of Essential Oils

9.3.1 General Overview

Despite their history of being regarded as secondary, non-essential plant metabolites, it is becoming clear that essential oils and their components have specific biological functions [14,17,121], many of which lend themselves to commercial exploitation. Given the range and complexity of the compounds present in essential oils it is hardly surprising that they have the capacity to affect many biological systems. The biological activities of greatest interest centre around applications in health, agriculture and the cosmetic and food industries. In the arena of health and medicine the diverse array of biological properties now being characterized includes antimicrobial, anticancer, analgesic, antioxidant, antiinflammatory, other immunomodulatory and antiplatelet, and antithrombotic [122–126] activities. Along with fragrance and solvent properties, several of these activities also find application in the cosmetic and food industries. Of greatest interest in agriculture is the antimicrobial and insecticidal potential of essential oils and their components.

9.3.2   Antimicrobial Activity

Microorganisms such as bacteria, fungi, viruses and protozoa are the aetiological agents of many infectious diseases, and compounds with specific activity against these microorganisms, that is antimicrobial activity, are our best weapon for treating these diseases. Even before the role of microorganisms in disease pathogenesis was appreciated or understood, attempts at treating such illnesses often utilized plant-based medicines that contained compounds with antimicrobial activity. These plant-based medicines included essential oils. Interest in their use for the treatment of bacterial infections only seemed to wane significantly with the advent of first the antibacterial sulfur drugs in the early twentieth century and then β-lactam antibiotics and others beginning in the 1940s. Given that modern antibiotics had the advantage of selective toxicity and the capacity to be administered systemically and that many important pathogenic bacteria were exquisitely susceptible to them, it is no wonder that they soon became the primary means of treating bacterial infections and that the use of essential oils and other plant-based medicines diminished. Renewed recent interest in their use has been attributed to several factors, including a general renaissance in the appeal of 'natural' products, the desire for antimicrobial compounds with even better safety and toxicity profiles, and more importantly, the need for alternative agents due to the reduced susceptibility to conventional antimicrobial agents shown by many important pathogens.

Whatever the reasons for the apparently renewed interest, there are now hundreds of reports of the *in vitro* antimicrobial activity of essential oils in the scientific and medical literature, including reviews of the medicinal properties of some of the more popular oils such as clove [90], lavender [127], *Lippia* spp. [128] and tea tree [30]. This antimicrobial activity includes activity against bacteria, fungi, viruses and protozoa. Whereas typically these reports used to describe the activity of a single compositionally unspecified essential oil against one or two isolates of the microorganism of interest using nonstandard or *ad hoc* methods, increasingly they report on the activity of well-characterized essential-oil samples [129] or individual components against a wider range of genera and species [130], often testing larger number of isolates [131–133] using widely-applied or standardized methods. Reports of antibacterial and antifungal activity seem to dominate these reports, perhaps because of greater access to and simplicity of these methods. However, data on the activity of essential oils against viruses and protozoa are becoming more available.

The antimicrobial activity of essential oils can be attributed largely to the major groups of compounds found in them: monoterpenes, sesquiterpenes and nonterpenaceous components such as phenylpropanoids. Where they are present in significant proportions, sulfur compounds such as those found in *Allium* spp. are often the main antimicrobial compounds.

9.3.2.1   Antibacterial Activity

Both major groups of bacteria, Gram-positive and Gram-negative, have demonstrated susceptibility *in vitro* to essential oils and components. The methods used are usually disc-diffusion methods or agar- or broth-dilution methods [2, 5, 134–136]. In disc-diffusion methods, a paper disc impregnated with essential oil is laid on an inoculated agar medium and after incubation the diameter of the area around the disc in which bacteria were unable to grow is measured. Although disc-diffusion methods are popular the data they offer, in the form of zones of inhibition, are less useful than data from agar- and broth-dilution methods. Furthermore, the diffusion of essential oils through agar, a fundamental aspect of disc-diffusion tests, is greatly compromised by the hydrophobic nature and limited aqueous solubility of essential oils. Agar- or broth-dilution methods, in which serial dilutions of the test oil in agar or broth media are inoculated with a known concentration of test organism, allow minimum inhibitory concentrations (MICs) to be determined. The MIC is generally defined as the lowest concentration of essential oil that inhibits growth of the test organism. Although solubilization of essential oils in these systems is still problematic, adequate solubilization or dispersion may be achieved through the use of low concentrations of surfactants or solvents. MICs are more useful than zones of growth inhibition since they can help establish safe and effective final concentrations in formulated products.

Most essential oils possess at least some degree of antibacterial activity. However, those attracting the most attention are the ones which inhibit or kill bacteria *in vitro* at concentrations below 1% vol/vol (10 000 ppm). Oregano (*Origanum* spp.), tea-tree (*Melaleuca alternifolia*), lemongrass (*Cymbopogon citratus*), lemon-myrtle (*Backhousia citriodora*) and clove (*Syzigium aromaticum*) oils are examples of essential oils that have activity against a wide range of Gram-positive and Gram-negative bacteria (with the exception of *Pseudomonas aeruginosa*) with MICs of less than 1% or approximately 10 mg/ml. It is worth noting here that this level of activity, while still potentially useful, is about 1000-fold lower than the activity of conventional antibiotics, for which MICs of susceptible bacteria are expressed in µg/ml quantities.

The *in vitro* susceptibility to essential oils of a wide range of bacteria has been tested, but those bacteria important in human health care [2,30,134] and the food industry [2,135,137] have been the focus of most investigations. In the human health-care field there has been particular interest in the susceptibility to essential oils of multidrug-resistant bacteria, such as the Gram-positive methicillin-resistant *Staphylococcus aureus* (MRSA) and vancomycin-resistant enterococci, which cause serious infections. Table 9.1 summarizes the antimicrobial activity determined by broth- or agar-dilution methods of essential oils against MRSA. The number of publications and the range of essential oils tested typify the interest in the susceptibility of multidrug-resistant bacteria and bacteria in general. Although many essential oils may not be taken internally for the treatment of frank infections due to their systemic toxicity at the doses required to be antimicrobially effective, there is still interest in using them topically to prevent handborne transmission of pathogenic bacteria in skin antisepsis products.

Many investigations of the antibacterial activity of essential oils report greater antibacterial activity against one or other of the two major divisions of bacteria, namely Gram-positive and Gram-negative. In most cases greater essential oil activity against Gram-positive bacteria is claimed [138–144] and this has led to the notion that in general essential oils have greater activity against Gram-positive bacteria. Such generalizations are without sound basis [145] and compelling evidence for this apparent bias in activity is lacking since in many cases the differences in activity between the groups are insufficient to support such claims. In addition, the sample of bacterial genera tested often skews the results. For example, many investigations include in their data for Gram-negative bacteria the susceptibility of the Gram-negative bacterium *P. aeruginosa*. This bacterium is widely acknowledged as being highly resistant to many antimicrobial agents and its inclusion unfairly skews data in favour of the greater susceptibility of Gram-positive bacteria. Studies testing a larger number of essential oils against a wider variety of bacteria tend to identify no such pattern

Table 9.1 The minimum inhibitory concentrations (MICs) of essential oils against methicillin-resistant Staphylococcus aureus (MRSA).

Plant source	Essential oil (common name)	No. isolates	Method	MIC range	Reference
Allium odorum L.	Chinese leek	60	Broth microdilution	48 mg/l	[132]
Allium sativum L.	Garlic	60	Broth microdilution	32 mg/l	[132]
Backhousia citriodora	Lemon myrtle	1	Agar dilution	0.2%	[146]
Centaurea aladagensis	—	1	Broth microdilution	0.22 mg/ml	[63]
	—	3	Broth microdilution	8 μl/ml	[19]
Juniperus communis	Juniper berry	15	Broth microdilution	>2%	[147]
		1	Agar dilution	1.2 μl/ml	[148]
Lavandula angustifolia	Lavender	15	Broth microdilution	0.5%	[147]
Melaleuca alternifolia	Tea tree	64	Broth microdilution	0.25–0.312%	[149]
		100	Broth microdilution	0.32%, median	[131]
		28	Broth microdilution	0.25–0.5%	[150]
		15	Broth microdilution	0.25%	[147]
		30	Broth microdilution	0.25–2%	[151]
		1	Agar dilution	0.3%	[146]
		30	Broth microdilution	0.125–1%	[152]
		98	Broth dilution	512–2048 mg/l	[133]
		1	Broth dilution	0.125%	[153]
		4	Agar dilution	1.05 ± 0.29%	[143]
		4	Broth microdilution	1%	[154]

[57, 130, 155]. While the phenomenon may occur, there are only sufficient, convincing data in very few cases, such as manuka oil from the New Zealand native *Leptospermum scoparium* (Myrtaceae), which has activity against Gram-negative bacteria that is 32- to 64-fold lower than that against Gram-positive bacteria [156].

The antimicrobial activity of the complex chemical mixtures that constitute essential oils has led to attempts to identify and isolate the antimicrobially active components of these oils. In many cases, the component(s) or fraction(s) responsible for the antibacterial activity or for a large part of it have been identified, such as terpinen-4-ol in *M. alternifolia* (tea-tree) oil [157], carvacrol and thymol in oregano oil [158], and carvacrol and eugenol in *Eugenia caryophyllata* (clove) oil [90].

Most essential oils possess at least limited antibacterial activity, with some oils and components exhibiting a much greater degree of activity. Surveys of the antibacterial activity of essential-oil constituents have consistently indicated that aldehydes and phenolics tend to exhibit greater antibacterial activity [159] than other types of constituents, often followed by the nonphenolic alcohols, with oxides and hydrocarbons having the least antibacterial activity [140, 143, 160–162]. More comprehensive analyses of structure–activity relationships have confirmed and expanded our knowledge of this trend. Griffin *et al.* [22] found that terpene acetates and hydrocarbons tended to exhibit the lowest levels of antimicrobial activity and were able to relate this to their limited hydrogen-bonding capacity and lower water-solubility. Higher levels of antimicrobial activity were associated with hydrogen-bonding parameters and, in the case of Gram-negative bacteria, smaller molecular size [22]. Subgroups such as the phenolics have also been subjected to such analyses [163], with hydrophobic factors being identified as the main determinant of antibacterial activity at least for the bacteria investigated [38].

9.3.2.2 Antifungal Activity

Essential oils and components also exhibit activity against fungi, activity that is becoming increasingly well described. A wide range of human, animal and agricultural fungal pathogens have been shown *in vitro* to be inhibited and/or killed by essential oils, heightening interest in their therapeutic or industrial application. Amongst the human and animal pathogens of interest, yeasts in the genus *Candida* and the dermatophytes *Epidermophyton*, *Microsporum* and *Trichophyton* have attracted the greatest interest, perhaps because the limited range and effectiveness of conventional antifungal agents fuels the search for novel therapies. In contrast to the pattern seen with bacteria, in which minimum inhibitory and cidal concentrations of oils are frequently the same or only one or two serial dilutions different [31, 130, 164, 165], the oil concentrations necessary to kill fungi are often much higher than those required to merely inhibit their growth [166].

There has been particular interest in the activity of essential oils and their components against food-spoilage fungi and essential oils and their components have been shown to inhibit the growth of many of them, including species of *Aspergillus*, *Microsproum*, *Mucor*, *Penicillium*, *Eurotium*, *Debaryomyces*, *Pichia*, *Zygosaccharomyces* and *Candida* [41, 135, 167–178]. However, one of the key issues with agents intended to preserve food is maintenance of the aroma, taste, colour and texture of the food. An undesirable effect of using essential oils or their components as food-preservation agents is that these organoleptic properties may be compromised at the concentrations required to inhibit microbial growth [135, 179–183]. More highly flavoured foods lend themselves better to preservation in this manner [2]. Alternatively, the use of lower concentrations of essential oils or components may be possible if multiple food-preservation strategies that result in additive or synergistic effects on antimicrobial activity are involved [184–186]. This approach fits well with the concepts of hurdle technology, in which multiple simultaneous preservation strategies are applied [187].

One application of the antibacterial and antifungal activity of essential oils and components in food preservation that has received particular interest is active packaging, in which oils or components are incorporated into the packaging. They may be included in the plastic or paper-based packaging itself or in the atmosphere contained within it. When incorporated into the atmosphere around bread, mustard oil and its primary antimicrobial component, allyl isothiocyanate, have been shown to effectively inhibit the growth of several bread-spoilage fungi, including *Aspergillus flavus*, *Endomyces fibuliger*, *Penicillium* spp. and *Pichia anomala* [188, 189]. Mould and yeast counts on sweet cherries were significantly reduced in modified atmospheres containing eugenol, thymol or menthol but not eucalyptol [190].

9.3.2.3   *Mechanisms of Antibacterial and Antifungal Action*

While the spectrum and scale of the antimicrobial activity of essential oils are becoming better characterized, much less is known about the means by which they exert their activity. For many years, the precise mechanisms by which microorganisms were inhibited and/or killed remained unclear and were attributed largely to unspecified effects on microbial membranes or envelopes. Over the last decade or two, a deeper understanding has been gained of the precise effects of essential oils and their components on microorganisms. As long believed, many of the described effects involve interactions with biological membranes. However, the specificity and subtlety of many of these interactions is only beginning to be fully appreciated. Where the mechanisms of antimicrobial action of essential oils have been investigated, many varied and specific effects have been described. For example, in bacteria, carvacrol has been shown to cause collapse of the proton-motive force and depletion of the ATP pool, leading to death [159, 191–193], while tea-tree oil (*M. alternifolia*) and its major component terpinen-4-ol increase membrane permeability to potassium ions [194] and 260 nm-absorbing materials presumed to be nucleotides [157], and the phenylpropanoid cinnamaldehyde has been shown to interfere with the crucial bacterial division protein FtsZ, thereby preventing cell division [195]. Carvacrol has also been shown to inhibit the synthesis of flagellin, the protein that makes up flagella used for bacterial motility, in the important foodborne pathogen *Escherichia coli* O157 : H7 [196]. Specific effects on bacterial virulence factors, that is the products by which bacteria establish infection and produce disease, have also been identified. Examples include that cinnamaldehyde interferes with quorum sensing communication processes mediated by two different types of signalling compounds, acyl homoserine lactones and a group known collectively as autoinducer-2 (AI-2) [197, 198]; *Ocimum gratissimum* essential oil inhibits extracellular protease activity and cell-wall lipopolysaccharide expression [199]; eugenol inhibits listeriolysin O production [200]; and mint (*Mentha piperita*) essential oil reduces levels of staphylococcal enterotoxin B [201]. In fungi too, specific effects have been identified that compromise cell integrity [202], viability or virulence [203, 204]. These examples serve to illustrate that where specific effects of essential oils and components have been investigated, many have been identified, greatly enhancing our understanding of their mechanisms of action and undermining previous assumptions of generic 'cytotoxic effects'.

9.3.2.4   *Antiviral Activity*

For many years data on the antiviral properties of essential oils and their constituents lagged behind those for other microorganisms with respect to the range of oils and viruses tested and characterization of the mechanisms of action. This was reflected in the relatively few publications covering the subject. More recently, numerous publications have described the *in vitro* activity of a wide range of essential oils. The majority of *in vitro* studies have been conducted using the enveloped influenza or herpes simplex viruses 1 or 2 (HSV-1 or -2). Essential oils from *Artemisia glabella* [205], *Cynanchum stauntonii* [206], *Houttuynia cordata* [207], *Oenanthe crocata* [208], *Origanum acutidens* [176], *Salvia limbata* and *S. sclarea* [209] and the component cinnamaldehyde [210] have been tested against influenza viruses. Oils from tea tree and eucalyptus [211], anise (*Illicium verum*), hyssop (*Hyssopus officinalis*), thyme (*Thymus vulgaris*), ginger (*Zingiber officinale*), chamomile (*Matricaria recutita*) and sandalwood (*S. album*) [212], *A. aborescens* [53,213], *H. cordata* [207], *L. scoparium* [214], *Melaleuca ericifolia*, *M. leucadendron* and *M. armillaris* [215], *Melissa officinalis* [216,217], *M. piperita* [218], *O. crocata* [208], *Salvia fructicosa* [219], *S. limbata* and *S. sclarea* [209], *S. album* [220], *Santolina insularis* [221,222], a range of South American plants including *Aloysia*, *Artemisia* and *Lippia* spp. [223, 224], and the components eugenol [225, 226] and isoborneol [227] have been tested against HSV-1 and/or -2. Minami *et al.* [228] tested oils from *Cupressus sempervirens* (cypress), *Juniperus communis* (juniper), *M. alternifolia* (tea tree), *Ocimum basilicum album* (tropical basil), *M. piperita* (peppermint), *Origanum majorana* (marjoram), *Eucalyptus globulus* (eucalyptus), *Ravensara aromatica* (ravensara), *Lavandula latifolia* (lavender), *Citrus limonum* (lemon), *Rosmarinus officinalis* (rosemary) and *Cymbopogon citrates* (lemongrass) against HSV-1. Most of these essential oils have been evaluated by measuring the inhibition of plaque formation in tissue cultures of appropriate host cells, a widely used method of measuring antiviral activity. In general, the concentration of oils or components that reduced plaque formation by 50% ranged from 0.000 06 to 1%. Often the concentrations of oil that inhibit plaque formation are only marginally lower than the concentrations that prove cytotoxic to the tissue-culture cells, resulting in a comparatively low therapeutic index. In some instances, no antiviral activity is seen at concentrations that are noncytotoxic to the host cell line, such as when *Juniperus oxycedrus* ssp. *badia* oil was tested against two strains of human immunodeficiency virus (HIV) [229]. Nevertheless, interest in the potential of oils as antiviral agents persists, particularly for topical use such as hand and skin antisepsis.

Apart from the more widely tested influenza virus and HSV, dengue virus type 2 and Junin virus [223, 224], adenovirus and mumps virus [230], human respiratory syncytial virus [231], HIV [232], Newcastle disease virus [205], poliovirus [208], tobacco mosaic virus [233], two bacteriophage [234], yellow fever virus [235] and the viral aetiological agent of severe acute respiratory syndrome, a novel coronavirus [236], have also been tested against a range of oils and components with similar results.

With regard to the mechanism of antiviral action, in most cases where antiviral effects have been assessed before and after host-cell adsorption, the antiviral effect has occurred largely upon treatment of virus particles with oil prior to their adsorption or addition to cell monolayers. This suggests a direct effect of oil on free virus particles rather than an intracellular antiviral effect [211,217,218,221,228]. The site of action has not been identified but most of the viruses tested have been enveloped viruses, with the exception of adenovirus, poliovirus and tobacco mosaic virus. Viral envelopes are typically derived from the membrane of the host cell and have a phospholipid bilayer structure. Since many essential oils have the capacity to disrupt biological membranes, it follows that viral envelopes may also be disrupted by essential oils, a contention supported by electron micrographs of HSV-1 after treatment with oregano or clove oils showing envelope disruption [237].

In contrast to the growing body of *in vitro* data, there are very few reports of *in vivo* activity. Black-seed oil (*Nigella sativa*) was tested against murine cytomegalovirus [238] and *C. stauntonii* oil [206], *Heracleum* spp. oils [239] and cinnamaldehyde [210] were tested against influenza in mouse models. Studies in humans are also very limited. Tea-tree oil has been evaluated for the treatment of herpes labialis in a small pilot study with promising results, suggesting a reduction in the time to complete healing with the use of tea-tree-oil ointment compared to control [240], and *B. citriodora* essential oil was trialled in the treatment of molluscum contagiosum, a cutaneous viral infection caused by the virus of the same name, in which lesions were resolved completely in 5 of 16 participants, reduced in number by more than 90% in 4 of 16 and reduced in number less than 90% in 6 of 16. One participant was lost to follow-up. In the vehicle placebo group 0 of 15 met the criteria of a 90% reduction in lesion number while 12 of 15 had no change or an increase in lesion number [241].

9.3.2.5 Antiprotozoal Activity

As with studies investigating the antiviral properties of essential oils and their components, data on the activity of essential oils against parasites such as protozoa have become increasingly available in the last decade. Protozoa are single-celled eukaryotic microorganisms and many oils have now been evaluated as antiprotozoal agents [242] with a view to applications in human and animal health care. The more complex life cycle of protozoa compared to bacteria and fungi complicates the determination of their susceptibility and most methods assess the susceptibility of one or two life cycle stages, such as the promastigotes and amastigotes of trypanosomal protozoa.

Leishmania spp. are the causative agents of leishmaniasis, a disease that manifests itself in a variety of presentations, and their susceptibility to several essential oils has been investigated, including *O. gratissimum* oil [243], *T. vulgaris* oil [244] and the components limonene [245], linalool [246], nerolidol [245] and terpinen-4-ol [244]. *Leishmania amazonensis* proved susceptible to the linalool-rich essential oil of *Croton cajucara* and to linalool with MICs of 85 and 22 pg/ml, respectively [246], and to *Chenopodium ambrosioides* oil with a promastigote MIC of 27.82 μg/ml [247]. The MIC of *C. ambrosioides* oil for *Leishmania donovani* promastigotes was 25 μg/ml [248].

Lemongrass (*C. citratus*) oil showed *in vitro* antitrypanosomal activity against *Crithidia deanei* [249], as did *O. gratissimum* oil against *Herpetomonas samuelpessoai* [250]. While both test organisms are considered nonpathogenic in vertebrates, they have proved to be useful models of trypanosomal infections important in humans. *Trypanosoma cruzi* were inhibited by lemongrass oil and citral [251], *Origanum vulgare* and *T. vulgaris* oils, and thymol [252] and *Achillea millefolium, Syzygium aromaticum* and *O. basilicum* oils, as well as eugenol and linalool [253]. Essential oil from the leaves of *Strychnos spinosa* inhibited *Trypanosoma brucei brucei*, but when tested alone two of its components, (*E*)-nerolidol and linalool, showed more potent and selective activity against the test organism [62].

The susceptibility of *Plasmodium* spp. has also been evaluated. A selection of oils from the Cameroonian medicinal plants *Xylopia phloiodora, Pachypodanthium confine, Antidesma laciniatum, Xylopia aethiopica* and *Hexalobus crispiflorus* inhibited the growth of *Plasmodium falciparum* [254] and farnesol, nerolidol, limonene and linalool inhibited parasite development and isoprenoid synthesis in *P. falciparum* [255]. Two *Lavandula* essential oils, including *L. angustifolia*, inhibited the human protozoan pathogens *Giardia duodenalis* and *Trichomonas vaginalis* and the fish pathogen *Hexamita inflata* [256]. *O. basilicum* oil, as well as linalool and to a lesser extent eugenol, two of its major components, demonstrated cidal activity against *Giardia lamblia* [257]. The susceptibility of *Histomonas meleagridis, Tetratrichomonas gallinarum* and *Blastocystis* sp.

to carvacrol, *Cassia* oil and an essential oil mixture containing thyme and rosemary oil was evaluated with minimal lethal concentrations of 0.1 to 0.75 μl/ml [258]. The minimum lethal concentrations of oils from *Cinnamomum aromaticum, Citrus limon* and *A. sativum* were determined for *T. gallinarum* and *H. meleagridis* and ranged from 0.125 to 1 μl/ml [259].

The limited *in vivo* work includes the assessment of *C. citratus* and *O. gratissimum* oils in a murine model of *Plasmodium berghei* infection in which both oils significantly suppressed parasitaemia [260]. In another mouse study, the intraperitoneal and oral routes of administration of *C. ambrosioides* oil prevented lesion development and retarded the course of infection with *L. amazonensis* compared to untreated mice [261]. Poultry experimentally infected with *Eimeria tenella*, the aetiological agent of caecal coccidiosis, whose diet was supplemented with oregano oil fared no differently from uninfected control group [262]. Human studies are rare but in one 14 adults with positive stool tests for one or more of the enteric parasites *Blastocystis hominis, Entamoeba hartmanni* and *Endolimax nana* received oregano oil daily for six weeks, eliminating the organisms in eight, four and one cases, respectively [263].

The mechanisms of antiprotozoal action are believed to be twofold. Direct effects of the essential oil on protozoa have been described, such as the disruption of flagellar membranes, mitochondrial swelling, and gross alterations in the organization of the nuclear and kinetoplast chromatins seen by electron microscopy after *L. amazonensis* promastigotes were treated with *C. cajucara* oil [246]. Monoterpenoids and sesquiterpenoids have been shown to exert antiprotozoal activity by interfering with the isoprenoid pathway present in protozoa, providing another rationale for their antiprotozoal effects [255].

Indirect effects of essential oils on the host immune system have also been described. *C. cajucara* oil can increase nitric oxide production by infected peritoneal macrophages, an important mechanism of intracellular parasite killing, but alone the major component of this oil, linalool (approximately 40% of the oil) [264], cannot [246]. Immunomodulatory effects that seem disadvantageous to the intracellular killing of parasites, such as the reduction of nitric oxide production by infected peritoneal macrophages, have also been described but have been shown to induce a cascade of events that results in parasite death [242].

9.3.3   Anticancer Activity

The sheer complexity of the processes involved in carcinogenesis not only confounds their elucidation and hampers attempts to find effective preventive or therapeutic interventions, but also provides a myriad of potential targets for chemopreventive or chemotherapeutic action. Key amongst these potential targets is the mevalonic-acid pathway in mammalian cells [265], the functioning of which is known to be modified by many isoprenoids found in essential oils. This is reflected in the diverse ways in which essential oils and their components have been demonstrated to possess anticancer properties. For example, one function of the mevalonic-acid pathway in mammalian cells is cholesterol synthesis. Cancer cells synthesize and accumulate cholesterol faster than normal cells and isoprenoid influences on the enzyme 3-hydroxy-3-methylglutaryl-CoA (HMG-CoA) reductase early in this pathway can prevent or inhibit tumour growth [266, 267]. Monoterpenes such as cineole, farnesol, geraniol, d-limonene, menthol and perillyl alcohol have been shown to significantly reduce the activity [268, 269], synthesis [270] and degradation of HMG-CoA reductase [271, 272].

In vitro, many essential oils and their components have been shown to inhibit the proliferation of numerous cell lines representative of different cancers. Oils from *Casaeria sylvestris* and *Zanthoxylum rhoifolium* have been tested against cervical-, colon- and lung-cancer cell lines [273, 274], *Comptonia peregrina* against colon- and lung-cancer cell lines [275], *Curcuma wenyujin* against liver-cancer cells [276], *Cyperus rotundus* against a leukaemia cell line [277], *Eugenia zuchowskiae* against breast- and melanoma-cancer cell lines [278], *Juniperus excelsa* against breast-, colon-, epidermal-, lung- and prostate-cancer cell lines [279], *Photinia serrulata* against cervical-, lung- and liver-cancer cell lines [280], *Salvia libanotica* against colon-cancer cell lines [281], *Schefflera heptaphylla* against breast-, liver- and melanoma-cancer cell lines [282], *Schinus molle* against breast-cancer and leukaemia cell lines [283], *Tetraclinis articulata* against human melanoma-, breast- and ovarian-cancer cell lines [284], *Thymus broussonettii* against ovarian-cancer cell lines [285] and *Photinia serrula* against cervical-, liver- and lung-cancer cell lines [280].

Many isoprenoids that occur in essential oils have been shown to suppress the proliferation of cancer cell lines *in vitro*, including carvacrol, citral, *p*-cymene, farnesol, geraniol, *d*-limonene, nerolidol, perillyl alcohol, α-pinene, α-terpineol, thymol, verbenone and α- and β-ionone [267, 286–288]. Mechanistic studies of the effects of essential-oil compounds highlight the numerous ways in which these compounds have chemotherapeutic potential. For instance, the cyclic monoterpene perillyl alcohol, one of the most studied and most promising anticancer terpenoids, has been shown to arrest cell proliferation in the G1 phase at numerous different points [289–294], induce apoptosis (the programmed cell death considered desirable in cancer chemotherapeutic agents) [293, 295–299], suppress prenyl transferase activities [300] (preventing protein prenylation and in turn the signalling and oncogenic activities of proteins such as Ras that play a role in cell-growth-promoting signal transduction) [301, 302] and inhibit synthesis and activity of 3-hydroxy-3-methylglutaryl-CoA reductase, the rate-limiting enzyme in mevalonate biosynthesis in mammalian cells [270], which is thought to suppress tumour cell growth and induce apoptotic cell death [267, 286]. Perillyl alcohol has also been demonstrated to suppress the statin-mediated upregulation of Ras protein and other G proteins [303–305], suppress the synthesis of G proteins [306] and affect signal transduction [307].

Isothiocyanates [100], thiosulfinates and other oil components derived from *Allium* spp. [308, 309] and to a lesser extent phenylpropanoids such as myristicin [310] or oils rich in them [89, 92] have also been investigated for their anticancer properties [311]. The anticancer properties of thiosulfinates are thought to be mediated by a number of mechanisms, including stimulation of the important hepatic detoxifying glutathione transferase enzymes, downregulation of cancer-promoting enzymes [312, 313], inhibition of proliferation [308, 314] and induction of apoptosis [315, 316].

Amongst the phenylpropanoids, eugenol has been identified as having anticancer effects [317–320], along with oils from *Pimpinella* spp. [321]. Cinnamaldehyde has received considerable attention, showing apoptotic activity mediated by the generation of reactive oxygen species and a caspase-dependent mechanism [322] and the ability to inhibit the development of mutagen-induced lung carcinogenesis in mice [323]. More recently, the activation of pro-apoptotic proteins and the family of mitogen-activated protein kinases that play a role in cell signalling, particularly pathways involved in regulating cell survival, differentiation and apoptosis, have been implicated in the apoptotic activity of cinnalmaldehyde [324, 325]. The release of proteins from the mitochondrial membrane bilayer is known to play a crucial role in apoptosis, and permeabilization of the outer membrane can influence this [326]. Cinnamaldehyde has been shown to adversely influence several mitochondrial membrane functions [327] and may also exert its apoptotic effects in this manner.

Apart from their direct effects on the proliferation of cells, essential oils and their components may also exert anticancer effects through their antimutagenic, antiangiogenic, antiinflammatory [321, 328] or antioxidant properties. These properties may be effected by multiple mechanisms, once again highlighting the complexity of unravelling the effects of essential oils and components *in vitro* and *in vivo*. Antimutagenic properties can be due to altered cell permeability which prevents or inhibits mutagen penetration into cells, extracellular interaction between the antimutagen and the mutagen that results in physical, chemical or enzymatically-catalysed changes in mutagenicity, alteration of cellular mechanisms that results in mutagenicity, or effects on DNA repair [1, 329]. The phenylpropanoid cinnamaldehyde has been shown to possess modest antimutagenic activity through its influence on DNA repair processes [330]. It has also been shown to inhibit chemical and physical mutagenesis in bacterial and mammalian models and to reduce chromosomal aberrations in Chinese hamster ovary cells exposed to UV light and X-rays, possibly through the activation of recombinatorial repair [331, 332].

The formation of new blood vessels, or angiogenesis, is crucial for the progression of solid tumours, and inhibitors of this process can have potent chemotherapeutic effects in the treatment of cancer [333, 334]. In addition to other effects, perillyl alcohol has also been shown to be antiangiogenic in that it inhibited the growth of new vessels in the chorioallantoic membrane assay, inhibited endothelial cell proliferation and their organization into tube-like structures, and altered the production of angiogenic growth factors in a manner that encouraged vessel regression [335]. Inhibition of angiogenesis has also been reported for *d*-limonene [336], perillyl alcohol, farnesol, and geraniol [298] and mastic (*P. lentiscus* var. *chia*) oils [337].

Despite all the promise of *in vitro* work and *in vivo* animal models, early trials in humans using compounds such as perillyl alcohol to treat breast [338], colorectal [339], ovarian [340] and prostate [341] cancer have been disappointing. However, results from the treatment of some malignancies such as gliomas are worth pursuing [342] and interest in the therapeutic potential of this and other essential oil compounds remains, as does the chemopreventive possibilities for such compounds [343].

9.4   Uses of Essential Oils

9.4.1   Pharmaceutical Products

Combined with their established historical use as medicines, the range of biological activities demonstrated by essential oils and their components has naturally aroused great interest in their use as medicinal products. Consequently, essential oils are ingredients in medicinal products sold for a wide range of therapeutic applications. The range of products into which essential oils are formulated, their indications and the claims made for these products vary considerably throughout the world, as do the laws governing their sale. In the United States herbal products designed for ingestion, including those containing essential oils, are classified as dietary supplements. As such they are usually exempt from the stricter regulatory requirements applicable to drugs and foods. In Australia, medicinal products containing essential oils for which therapeutic claims are made are regarded as pharmaceutical products. Under the Australian pharmaceutical regulatory framework, they may be classified as registered or listed products. Listed products must meet less stringent criteria than those in the registered products category and may make concomitantly smaller therapeutic claims [344, 345]. In the European Union, different regulatory approaches are taken in different member states, although the goal is harmonization across member states [345, 346]. It is hoped that once harmonized, essential oils formulated into therapeutic products may be approved for sale under a simplified procedure, without needing to fulfil the criteria for a full product license [346].

Over-the-counter (OTC) medicines in most Western countries are designed for the treatment of minor, self-limiting conditions and their symptoms. Essential oils are active ingredients or excipients in many OTC medicines. For example, eucalyptus oil is found in more than 100 OTC products [347] intended mainly for use in the treatment or management of symptoms of upper respiratory tract infections. Since these infections are usually mild and self-limiting, the therapeutic claims made for these OTC medicines are generally more modest than those that can be made for prescription products.

Worldwide, few if any pharmaceutical agents containing essential oils as the active ingredient have been fully licensed as drugs by fulfilling the requirements for safety and efficacy data that must be met by new, conventional pharmaceutical agents. The reasons for this are likely to be manifold, including limitations on the resources available to adequately research and document the safety and efficacy of such products. In most cases the intellectual property arising from such efforts could not be protected sufficiently to warrant the investment by private enterprise. Publicly funded research seems the most likely mechanism by which some essential-oil therapeutic products might eventually be evaluated and registered in the same manner as drugs.

9.4.2   Foods and Beverages

Essential oils and their constituents are widely used in many foods and beverages, primarily as flavouring agents [348, 349]. Citrus-peel essential oils are amongst the most important of these, including orange, lemon, mandarin, tangerine and grapefruit oils [350, 351]. Peppermint, cornmint, eucalyptus and citronella oils are other leading oils in terms of volume [351]. Amongst single constituents, one of the most important to the flavour industry is menthol [352].

The concentrations used in foods and beverages are generally low; in beverages levels are typically at or below 0.1% [349]. In foods in Europe, for example, eucalyptus oil is approved for use as a flavouring agent at 5 mg/kg or less and in confectionery at 15 mg/kg [353]. As discussed previously, the levels of essential oils that are used in foods are governed largely by their effect on the organoleptic properties of the food. Their presence in food may also contribute to preservation of the products, depending on the concentrations used and the interaction they have with other ingredients and preservation factors in the product [135].

9.5   Conclusions

The diversity of compounds that make up essential oils is becoming increasingly well characterized. Similarly, the spectrum of biological activity of essential oils and their components is beginning to be fully appreciated and understood. The challenge remains to further explore the range of biological effects of essential oils and their potential applications.

References

1. Bakkali, F., Averbeck, S., Averbeck, D. and Idaomar, M. (2008) Biological effects of essential oils: a review. *Food Chem. Toxicol.*, **46**, 446–475.
2. Burt, S. (2004) Essential oils: their antibacterial properties and potential applications in foods: a review. *Int. J. Food Microbiol.*, **94**, 223–253.
3. Pengelly, A. (2004) *The Constituents of Medicinal Plants*, 2nd ed, Allen & Unwin, Sydney, Australia.
4. Başer, K.H.C. and Demirci, F. (2007) Chemistry of essential oils, in *Flavours and Fragrances Chemistry, Bioprocessing and Sustainability* (ed R.G. Berger), Springer, Berlin, Germany, pp. 43–86.
5. Lahlou, M. (2004) Methods to study the phytochemistry and bioactivity of essential oils. *Phytother. Res.*, **18**, 435–448.
6. Langenheim, J.H. (1994) Higher plant terpenoids: a phytocentric overview of their ecological roles. *J. Chem. Ecol.*, **20**, 1223–1280.
7. Croteau, R. (1986) Biochemistry of monoterpenes and sesquiterpenes of the essential oils, in *Herbs, Spices, and Medicinal Plants: Recent Advances in Botany, Horticulture, and Pharmacology* (eds L.E. Craker and J.E. Simon), Haworth Press, New York, NY, USA, pp. 81–133.
8. Figueiredo, A.C., Barroso, J.G., Pedro, L.G. and Scheffer, J.J.C. (2008) Factors affecting secondary metabolite production in plants: volatile components and essential oils. *Flavour Fragrance J.*, **23**, 213–226.
9. Lee, K.-W., Everts, H. and Beynen, A.C. (2004) Essential oils in broiler nutrition. *Int. J. Poult. Nutr.*, **3**, 738–752.
10. Edris, A.E. (2007) Pharmaceutical and therapeutic potentials of essential oils and their individual volatile constituents: a review. *Phytother. Res.*, **21**, 308–323.
11. Obst, J.R. (1998) Special (secondary) metabolites from wood, in *Forest Products Biotechnology* (eds A. Bruce and J.W. Palfreyman), Taylor & Francis, London, pp. 151–165.
12. Croteau, R., Kutchan, T.M. and Lewis, N.G. (2000) Natural products (secondary metabolites), in *Biochemistry and Molecular Biology of Plants* (eds B. Buchanan, W. Gruissem, and R. Jones), American Society of Plant Biologists, Rockville, MD, USA, pp. 1250–1268.
13. Theis, N. and Lerdau, M. (2003) The evolution of function in plant secondary metabolites. *Int. J. Plant Sci.*, **164**, S93–S102.
14. Pichersky, E., Sharkey, T.D. and Gershenzon, J. (2006) Plant volatiles: a lack of function or a lack of knowledge? *Trends Plant Sci.*, **11**, 421–421.
15. McCaskill, D. and Croteau, R. (1998) Some caveats for bioengineering terpenoid metabolism in plants. *Trends Biotechnol.*, **16**, 349–355.
16. Gang, D.R., Lavid, N., Zubieta, C. *et al.* (2002) Characterization of phenylpropene O-methyltransferases from sweet basil: facile change of substrate specificity and convergent evolution within a plant O-methyltransferase family. *Plant Cell*, **14**, 505–519.
17. Gershenzon, J. and Dudareva, N. (2007) The function of terpene natural products in the natural world. *Nat. Chem. Biol.*, **3**, 408–414.
18. İşcan, G., Kirimer, N., Kurkcuoglu, M. and Başer, K.H.C. (2005) Composition and antimicrobial activity of the essential oils of two endemic species from Turkey: *Sideritis cilicica* and *Sideritis bilgerana*. *Chem. Nat. Compd.*, **41**, 679–682.

19. Dung, N.T., Kim, J.M. and Kang, S.C. (2008) Chemical composition, antimicrobial and antioxidant activities of the essential oil and the ethanol extract of *Cleistocalyx operculatus* (Roxb.) Merr and Perry buds. *Food Chem. Toxicol.*, **46**, 3632–3639.

20. Friedrich, H. (1976) Phenylpropanoid constituents of essential oils. *Lloydia*, **39**, 1–7.

21. Weidenhamer, J.D., Macias, F.A., Fischer, N.H. and Williamson, G.B. (1993) Just how insoluble are monoterpenes? *J. Chem. Ecol.*, **19**, 1799–1807.

22. Griffin, S.G., Wyllie, S.G., Markham, J.L. and Leach, D.N. (1999) The role of structure and molecular properties of terpenoids in determining their antimicrobial activity. *Flavour Fragrance J.*, **14**, 322–332.

23. Breitmaier, E. (2006) *Terpenes. Flavors, Fragrances, Pharmaca, Pheromones*, Wiley-VCH, Weinheim, Germany.

24. Dubey, V.S., Bhalla, R. and Luthra, R. (2003) An overview of the non-mevalonate pathway for terpenoid biosynthesis in plants. *J. Biosci.*, **28**, 637–646.

25. Glaeske, K.W. and Boehlke, P.R. (2002) Making sense of terpenes: an exploration into biological chemistry. *Am. Biol. Teach.*, **64**, S208–211.

26. Bouvier, F., Rahier, A. and Camara, B. (2005) Biogenesis, molecular regulation and function of plant isoprenoids. *Prog. Lipid Res.*, **44**, 357–429.

27. Ruzicka, L. (1953) The isoprene rule and the biogenesis of terpenic compounds. *Cell. Mol. Life Sci.*, **9**, 357–367.

28. Little, D.B. and Croteau, R.B. (1999) Biochemistry of essential oil terpenes: a thirty year overview, in *Flavor Chemistry: Thirty Years of Progress* (eds R. Teranishi, E.L. Wick and I. Hornstein), Springer, pp. 239–254.

29. Chinou, I. (2005) Labdanes of natural origin: biological activities (1981–2004). *Curr. Med. Chem.*, **12**, 1295–1317.

30. Carson, C.F., Hammer, K.A. and Riley, T.V. (2006) *Melaleuca alternifolia* (tea tree) oil: a review of antimicrobial and other properties. *Clin. Microbiol. Rev.*, **19**, 50–62.

31. Yu, J., Lei, J., Yu, H. *et al.* (2004) Chemical composition and antimicrobial activity of the essential oil of *Scutellaria barbata*. *Phytochemistry*, **65**, 881–884.

32. Nostro, A., Blanco, A.R., Cannatelli, M.A. *et al.* (2004) Susceptibility of methicillin-resistant staphylococci to oregano essential oil, carvacrol and thymol. *FEMS Microbiol. Lett.*, **230**, 191–195.

33. Skočibušić, M. and Bezić, N. (2004) Phytochemical analysis and in vitro antimicrobial activity of two *Satureja* species essential oils. *Phytother. Res.*, **18**, 967–970.

34. Bohlmann, J. and Keeling, C.I. (2008) Harnessing plant biomass for biofuels and biomaterials. Terpenoid biomaterials. *Plant J.*, **54**, 656–669.

35. Keszei, A., Brubaker, C.L. and Foley, W.J. (2008) A molecular perspective on terpene variation in Australian Myrtaceae. *Aust. J. Bot.*, **56**, 197–213.

36. McGarvey, D.J. and Croteau, R. (1995) Terpenoid metabolism. *Plant Cell*, **7**, 1015–1026.

37. Koparal, A.T. and Zeytinoglu, M. (2003) Effects of carvacrol on a human non-small cell lung cancer (NSCLC) cell line, A549. *Cytotechnology*, **43**, 149–154.

38. Shapiro, S. and Guggenheim, B. (1998) Inhibition of oral bacteria by phenolic compounds. Part 1. QSAR analysis using molecular connectivity. *Quant. Struct-Act. Relat.*, **17**, 327–337.

39. McNaught, A.D. and Wilkinson, A. (1997) *Compendium of Chemical Terminology*, 2nd ed., Blackwell Science, Oxford, UK.

40. Castola, V., Bighelli, A. and Casanova, J. (2000) Intraspecific chemical variability of the essential oil of *Pistacia lentiscus* L. from Corsica. *Biochem. Syst. Ecol.*, **28**, 79–88.

41. Cosentino, S., Barra, A., Pisano, B. *et al.* (2003) Composition and antimicrobial properties of Sardinian *Juniperus* essential oils against foodborne pathogens and spoilage microorganisms. *J. Food Protect.*, **66**, 1288–1291.

42. Zhao, J. (2007) Plant troponoids: chemistry, biological activity, and biosynthesis. *Curr. Med. Chem.*, **14**, 2597–2621.

43. Bentley, R. (2008) A fresh look at natural tropolonoids. *Nat. Prod. Rep.*, **25**, 118–138.

44. Sy, L.-K. and Brown, G.D. (2001) Deoxyarteannuin B, dihydro-deoxyarteannuin B and trans-5-hydroxy-2-isopropenyl-5-methylhex-3-en-1-ol from *Artemisia anuua*. *Phytochemistry*, **58**, 1159–1166.

45. Umlauf, D., Zapp, J., Becker, H. and Adam, K.P. (2004) Biosynthesis of the irregular monoterpene artemisia ketone, the sesquiterpene germacrene D and other isoprenoids in Tanacetum vulgare L. (Asteraceae). *Phytochemistry*, **65**, 2463–2470.

46. Salido, S., Valenzuela, L.R., Altarejos, J. *et al.* (2004) Composition and infraspecific variability of *Artemisia herba-alba* from southern Spain. *Biochem. Syst. Ecol.*, **32**, 265–277.

47. Fugh-Berman, A. and Myers, A. (2004) *Citrus aurantium*, an ingredient of dietary supplements marketed for weight loss: current status of clinical and basic research. *Exp. Biol. Med.*, **229**, 698–704.

48. Limberger, R.P., Sobral, M. and Henriques, A.T. (2005) Intraspecific volatile oil variation in *Myrceugenia cucullata* (Myrtaceae). *Biochem. Syst. Ecol.*, **33**, 287–293.

49. Piggott, M.J., Ghisalberti, E.L. and Trengove, R.D. (1997) Western Australian sandalwood oil: extraction by different techniques and variations of the major components in different sections of a single tree. *Flavour Fragrance J.*, **12**, 43–46.

50. Ferrari, B., Tomi, F., Richomme, P. and Casanova, J. (2005) Two new irregular acyclic sesquiterpenes aldehydes from *Santolina corsica* essential oil. *Magn. Reson. Chem.*, **43**, 73–74.

51. Liu, K., Rossi, P.G., Ferrari, B. *et al.* (2007) Composition, irregular terpenoids, chemical variability and antibacterial activity of the essential oil from *Santolina corsica* Jordan et Fourr. *Phytochemistry*, **68**, 1698–1705.

52. Ganzera, M., Schneider, P. and Stuppner, H. (2006) Inhibitory effects of the essential oil of chamomile (*Matricaria recutita* L.) and its major constituents on human cytochrome P450 enzymes. *Life Sci.*, **78**, 856–861.

53. Sinico, C., De Logu, A., Lai, F. *et al.* (2005) Liposomal incorporation of *Artemisia arborescens* L. essential oil and in vitro antiviral activity. *Eur. J. Pharm. Biopharm.*, **59**, 161–168.

54. Henriques, A.T., Sobral, M.E., Cauduro, A.D. *et al.* (1993) Aromatic plants from Brazil. II. The chemical composition of some *Eugenia* essential oils. *J. Essent. Oil Res.*, **5**, 501–505.

55. Sabulal, B., Dan, M., Aj, J. *et al.* (2006) Caryophyllene-rich rhizome oil of *Zingiber nimmonii* from South India: chemical characterization and antimicrobial activity. *Phytochemistry*, **67**, 2469–2473.

56. Adams, R.P. (1991) Cedar wood oil: Analyses and properties, in *Modern Methods of Plant Analysis* (eds H. F. Linskens and J. F. Jackson), Springer-Verlag, Berlin, Germany, pp. 159–173.

57. Lis-Balchin, M., Deans, S.G. and Eaglesham, E. (1998) Relationship between bioactivity and chemical composition of commercial essential oils. *Flavour Fragrance J.*, **13**, 98–104.

58. Sangwan, N.S., Farooqi, A.H.A., Shabih, F. and Sangwan, R.S. (2001) Regulation of essential oil production in plants. *Plant Growth Regul.*, **34**, 3–21.

59. Langenheim, J.H. (2003) *Plant Resins: Chemistry, Evolution, Ecology, and Ethnobotany*, Timber Press, Portland, OR, USA.

60. Reverchon, E. (1997) Supercritical fluid extraction and fractionation of essential oils and related products. *J. Supercrit. Fluids*, **10**, 1–37.

61. Wu, T.-S., Damu, A.G., Su, C.-R. and Kuo, P.-C. (2004) Terpenoids of *Aristolochia* and their biological activities. *Nat. Prod. Rep.*, **21**, 594–624.

62. Hoct, S., Stévigny, C., Hérent, M.F. and Quetin-Leclercq, J. (2006) Antitrypanosomal compounds from the leaf essential oil of *Strychnos spinosa*. *Planta Med.*, **72**, 480–482.

63. Köse, Y.B., İşcan, G., Demirci, B. *et al.* (2007) Antimicrobial activity of the essential oil of *Centaurea aladagensis*. *Fitoterapia*, **78**, 253–254.

64. Jahangir, M., Kim, H.K., Choi, Y.H. and Verpoorte, R. (2009) Health-affecting compounds in Brassicaceae. *Compr. Rev. Food Sci. Food Saf.*, **8**, 31–43.

65. Wungsintaweekul, J., Sirisuntipong, T., Kongduang, D. *et al.* (2008) Transcription profiles analysis of genes encoding 1-deoxy-D-xylulose 5-phosphate synthase and 2C-methyl-D-erythritol 4-phosphate synthase in plaunotol biosynthesis from *Croton stellatopilosus*. *Biol. Pharm. Bull.*, **31**, 852–856.

66. Lo Presti, M., Sciarrone, D., Crupi, M.L. *et al.* (2008) Evaluation of the volatile and chiral composition in *Pistacia lentiscus* L. essential oil. *Flavour Fragrance J.*, **23**, 249–257.

67. Boelens, M.H. and Jimenez, R. (1991) Chemical composition of the essential oils from the gum and from various parts of *Pistacia lentiscus* l. (mastic gum tree). *Flavour Fragrance J.*, **6**, 271–275.

68. Ulubelen, A., Topcu, G., Eriş, C. *et al.* (1994) Terpenoids from *Salvia sclarea*. *Phytochemistry*, **36**, 971–974.

69. Gökdil, G., Topcu, G., Sönmez, U. and Ulubelen, A. (1997) Terpenoids and flavonoids from *Salvia cyanescens*. *Phytochemistry*, **46**, 799–800.

70. Dimas, K., Kokkinopoulos, D., Demetzos, C. *et al.* (1999) The effect of sclareol on growth and cell cycle progression of human leukemic cell lines. *Leuk. Res.*, **23**, 217–234.

71. Brophy, J.J., Goldsack, R.J., Wu, M.Z. *et al.* (2000) The steam volatile oil of *Wollemia nobilis* and its comparison with other members of the Araucariaceae (*Agathis* and *Araucaria*). *Biochem. Syst. Ecol.*, **28**, 563–578.

72. Briggs, L.H. and White, G.W. (1975) Constituents of the essential oil of *Araucaria araucana*. *Tetrahedron*, **31**, 1311–1314.

73. Otsuka, M., Kenmoku, H., Ogawa, M. *et al.* (2004) Emission of ent-kaurene, a diterpenoid hydrocarbon precursor for gibberellins, into the headspace from plants. *Plant Cell Physiol.*, **45**, 1129–1138.

74. Hirschberg, J. (2001) Carotenoid biosynthesis in flowering plants. *Current Opinion in Plant Biology*, **4**, 210–218.

75. Auldridge, M.E., McCarty, D.R. and Klee, H.J. (2006) Plant carotenoid cleavage oxygenases and their apocarotenoid products. *Curr. Opin. Plant Biol.*, **9**, 315–321.

76. Kloer, D.P. and Schulz, G.E. (2006) Structural and biological aspects of carotenoid cleavage. *Cell. Mol. Life Sci.*, **63**, 2291–2303.

77. Rodríguez-Bustamante, E. and Sánchez, S. (2007) Microbial production of C13-norisoprenoids and other aroma compounds via carotenoid cleavage. *Crit. Rev. Microbiol.*, **33**, 211–230.

78. Plummer, J.A., Wann, J.M., Considine, J.A. and Spadek, Z.E. (1996) Selection of *Boronia* for essential oils and cut flowers, in *Progress in New Crops* (ed. J. Janick), ASHS Press, Arlington, VA, USA, pp. 602–609.

79. Ghisalberti, E.L. (1998) Phytochemistry of the Australian Rutaceae: *Boronia*, *Eriostemon* and *Phebalium* species. *Phytochemistry*, **47**, 163–176.

80. Babu, K.G.D., Singh, B., Joshi, V.P. and Singh, V. (2002) Essential oil composition of Damask rose (*Rosa damascena* Mill.) distilled under different pressures and temperatures. *Flavour Fragrance J.*, **17**, 136–140.

81. Herrmann, K.M. and Weaver, L.M. (1999) The shikimate pathway. *Annu. Rev. Plant Physiol. Plant Mol. Biol.*, **50**, 473–503.

82. Ferrer, J.L., Austin, M.B., Stewart, C. and Noel, J.P. (2008) Structure and function of enzymes involved in the biosynthesis of phenylpropanoids. *Plant Physiol. Biochem.*, **46**, 356–370.

83. Wu, S. and Chappell, J. (2008) Metabolic engineering of natural products in plants; tools of the trade and challenges for the future. *Curr. Opin. Biotechnol.*, **19**, 145–152.

84. Dudareva, N., Pichersky, E. and Gershenzon, J. (2004) Biochemistry of plant volatiles. *Plant Physiol.*, **135**, 1893–1902.

85. Nogueira, P.C.de.L., Bittrich, V., Shepherd, G.J. *et al.* (2001) The ecological and taxonomic importance of flower volatiles of *Clusia* species (Guttiferae). *Phytochemistry*, **56**, 443–452.

86. Jirovetz, L., Buchbauer, G., Shafi, M.P. and Kaniampady, M.M. (2003) Chemotaxonomical analysis of the essential oil aroma compounds of four different *Ocimum* species from southern India. *Eur. Food Res. Technol.*, **217**, 120–124.

87. Clifford, M.N. (2000) Miscellaneous phenols in foods and beverages-nature, occurrence and dietary burden. *J. Sci. Food Agric.*, **80**, 1126–1137.

88. Gang, D.R., Wang, J., Dudareva, N. *et al.* (2001) An investigation of the storage and biosynthesis of phenylpropenes in sweet basil. *Plant Physiol.*, **125**, 539–555.

89. Medina-Holguín, A.L., Holguín, F. O., Micheletto, S. *et al.* (2008) Chemotypic variation of essential oils in the medicinal plant, *Anemopsis californica*. *Phytochemistry*, **69**, 919–927.

90. Chaieb, K., Hajlaoui, H., Zmantar, T. *et al.* (2007) The chemical composition and biological activity of clove essential oil, *Eugenia caryophyllata* (*Syzigium aromaticum* L. Myrtaceae): a short review. *Phytother. Res.*, **21**, 501–506.

91. Martins, A.P., Salgueiro, L., Vila, R. *et al.* (1998) Essential oils from four *Piper* species. *Phytochemistry*, **49**, 2019–2023.

92. Siani, A.C., Ramos, M.F.S., Menezes-de-Lima, O. *et al.* (1999) Evaluation of anti-inflammatory-related activity of essential oils from the leaves and resin of species of *Protium. J. Ethnopharmacol.*, **66**, 57–69.

93. Gog, L., Berenbaum, M.R., DeLucia, E.H. and Zangerl, A.R. (2005) Autotoxic effects of essential oils on photosynthesis in parsley, parsnip, and rough lemon. *Chemoecology*, **15**, 115–119.

94. de Morais, S.M., Facundo, V.A., Bertini, L.M. *et al.* (2007) Chemical composition and larvicidal activity of essential oils from *Piper* species. *Biochem. Syst. Ecol.*, **35**, 670–675.

95. Mimica-Dukić, N., Kujundžić, S., Soković, M. and Couladis, M. (2003) Essential oil composition and antifungal activity of *Foeniculum vulgare* Mill. obtained by different distillation conditions. *Phytother. Res.*, **17**, 368–371.

96. Flamini, G., Tebano, M., Cioni, P.L. *et al.* (2007) Comparison between the conventional method of extraction of essential oil of *Laurus nobilis* L. and a novel method which uses microwaves applied in situ, without resorting to an oven. *J. Chromatogr. A*, **1143**, 36–40.

97. Boelens, M.H. (1996) Chemical and sensory evaluation of trace compounds in naturals. *Perfum. Flavor.*, **21**, 25–31.

98. Goeke, A. (2002) Sulfur-containing odorants in fragrance chemistry. *J. Sulfur Chem.*, **23**, 243–278.

99. Halkier, B.A. and Gershenzon, J. (2006) Biology and biochemistry of glucosinolates. *Ann. Rev. Plant Biol.*, **57**, 303–333.

100. Fahey, J.W., Zalcmann, A.T. and Talalay, P. (2001) The chemical diversity and distribution of glucosinolates and isothiocyanates among plants. *Phytochemistry*, **56**, 5–51.

101. Tellez, M.R., Khan, I.A., Kobaisy, M. *et al.* (2002) Composition of the essential oil of Lepidium meyenii (Walp.). *Phytochemistry*, **61**, 149–155.

102. Dhingra, O.D., Costa, M.L.N., Silva, J.G.J. and Mizubuti, E.S.G. (2004) Essential oil of mustard to control *Rhizoctonia solani* causing seedling damping off and seedling blight in nursery. *Fitopatol. Bras.*, **29**, 683–686.

103. Miyazawa, M., Nishiguchi, T. and Yamafuji, C. (2005) Volatile components of the leaves of *Brassica rapa* L. var. *perviridis* Bailey. *Flavour Fragrance J.*, **20**, 158–160.

104. Turgis, M., Han, J., Caillet, S. and Lacroix, M. (2009) Antimicrobial activity of mustard essential oil against *Escherichia coli* O157 : H7 and *Salmonella typhi. Food Control*, **20**, 1073–1079.

105. Kurose, K. and Yatagai, M. (2005) Components of the essential oils of *Azadirachta indica* A. Juss, *Azadirachta siamensis* Velton, and *Azadirachta excelsa* (Jack) Jacobs and their comparison. *J. Wood Sci.*, **51**, 185–188.

106. Merat, E. and Vogel, J. (1976) Separation and spectrofluorimetric determination of methyl anthranilate and methyl N-methyl anthranilate in various beverages and flavorants. *Mitt. Geb. Lebensmittelunters. Hyg.*, **66**, 496–501.

107. Rath, C.C., Devi, S., Dash, S.K. and Mishra, R.K. (2008) Antibacterial potential assessment of jasmine essential oil against *E. coli. Indian J. Pharm. Sci.*, **70**, 238–241.

108. El-Sakhawy, F.S., El-Tantawy, M.E., Ross, S.A. and El-Sohly, M.A. (1998) Composition and antimicrobial activity of the essential oil of *Murraya exotica* L. *Flavour Fragrance J.*, **13**, 59–62.

109. Verzera, A., Mondello, L., Trozzi, A. and Dugo, P. (1997) On the genuineness of citrus essential oils. Part LII. Chemical characterization of essential oil of three cultivars of *Citrus clementine* Hort. *Flavour Fragrance J.*, **12**, 163–172.

110. Stashenko, E.E., Torres, W. and Morales, J.R.M. (1995) A study of the compositional variation of the essential oil of ylang-ylang (*Cananga odorata* Hook Fil. et Thomson, *formagenuina*) during flower development. *J. High Resolut. Chromatogr.*, **18**, 101–104.

111. Rchid, H., Nmila, R., Bessiere, J.M. *et al.* (2004) Volatile components of *Nigella damascena* L. and *Nigella sativa* L. seeds. *J. Essent. Oil Res.*, **16**, 585–587.

112. Mondello, L., Basile, A., Previti, P. and Dugo, G. (1997) Italian Citrus petitgrain oils. II. Composition of mandarin petitgrain oil. *J. Essent. Oil Res.*, **9**, 255–266.

113. El-Ghorab, A.H., El-Massry, K.F. and Mansour, A.F. (2003) Chemical composition, antifungal and radical scavenging activities of Egyptian mandarin petitgrain essential oil. *Bull. Natl. Res. Cent.*, **28**, 535–549.

114. Clery, R.A., Hammond, C.J. and Wright, A.C. (2005) Nitrogen compounds from Haitian vetiver oil. *J. Essent. Oil Res.*, **17**, 591–592.

115. Clery, R.A., Hammond, C.J. and Wright, A.C. (2006) Nitrogen-containing compounds in black pepper oil (*Piper nigrum* L.). *J. Essent. Oil Res.*, **18**, 1–3.

116. Krest, I., Glodek, J. and Keusgen, M. (2000) Cysteine sulfoxides and alliinase activity of some *Allium* species. *J. Agric. Food Chem.*, **48**, 3753–3760.

117. Štajner, D., Milić, N., Čanadanović-Brunet, J. *et al.* (2006) Exploring *Allium* species as a source of potential medicinal agents. *Phytother. Res.*, **20**, 581–584.

118. Amagase, H., Petesch, B.L., Matsuura, H. *et al.* (2001) Intake of garlic and its bioactive components. *J. Nutr.*, **131**, 955S–962S.

119. Omata, A., Yomogida, K., Nakamura, S. *et al.* (1991) New sulphur components of rose oil. *Flavour Fragrance J.*, **6**, 149–152.

120. Takahashi, K., Muraki, S. and Yoshida, T. (1981) Synthesis and distribution of (−)-mintsulfide, a novel sulfur-containing sesquiterpene. *Agric. Biol. Chem.*, **45**, 129–132.

121. Vickers, C.E., Gershenzon, J., Lerdau, M.T. and Loreto, F. (2009) A unified mechanism of action for volatile isoprenoids in plant abiotic stress. *Nat. Chem. Biol.*, **5**, 283–291.

122. Ballabeni, V., Tognolini, M., Chiavarini, M. *et al.* (2004) Novel antiplatelet and antithrombotic activities of essential oil from *Lavandula hybrida* Reverchon 'grosso'. *Phytomedicine*, **11**, 596–601.

123. Tognolini, M., Ballabeni, V., Bertoni, S. *et al.* (2007) Protective effect of *Foeniculum vulgare* essential oil and anethole in an experimental model of thrombosis. *Pharmacol. Res.*, **56**, 254–260.

124. Tognolini, M., Barocelli, E., Ballabeni, V. *et al.* (2006) Comparative screening of plant essential oils: phenylpropanoid moiety as basic core for antiplatelet activity. *Life Sci.*, **78**, 1419–1432.

125. Huang, J., Wang, S., Luo, X. *et al.* (2007) Cinnamaldehyde reduction of platelet aggregation and thrombosis in rodents. *Thromb. Res.*, **119**, 337–342.

126. Liao, B.C., Hsieh, C.W., Liu, Y.C. *et al.* (2008) Cinnamaldehyde inhibits the tumor necrosis factor-a-induced expression of cell adhesion molecules in endothelial cells by suppressing NF-kB activation: effects upon IkB and Nrf2. *Toxicol. Appl. Pharmacol.*, **229**, 161–171.

127. Cavanagh, H.M.A. and Wilkinson, J.M. (2002) Biological activities of lavender essential oil. *Phytother. Res.*, **16**, 301–308.

128. Pascual, M.E., Slowing, K., Carretero, E. *et al.* (2001) Lippia: traditional uses, chemistry and pharmacology: a review. *J. Ethnopharmacol.*, **76**, 201–214.

129. Jirovetz, L., Buchbauer, G., Denkova, Z. *et al.* (2006) Comparative study on the antimicrobial activities of different sandalwood essential oils of various origin. *Flavour Fragrance J.*, **21**, 465–468.

130. Hammer, K.A., Carson, C.F. and Riley, T.V. (1999) Antimicrobial activity of essential oils and other plant extracts. *J. Appl. Microbiol.*, **86**, 985–990.

131. Elsom, G.K.F. and Hide, D. (1999) Susceptibility of methicillin-resistant *Staphylococcus aureus* to tea tree oil and mupirocin. *J. Antimicrob. Chemother.*, **43**, 427–428.

132. Tsao, S.-M. and Yin, M.-C. (2001) In-vitro antimicrobial activity of four diallyl sulphides ocurring naturally in garlic and Chinese leek oils. *J. Med. Microbiol.*, **50**, 646–649.

133. LaPlante, K.L. (2007) In vitro activity of lysostaphin, mupirocin, and tea tree oil against clinical methicillin-resistant *Staphylococcus aureus*. *Diagn. Microbiol. Infect. Dis.*, **57**, 413–418.

134. Kalemba, D. and Kunicka, A. (2003) Antibacterial and antifungal properties of essential oils. *Curr. Med. Chem.*, **10**, 813–829.

135. Holley, R.A. and Patel, D. (2005) Improvement in shelf-life and safety of perishable foods by plant essential oils and smoke antimicrobials. *Food Microbiol.*, **22**, 273–292.

136. Wilkinson, J.M. (2006) Methods for testing the antimicrobial activity of extracts, in *Modern Phytomedicine. Turning Medicinal Plants into Drugs* (eds I. Ahmad, F. Aqil and M. Owais), Wiley-VCH, Weinheim, Germany, pp. 157–171.

137. Lanciotti, R., Gianotti, A., Patrignani, F. *et al.* (2004) Use of natural aroma compounds to improve shelf-life and safety of minimally processed fruits. *Trends Food Sci. Technol.*, **15**, 201–208.

138. Magiatis, P., Melliou, E., Skaltsounis, A.L. *et al.* (1999) Chemical composition and antimicrobial activity of the essential oils of *Pistacia lentiscus* var. *chia*. *Planta Med.*, **65**, 749–752.

139. Marino, M., Bersani, C. and Comi, G. (2001) Impedance measurements to study the antimicrobial activity of essential oils from Lamiaceae and Compositae. *Int. J. Food Microbiol.*, **67**, 187–195.

140. Delaquis, P.J., Stanich, K., Girard, B. and Mazza, G. (2002) Antimicrobial activity of individual and mixed fractions of dill, cilantro, coriander and eucalyptus essential oils. *Int. J. Food Microbiol.*, **74**, 101–109.

141. Pintore, G., Usai, M., Bradesi, P. *et al.* (2002) Chemical composition and antimicrobial activity of *Rosmarinus officinalis* L. oils from Sardinia and Corsica. *Flavour Fragrance J.*, **17**, 15–19.

142. Ali, B.H. and Blunden, G. (2003) Pharmacological and toxicological properties of *Nigella sativa*. *Phytother. Res.*, **17**, 299–305.

143. Mayaud, L., Carricajo, A., Zhiri, A. and Aubert, G. (2008) Comparison of bacteriostatic and bactericidal activity of 13 essential oils against strains with varying sensitivity to antibiotics. *Lett. Appl. Microbiol.*, **47**, 167–173.

144. Di Pasqua, R., De Feo, V., Villani, F. and Mauriello, G. (2005) In vitro antimicrobial activity of essential oils from Mediterranean Apiaceae, Verbenaceae and Lamiaceae against foodborne pathogens and spoilage bacteria. *Ann. Microbiol.*, **55**, 139–143.

145. Lis-Balchin, M., Steyrl, H. and Krenn, E. (2003) The comparative effect of novel *Pelargonium* essential oils and their corresponding hydrosols as antimicrobial agents in a model food system. *Phytother. Res.*, **17**, 60–65.

146. Hayes, A.J. and Markovic, B. (2002) Toxicity of Australian essential oil *Backhousia citriodora* (Lemon myrtle). Part 1. Antimicrobial activity and in vitro cytotoxicity. *Food Chem. Toxicol.*, **40**, 535–543.

147. Nelson, R.R.S. (1997) In-vitro activities of five plant essential oils against methicillin-resistant *Staphylococcus aureus* and vancomycin-resistant *Enterococcus faecium*. *J. Antimicrob. Chemother.*, **40**, 305–306.

148. Filipowicz, N., Kamiński, M., Kurlenda, J. *et al.* (2003) Antibacterial and antifungal activity of juniper berry oil and its selected components. *Phytother. Res.*, **17**, 227–231.

149. Carson, C.F., Cookson, B.D., Farrelly, H.D. and Riley, T.V. (1995) Susceptibility of methicillin-resistant *Staphylococcus aureus* to the essential oil of *Melaleuca alternifolia*. *J. Antimicrob. Chemother.*, **35**, 421–424.

150. Chan, C.H. and Loudon, K.W. (1998) Activity of tea tree oil on methicillin-resistant *Staphylococcus aureus* (MRSA) [letter]. *J. Hosp. Infect.*, **39**, 244–245.

151. Brady, A., Loughlin, R., Gilpin, D. *et al.* (2006) In vitro activity of tea-tree oil against clinical skin isolates of meticillin-resistant and -sensitive *Staphylococcus aureus* and coagulase-negative staphylococci growing planktonically and as biofilms. *J. Med. Microbiol.*, **55**, 1375–1380.

152. Loughlin, R., Gilmore, B.F., McCarron, P.A. and Tunney, M.M. (2008) Comparison of the cidal activity of tea tree oil and terpinen-4-ol against clinical bacterial skin isolates and human fibroblast cells. *Lett. Appl. Microbiol.*, **46**, 428–433.

153. Davis, A.O., O'Leary, J.O., Muthaiyan, A. *et al.* (2005) Characterization of *Staphylococcus aureus* mutants expressing reduced susceptibility to common house-cleaners. *J. Appl. Microbiol.*, **98**, 364–372.

154. Banes-Marshall, L., Cawley, P. and Phillips, C.A. (2001) *In vitro* activity of *Melaleuca alternifolia* (tea tree) oil against bacterial and *Candida* spp. isolates from clinical specimens. *Br. J. Biomed. Sci.*, **58**, 139–145.

155. Deans, S.G. and Ritchie, G. (1987) Antibacterial properties of plant essential oils. *Int. J. Food Microbiol.*, **5**, 165–180.

156. Porter, N.G. and Wilkins, A.L. (1999) Chemical, physical and antimicrobial properties of essential oils of *Leptospermum scoparium* and *Kunzea ericoides*. *Phytochemistry*, **50**, 407–415.

157. Carson, C.F., Mee, B.J. and Riley, T.V. (2002) Mechanism of action of *Melaleuca alternifolia* (tea tree) oil on *Staphylococcus aureus* determined by time-kill, lysis, leakage, and salt tolerance assays and electron microscopy. *Antimicrob. Agents Chemother.*, **46**, 1914–1920.

158. Lambert, R.J.W., Skandamis, P.N., Coote, P.J. and Nychas, G.-J.E. (2001) A study of the minimum inhibitory concentration and mode of action of oregano essential oil, thymol and carvacrol. *J. Appl. Microbiol.*, **91**, 453–462.

159. Ultee, A., Bennik, M.H.J. and Moezelaar, R. (2002) The phenolic hydroxyl group of carvacrol is essential for action against the food-borne pathogen *Bacillus cereus*. *Appl. Environ. Microbiol.*, **68**, 1561–1568.

160. Carson, C.F. and Riley, T.V. (1995) Antimicrobial activity of the major components of the essential oil of *Melaleuca alternifolia*. *J. Appl. Bacteriol.*, **78**, 264–269.

161. Inouye, S., Takizawa, T. and Yamaguchi, H. (2001) Antibacterial activity of essential oils and their major constituents against respiratory tract pathogens by gaseous contact. *J. Antimicrob. Chemother.*, **47**, 565–573.

162. Peñalver, P., Huerta, B., Borge, C. *et al.* (2005) Antimicrobial activity of five essential oils against origin strains of the Enterobacteriaceae family. *Acta Pathol. Microbiol. Immunol. Scand.*, **113**, 1–6.

163. Veldhuizen, E.J.A., Tjeerdsma-van Bokhoven, J.L.M., Zweijtzer, C. *et al.* (2006) Structural requirements for the antimicrobial activity of carvacrol. *J. Agric. Food Chem.*, **54**, 1874–1879.

164. Hammer, K.A., Carson, C.F. and Riley, T.V. (1996) Susceptibility of transient and commensal skin flora to the essential oil of *Melaleuca alternifolia* (tea tree oil). *Am. J. Infect. Control*, **24**, 186–189.

165. Preuss, H.G., Echard, B., Enig, M. *et al.* (2005) Minimum inhibitory concentrations of herbal essential oils and monolaurin for Gram-positive and Gram-negative bacteria. *Mol. Cell. Biochem.*, **272**, 29–34.

166. Hammer, K.A., Carson, C.F. and Riley, T.V. (2003) Antifungal activity of the components of *Melaleuca alternifolia* (tea tree) oil. *J. Appl. Microbiol.*, **95**, 853–860.

167. Matan, N., Rimkeeree, H., Mawson, A.J. *et al.* (2006) Antimicrobial activity of cinnamon and clove oils under modified atmosphere conditions. *Int. J. Food Microbiol.*, **107**, 180–185.

168. Thompson, D.P. (1989) Fungitoxic activity of essential oil components on food storage fungi. *Mycologia*, **81**, 151–153.

169. Araújo, C., Sousa, M.J., Ferreira, M.F. and Leão, C. (2003) Activity of essential oils from Mediterranean *Lamiaceae* species against food spoilage yeasts. *J. Food Protect.*, **66**, 625–632.

170. Conner, D.E. and Beuchat, L.R. (1984) Effects of essential oils from plants on growth of food spoilage yeasts. *J. Food Sci.*, **49**, 429–434.

171. López, P., Sánchez, C., Batlle, R. and Nerín, C. (2005) Solid- and vapor-phase antimicrobial activities of six essential oils: susceptibility of selected foodborne bacterial and fungal strains. *J. Agric. Food Chem.*, **53**, 6939–6946.

172. Nguefack, J., Leth, V., Amvam, P.H. and Mathur, S.B. (2004) Evaluation of five essential oils from aromatic plants of Cameroon for controlling food spoilage and mycotoxin producing fungi. *Int. J. Food Microbiol.*, **94**, 329–334.

173. Benkeblia, N. (2004) Antimicrobial activity of essential oil extracts of various onions (*Allium cepa*) and garlic (*Allium sativum*). *LWT Food Sci. Technol.*, **37**, 263–268.

174. Şahin, F., Güllüce, M., Daferera, D. *et al.* (2004) Biological activities of the essential oils and methanol extract of *Origanum vulgare* ssp. vulgare in the Eastern Anatolia region of Turkey. *Food Control*, **15**, 549–557.

175. Sacchetti, G., Maietti, S., Muzzoli, M. *et al.* (2005) Comparative evaluation of 11 essential oils of different origin as functional antioxidants, antiradicals and antimicrobials in foods. *Food Chem.*, **91**, 621–632.

176. Sökmen, M., Serkedjieva, J., Daferera, D. *et al.* (2004) In vitro antioxidant, antimicrobial, and antiviral activities of the essential oil and various extracts from herbal parts and callus cultures of *Origanum acutidens*. *J. Agric. Food Chem.*, **52**, 3309–3312.

177. Kumar, R., Mishra, A.K., Dubey, N.K. and Tripathi, Y.B. (2007) Evaluation of *Chenopodium ambrosioides* oil as a potential source of antifungal, antiaflatoxigenic and antioxidant activity. *Int. J. Food Microbiol.*, **115**, 159–164.

178. Suppakul, P., Miltz, J., Sonneveld, K. and Biggers, S.W. (2003) Antimicrobial properties of basil and its possible application in food packaging. *J. Agric. Food Chem.*, **51**, 3197–3207.

179. Cerrutti, P. and Alzamora, S.M. (1996) Inhibitory effects of vanillin on some food spoilage yeasts in laboratory media and fruit purees. *Int. J. Food Microbiol.*, **29**, 379–386.

180. Skandamis, P., Tsigarida, E. and Nychas, G.-J.E. (2002) The effect of oregano essential oil on survival/death of *Salmonella typhimurium* in meat stored at 5 °C under aerobic,VP/MAP conditions. *Food Microbiol.*, **19**, 97–103.
181. Gutierrez, J., Bourke, P., Lonchamp, J. and Barry-Ryan, C. (2009) Impact of plant essential oils on microbiological, organoleptic and quality markers of minimally processed vegetables. *Innov. Food Sci. Emerg. Technol.*, **10**, 195–202.
182. Gutierrez, J., Barry-Ryan, C. and Bourke, P. (2009) Antimicrobial activity of plant essential oils using food model media: efficacy, synergistic potential and interactions with food components. *Food Microbiol.*, **26**, 142–150.
183. Solomakos, N., Govaris, A., Koidis, P. and Botsoglou, N. (2008) The antimicrobial effect of thyme essential oil, nisin, and their combination against *Listeria monocytogenes* in minced beef during refrigerated storage. *Food Microbiol.*, **25**, 120–127.
184. Gutierrez, J., Barry-Ryan, C. and Bourke, P. (2008) The antimicrobial efficacy of plant essential oil combinations and interactions with food ingredients. *Int. J. Food Microbiol.*, **124**, 91–97.
185. Burt, S.A., Vlielander, R., Haagsman, H.P. and Veldhuizen, E.J. (2005) Increase in activity of essential oil components carvacrol and thymol against *Escherichia coli* O157 : H7 by addition of food stabilizers. *J. Food Protect.*, **68**, 919–926.
186. Goñi, P., López, P., Sánchez, C. *et al.* (2009) Antimicrobial activity in the vapour phase of a combination of cinnamon and clove essential oils. *Food Chem.*, **116**, 982–989.
187. Leistner, L. (2000) Basic aspects of food preservation by hurdle technology. *Int. J. Food Microbiol.*, **55**, 181–186.
188. Nielsen, P.V. and Rios, R. (2000) Inhibition of fungal growth on bread by volatile components from spices and herbs, and the possible application in active packaging, with special emphasis on mustard essential oil. *Int. J. Food Microbiol.*, **60**, 219–229.
189. Suhr, K.I. and Nielsen, P.V. (2005) Inhibition of fungal growth on wheat and rye bread by modified atmosphere packaging and active packaging using volatile mustard essential oil. *J. Food Sci.*, **70**, 37–44.
190. Serrano, M., Martínez-Romero, D., Castillo, S. *et al.* (2005) The use of natural antifungal compounds improves the beneficial effect of MAP in sweet cherry storage. *Innov. Food Sci. Emerg. Technol.*, **6**, 115–123.
191. Helander, I.M., Alakomi, H.L., Latva-Kala, K. *et al.* (1998) Characterization of the action of selected essential oil components on Gram-negative bacteria. *J. Agric. Food Chem.*, **46**, 3590–3595.
192. Ultee, A., Kets, E.P.W. and Smid, E.J. (1999) Mechanisms of action of carvacrol on the food-borne pathogen *Bacillus cereus*. *Appl. Environ. Microbiol.*, **65**, 4606–4610.
193. Gill, A.O. and Holley, R.A. (2006) Disruption of *Escherichia coli*, *Listeria monocytogenes* and *Lactobacillus sakei* cellular membranes by plant oil aromatics. *Int. J. Food Microbiol.*, **108**, 1–9.
194. Cox, S.D., Mann, C.M., Markham, J.L. *et al.* (2000) The mode of antimicrobial action of the essential oil of *Melaleuca alternifolia* (tea tree oil). *J. Appl. Microbiol.*, **88**, 170–175.
195. Domadia, P., Swarup, S., Bhunia, A. *et al.* (2007) Inhibition of bacterial cell division protein FtsZ by cinnamaldehyde. *Biochem. Pharmacol.*, **74**, 831–840.
196. Burt, S.A., Van Der Zee, R., Koets, A.P. *et al.* (2007) Carvacrol induces heat shock protein 60 and inhibits synthesis of flagellin in *Escherichia coli* O157: H7. *Appl. Environ. Microbiol.*, **73**, 4484–4490.
197. Niu, C., Afre, S. and Gilbert, E.S. (2006) Subinhibitory concentrations of cinnamaldehyde interfere with quorum sensing. *Lett. Appl. Microbiol.*, **43**, 489–494.
198. Brackman, G., Defoirdt, T., Miyamoto, C. *et al.* (2008) Cinnamaldehyde and cinnamaldehyde derivatives reduce virulence in *Vibrio* spp. by decreasing the DNA-binding activity of the quorum sensing response regulator LuxR. *BMC Microbiol.*, **8**, 149.
199. Iwalokun, B.A., Gbenle, G.O., Adewole, T.A. *et al.* (2003) Effects of *Ocimum gratissimum* L. essential oil at subinhibitory concentrations on virulent and multidrug-resistant *Shigella* strains from Lagos, Nigeria. *APMIS*, **111**, 477–482.
200. Filgueiras, C.T. and Vanetti, M.C.D. (2006) Effect of eugenol on growth and listeriolysin O production by *Listeria monocytogenes*. *Braz. Arch. Biol. Technol.*, **49**, 405–409.

201. Tassou, C., Koutsoumanis, K. and Nychas, G.J.E. (2000) Inhibition of *Salmonella enteritidis* and *Staphylococcus aureus* in nutrient broth by mint essential oil. *Food Res. Int.*, **33**, 273–280.

202. Hammer, K.A., Carson, C.F. and Riley, T.V. (2004) Antifungal effects of *Melaleuca alternifolia* (tea tree) oil and its components on *Candida albicans*, *Candida glabrata* and *Saccharomyces cerevisiae*. *J. Antimicrob. Chemother.*, **53**, 1081–1085.

203. Hammer, K.A., Carson, C.F. and Riley, T.V. (2000) *Melaleuca alternifolia* (tea tree) oil inhibits germ tube formation by *Candida albicans*. *Med. Mycol.*, **38**, 355–362.

204. Alviano, W.S., Mendonça-Filho, R.R., Alviano, D.S. *et al.* (2005) Antimicrobial activity of *Croton cajucara* Benth linalool-rich essential oil on artificial biofilms and planktonic microorganisms. *Oral Microbiol. Immunol.*, **20**, 101–105.

205. Seidakhmetova, R.B., Beisenbaeva, A.A., Atazhanova, G.A. *et al.* (2002) Chemical composition and biological activity of the essential oil from *Artemisia glabella*. *Pharm. Chem. J.*, **36**, 135–138.

206. Zai-Chang, Y., Bo-Chu, W., Xiao-Sheng, Y. and Qiang, W. (2005) Chemical composition of the volatile oil from *Cynanchum stauntonii* and its activities of anti-influenza virus. *Colloids Surf. B Biointerfaces*, **43**, 198–202.

207. Hayashi, K., Kamiya, M. and Hayashi, T. (1995) Virucidal effects of the steam distillate from *Houttuynia cordata* and its components on HSV-1, influenza virus, and HIV. *Planta Med.*, **61**, 237–241.

208. Bonsignore, L., Casu, L., Loy, G. *et al.* (2004) Analysis of the essential oil of *Oenanthe crocata* L. and its biological activity. *J. Essent. Oil Res.*, **16**, 266–269.

209. Öğütçü, H., Sökmen, A., Sökmen, M. *et al.* (2008) Bioactivities of the various extracts and essential oils of *Salvia limbata* CA Mey. and *Salvia sclarea* L. *Turk. J. Biol.*, **32**, 181–192.

210. Hayashi, K., Imanishi, N., Kashiwayama, Y. *et al.* (2007) Inhibitory effect of cinnamaldehyde, derived from *Cinnamomi cortex*, on the growth of influenza A/PR/8 virus in vitro and in vivo. *Antiviral Res.*, **74**, 1–8.

211. Schnitzler, P., Schön, K. and Reichling, J. (2001) Antiviral activity of Australian tea tree oil and eucalyptus oil against herpes simplex virus in cell culture. *Pharmazie*, **56**, 343–347.

212. Koch, C., Reichling, J., Schneele, J. and Schnitzler, P. (2007) Inhibitory effect of essential oils against herpes simplex virus type 2. *Phytomedicine*, **15**, 71–78.

213. Saddi, M., Sanna, A., Cottiglia, F. *et al.* (2007) Antiherpevirus activity of *Artemisia arborescens* essential oil and inhibition of lateral diffusion in Vero cells. *Ann. Clin. Microbiol. Antimicrob.*, **6**, 10.

214. Reichling, J., Koch, C., Stahl-Biskup, E. *et al.* (2005) Virucidal activity of a β-triketone-rich essential oil of *Leptospermum scoparium* (manuka oil) against HSV-1 and HSV-2 in cell culture. *Planta Med.*, **71**, 1123–1127.

215. Farag, R.S., Shalaby, A.S., El-Baroty, G.A. *et al.* (2004) Chemical and biological evaluation of the essential oils of different *Melaleuca* species. *Phytother. Res.*, **18**, 30–35.

216. Allahverdiyev, A., Duran, N., Ozguven, M. and Koltas, S. (2004) Antiviral activity of the volatile oils of *Melissa officinalis* L. against Herpes simplex virus type-2. *Phytomedicine*, **11**, 657–661.

217. Schnitzler, P., Schuhmacher, A., Astani, A. and Reichling, J. (2008) *Melissa officinalis* oil affects infectivity of enveloped herpesviruses. *Phytomedicine*, **15**, 734–740.

218. Schuhmacher, A., Reichling, J. and Schnitzler, P. (2003) Virucidal effect of peppermint oil on the enveloped viruses herpes simplex virus type 1 and type 2 in vitro. *Phytomedicine*, **10**, 504–510.

219. Sivropoulou, A., Nikolaou, C., Papanikolaou, E. *et al.* (1997) Antimicrobial, cytotoxic, and antiviral activities of *Salvia fructicosa* essential oil. *J. Agric. Food Chem.*, **45**, 3197–3201.

220. Benencia, F. and Courrèges, M.C. (1999) Antiviral activity of sandalwood oil against herpes simplex viruses-1 and-2. *Phytomedicine*, **6**, 119–123.

221. De Logu, A., Loy, G., Pellerano, M.L. *et al.* (2000) Inactivation of HSV-1 and HSV-2 and prevention of cell-to-cell virus spread by *Santolina insularis* essential oil. *Antiviral Res.*, **48**, 177–185.

222. Valenti, D., De Logu, A., Loy, G. *et al.* (2001) Liposome-incorporated *Santolina insularis* essential oil: preparation, characterization and in vitro antiviral activity. *J. Liposome Res.*, **11**, 73–90.

223. García, C.C., Talarico, L., Almeida, N. *et al.* (2003) Virucidal activity of essential oils from aromatic plants of San Luis, Argentina. *Phytother. Res.*, **17**, 1073–1075.

224. Duschatzky, C.B., Possetto, M.L., Talarico, L.B. *et al.* (2005) Evaluation of chemical and antiviral properties of essential oils from South American plants. *Antivir. Chem. Chemother.*, **16**, 247–251.

225. Benencia, F. and Courrèges, M.C. (2000) In vitro and in vivo activity of eugenol on human herpesvirus. *Phytother. Res.*, **14**, 495–500.

226. Tragoolpua, Y. and Jatisatienr, A. (2007) Anti-herpes simplex virus activities of *Eugenia caryophyllus* (Spreng.) Bullock & SG Harrison and essential oil, eugenol. *Phytother. Res.*, **21**, 1153–1158.

227. Armaka, M., Papanikolaou, E., Sivropoulou, A. and Arsenakis, M. (1999) Antiviral properties of isoborneol, a potent inhibitor of herpes simplex virus type 1. *Antiviral Res.*, **43**, 79–92.

228. Minami, M., Kita, M., Nakaya, T. *et al.* (2003) The inhibitory effect of essential oils on herpes simplex virus type-1 replication in vitro. *Microbiol. Immunol.*, **47**, 681–684.

229. Salido, S., Altarejos, J., Nogueras, M. *et al.* (2002) Chemical studies of essential oils of *Juniperus oxycedrus* ssp. *badia*. *J. Ethnopharmacol.*, **81**, 129–134.

230. Cermelli, C., Fabio, A., Fabio, G. and Quaglio, P. (2008) Effect of *Eucalyptus* essential oil on respiratory bacteria and viruses. *Curr. Microbiol.*, **56**, 89–92.

231. Wang, K.C., Chang, J.S., Chiang, L.C. and Lin, C.C. (2009) 4-Methoxycinnamaldehyde inhibited human respiratory syncytial virus in a human larynx carcinoma cell line. *Phytomedicine*, **16**, 882–886.

232. Ross, S.A., El Sayed, K.A., El Sohly, M.A. *et al.* (1997) Phytochemical analysis of *Geigeria alata* and *Francoeria crispa* essential oils. *Planta Med.*, **63**, 479–482.

233. Bishop, C.D. (1995) Antiviral activity of the essential oil of *Melaleuca alternifolia* (Maiden & Betche) Cheel (tea tree) against tobacco mosaic virus. *J. Essent. Oil Res.*, **7**, 641–644.

234. Chao, S.C., Young, D.G. and Oberg, C.J. (2000) Screening for inhibitory activity of essential oils on selected bacteria, fungi and viruses. *J. Essent. Oil Res.*, **12**, 639–649.

235. Meneses, R., Ocazioncz, R.E., Martínez, J.R. and Stashenko, E.E. (2009) Inhibitory effect of essential oils obtained from plants grown in Colombia on yellow fever virus replication in vitro. *Ann. Clin. Microbiol. Antimicrob.*, **8**, 8.

236. Loizzo, M.R., Saab, A.M., Tundis, R. *et al.* (2008) Phytochemical analysis and in vitro antiviral activities of the essential oils of seven Lebanon species. *Chem. Biodivers.*, **5**, 461–470.

237. Siddiqui, Y.M., Ettayebi, M., Haddad, A.M. and Al-Ahdal, M.N. (1996) Effect of essential oils on the enveloped viruses: antiviral activity of oregano and clove oils on herpes simplex virus type 1 and Newcastle disease virus. *Med. Sci. Res.*, **24**, 185–186.

238. Salem, M.L. and Hossain, M.S. (2000) Protective effect of black seed oil from *Nigella sativa* against murine cytomegalovirus infection. *Int. J. Immunopharmacol.*, **22**, 729–740.

239. Tkachenko, K.G. (2006) Antiviral activity of the essential oils of some *Heracleum* L. species. *J. Herbs Spices Med. Plants*, **12**, 1–12.

240. Carson, C.F., Ashton, L., Dry, L. *et al.* (2001) *Melaleuca alternifolia* (tea tree) oil gel (6%) for the treatment of recurrent herpes labialis. *J. Antimicrob. Chemother.*, **48**, 450–451.

241. Burke, B.E., Baillie, J.E. and Olson, R.D. (2004) Essential oil of Australian lemon myrtle (*Backhousia citriodora*) in the treatment of molluscum contagiosum in children. *Biomed. Pharmacother.*, **58**, 245–247.

242. Anthony, J.P., Fyfe, L. and Smith, H. (2005) Plant active components: a resource for antiparasitic agents? *Trends Parasitol.*, **21**, 462–468.

243. Ueda-Nakamura, T., Mendonça-Filho, R.R., Morgado-Díaz, J.A. *et al.* (2006) Antileishmanial activity of eugenol-rich essential oil from *Ocimum gratissimum*. *Parasitol. Int.*, **55**, 99–105.

244. Mikus, J., Harkenthal, M., Steverding, D. and Reichling, J. (2000) In vitro effect of essential oils and isolated mono- and sesquiterpenes on *Leishmania major* and *Trypanosoma brucei*. *Planta Med.*, **66**, 366–368.

245. Arruda, D.C., D'Alexandri, F.L., Katzin, A.M. and Uliana, S.R.B. (2005) Antileishmanial activity of the terpene nerolidol. *Antimicrob. Agents Chemother.*, **49**, 1679–1687.

246. Rosa, M.S.S., Mendonca-Filho, R.R., Bizzo, H.R. *et al.* (2003) Antileishmanial activity of a linalool-rich essential oil from *Croton cajucara*. *Antimicrob. Agents Chemother.*, **47**, 1895–1901.

247. Monzote, L., Montalvo, A.M., Scull, R. *et al.* (2007) Combined effect of the essential oil from *Chenopodium ambrosioides* and antileishmanial drugs on promastigotes of *Leishmania amazonensis*. *Rev. Inst. Med. Trop. Sao Paulo*, **49**, 257–260.

248. Monzote, L., García, M., Montalvo, A.M. *et al.* (2007) In vitro activity of an essential oil against *Leishmania donovani*. *Phytother. Res.*, **21**, 1055–1058.

249. Pedroso, R.B., Ueda-Nakamura, T., Filho, B.P.D. *et al.* (2006) Biological activities of essential oil obtained from *Cymbopogon citratus* on *Crithidia deanei*. *Acta Protozool.*, **45**, 231–240.

250. Holetz, F.B., Ueda-Nakamura, T., Dias Filho, B.P. *et al.* (2003) Effect of essential oil of *Ocimum gratissimum* on the trypanosomatid *Herpetomonas samuelpessoai*. *Acta Protozool.*, **42**, 269–275.

251. Santoro, G.F., Cardoso, M.G., Guimarães, L.G.L. *et al.* (2007) Anti-proliferative effect of the essential oil of *Cymbopogon citratus* (DC) Stapf (lemongrass) on intracellular amastigotes, bloodstream trypomastigotes and culture epimastigotes of *Trypanosoma cruzi* (Protozoa: Kinetoplastida). *Parasitology*, **134**, 1649–1656.

252. Santoro, G.F., das Graças Cardoso, M., Guimarães, L.G.L. *et al.* (2007) Effect of oregano (*Origanum vulgare* L.) and thyme (*Thymus vulgaris* L.) essential oils on *Trypanosoma cruzi* (Protozoa: Kinetoplastida) growth and ultrastructure. *Parasitol. Res.*, **100**, 783–790.

253. Santoro, G.F., Cardoso, M.G., Guimarães, L.G.L. *et al.* (2007) *Trypanosoma cruzi*: activity of essential oils from *Achillea millefolium* L., *Syzygium aromaticum* L. and *Ocimum basilicum* L. on epimastigotes and trypomastigotes. *Exp. Parasitol.*, **116**, 283–290.

254. Boyom, F.F., Ngouana, V., Zollo, P.H.A. *et al.* (2003) Composition and anti-plasmodial activities of essential oils from some Cameroonian medicinal plants. *Phytochemistry*, **64**, 1269–1275.

255. RodriguesGoulart, H., Kimura, E.A., Peres, V.J. *et al.* (2004) Terpenes arrest parasite development and inhibit biosynthesis of isoprenoids in *Plasmodium falciparum*. *Antimicrob. Agents Chemother.*, **48**, 2502–2509.

256. Moon, T., Wilkinson, J.M. and Cavanagh, H.M.A. (2006) Antiparasitic activity of two *Lavandula* essential oils against *Giardia duodenalis*, *Trichomonas vaginalis* and *Hexamita inflata*. *Parasitol. Res.*, **99**, 722–728.

257. Almeida, I., Alviano, D.S., Vieira, D.P. *et al.* (2007) Antigiardial activity of *Ocimum basilicum* essential oil. *Parasitol. Res.*, **101**, 443–452.

258. Grabensteiner, E., Arshad, N. and Hess, M. (2007) Differences in the in vitro susceptibility of mono-eukaryotic cultures of *Histomonas meleagridis*, *Tetratrichomonas gallinarum* and *Blastocystis* sp. to natural organic compounds. *Parasitol. Res.*, **101**, 193–199.

259. Zenner, L., Callait, M.P., Granier, C. and Chauve, C. (2003) In vitro effect of essential oils from *Cinnamomum aromaticum*, *Citrus limon* and *Allium sativum* on two intestinal flagellates of poultry, *Tetratrichomonas gallinarum* and *Histomonas meleagridis*. *Parasite*, **10**, 153–157.

260. Tchoumbougnang, F., Zollo, P.H.A., Dagne, E. and Mekonnen, Y. (2005) In vivo antimalarial activity of essential oils from *Cymbopogon citratus* and *Ocimum gratissimum* on mice infected with *Plasmodium berghei*. *Planta Med.*, **71**, 20–23.

261. Monzote, L., Montalvo, A.M., Scull, R. *et al.* (2007) Activity, toxicity and analysis of resistance of essential oil from *Chenopodium ambrosioides* after intraperitoneal, oral and intralesional administration in BALB/c mice infected with *Leishmania amazonensis*: a preliminary study. *Biomed. Pharmacother.*, **61**, 148–153.

262. Giannenas, I., Florou-Paneri, P., Papazahariadou, M. *et al.* (2003) Effect of dietary supplementation with oregano essential oil on performance of broilers after experimental infection with *Eimeria tenella*. *Arch. Anim. Nutr.*, **57**, 99–106.

263. Force, M., Sparks, W.S. and Ronzio, R.A. (2000) Inhibition of enteric parasites by emulsified oil of oregano in vivo. *Phytother. Res.*, **14**, 213–214.

264. Lopes, D., Bizzo, H.R., Sobrinho, A.F.S. and Pereira, M.V.G. (2000) Linalool-rich essential oil from leaves of *Croton cajucara* Benth.. *J. Essent. Oil Res.*, **12**, 705–708.

265. Swanson, K.M. and Hohl, R.J. (2006) Anti-cancer therapy: targeting the mevalonate pathway. *Curr. Cancer Drug Targets*, **6**, 15–37.

266. Craig, W.J. (1999) Health-promoting properties of common herbs. *Am. J. Clin. Nutr.*, **70**, S491–S499.

267. Mo, H. and Elson, C.E. (2004) Studies of the isoprenoid-mediated inhibition of mevalonate synthesis applied to cancer chemotherapy and chemoprevention. *Exp. Biol. Med.*, **229**, 567–585.

268. Clegg, R.J., Middleton, B., Bell, G.D. and White, D.A. (1980) Inhibition of hepatic cholesterol synthesis and S-3-hydroxy-3-methylglutaryl-CoA reductase by mono and bicyclic monoterpenes administered in vivo. *Biochem. Pharmacol.*, **29**, 2125–2127.

269. Clegg, R.J., Middleton, B., Bell, G.D. and White, D.A. (1982) The mechanism of cyclic monoterpene inhibition of hepatic 3-hydroxy-3-methylglutaryl coenzyme A reductase in vivo in the rat. *J. Biol. Chem.*, **257**, 2294–2299.

270. Peffley, D.M. and Gayen, A.K. (2003) Plant-derived monoterpenes suppress hamster kidney cell 3-hydroxy-3-methylglutaryl coenzyme A reductase synthesis at the post-transcriptional level 1. *J. Nutr.*, **133**, 38–44.

271. Meigs, T.E., Roseman, D.S. and Simoni, R.D. (1996) Regulation of 3-hydroxy-3-methylglutaryl-coenzyme A reductase degradation by the nonsterol mevalonate metabolite farnesol in vivo. *J. Biol. Chem.*, **271**, 7916–7922.

272. Meigs, T.E. and Simoni, R.D. (1997) Farnesol as a regulator of HMG-CoA reductase degradation: characterization and role of farnesyl pyrophosphatase. *Arch. Biochem. Biophys.*, **345**, 1–9.

273. Silva, S.L., Chaar, J.S., Figueiredo, P.M.S. and Yano, T. (2008) Cytotoxic evaluation of essential oil from *Casearia sylvestris* Sw on human cancer cells and erythrocytes. *Acta Amazon.*, **38**, 107–112.

274. Silva, S.L., Figueiredo, P.M. and Yano, T. (2007) Cytotoxic evaluation of essential oil from *Zanthoxylum rhoifolium* Lam. leaves. *Acta Amazon.*, **37**, 281–286.

275. Sylvestre, M., Pichette, A., Lavoie, S. *et al.* (2007) Composition and cytotoxic activity of the leaf essential oil of *Comptonia peregrina* (L.) Coulter. *Phytother. Res.*, **21**, 536–540.

276. Xiao, Y., Yang, F.Q., Li, S.P. *et al.* (2008) Essential oil of *Curcuma wenyujin* induces apoptosis in human hepatoma cells. *World J. Gastroenterol.*, **14**, 4309–4318.

277. Kilani, S., Ledauphin, J., Bouhlel, I. *et al.* (2008) Comparative study of *Cyperus rotundus* essential oil by a modified GC/MS analysis method. Evaluation of its antioxidant, cytotoxic, and apoptotic effects. *Chem. Biodivers.*, **5**, 729–742.

278. Cole, R.A., Bansal, A., Moriarity, D.M. *et al.* (2007) Chemical composition and cytotoxic activity of the leaf essential oil of *Eugenia zuchowskiae* from Monteverde, Costa Rica. *J. Nat. Med.*, **61**, 414–417.

279. Topçu, G., Gören, A.C., Bilsel, G. *et al.* (2005) Cytotoxic activity and essential oil composition of leaves and berries of *Juniperus excelsa*. *Pharm. Biol.*, **43**, 125–128.

280. Hou, J., Sun, T., Hu, J. *et al.* (2007) Chemical composition, cytotoxic and antioxidant activity of the leaf essential oil of *Photinia serrulata*. *Food Chem.*, **103**, 355–358.

281. Itani, W.S., El-Banna, S.H., Hassan, S.B. *et al.* (2008) Anti colon cancer components from lebanese sage (*Salvia libanotica*) essential oil: mechanistic basis. *Cancer Biol. Ther.*, **7**, 1765–1773.

282. Li, Y.L., Yeung, C.M., Chiu, L.C.M. *et al.* (2009) Chemical composition and antiproliferative activity of essential oil from the leaves of a medicinal herb, *Schefflera heptaphylla*. *Phytother. Res.*, **23**, 140–142.

283. Díaz, C., Quesada, S., Brenes, O. *et al.* (2008) Chemical composition of *Schinus molle* essential oil and its cytotoxic activity on tumour cell lines. *Nat. Prod. Res.*, **22**, 1521–1534.

284. Buhagiar, J.A., Podesta, M.T., Wilson, A.P. *et al.* (1999) The induction of apoptosis in human melanoma, breast and ovarian cancer cell lines using an essential oil extract from the conifer *Tetraclinis articulata*. *Anticancer Res.*, **19**, 5435–5443.

285. Ait M'Barek, L., Ait Mouse, H., Jaâfari, A. *et al.* (2007) Cytotoxic effect of essential oil of thyme (*Thymus broussonettii*) on the IGR-OV1 tumor cells resistant to chemotherapy. *Braz. J. Med. Biol. Res.*, **40**, 1537–1544.

286. Elson, C.E., Peffley, D.M., Hentosh, P. and Mo, H. (1999) Isoprenoid-mediated inhibition of mevalonate synthesis: potential application to cancer. *Exp. Biol. Med.*, **221**, 294–311.

287. Tatman, D. and Mo, H. (2002) Volatile isoprenoid constituents of fruits, vegetables and herbs cumulatively suppress the proliferation of murine B16 melanoma and human HL-60 leukemia cells. *Cancer Lett.*, **175**, 129–140.

288. Joo, J.H. and Jetten, A.M. (2009) Molecular mechanisms involved in farnesol-induced apoptosis. *Cancer Lett.*, **287**, 123–135.

289. Bardon, S., Picard, K. and Martel, P. (1998) Monoterpenes inhibit cell growth, cell cycle progression, and cyclin D1 gene expression in human breast cancer cell lines. *Nutr. Cancer*, **32**, 1–7.

290. Ferri, N., Arnaboldi, L., Orlandi, A. *et al.* (2001) Effect of S (–) perillic acid on protein prenylation and arterial smooth muscle cell proliferation. *Biochem. Pharmacol.*, **62**, 1637–1645.

291. Bardon, S., Foussard, V., Fournel, S. and Loubat, A. (2002) Monoterpenes inhibit proliferation of human colon cancer cells by modulating cell cycle-related protein expression. *Cancer Lett.*, **181**, 187–194.

292. Shi, W. and Gould, M.N. (2002) Induction of cytostasis in mammary carcinoma cells treated with the anticancer agent perillyl alcohol. *Carcinogenesis*, **23**, 131–142.

293. Elegbede, J.A., Flores, R. and Wang, R.C. (2003) Perillyl alcohol and perillaldehyde induced cell cycle arrest and cell death in BroTo and A549 cells cultured in vitro. *Life Sci.*, **73**, 2831–2840.

294. Wiseman, D.A., Werner, S.R. and Crowell, P.L. (2007) Cell cycle arrest by the isoprenoids perillyl alcohol, geraniol, and farnesol is mediated by p21Cip1 and p27Kip1 in human pancreatic adenocarcinoma cells. *J. Pharmacol. Exp. Ther.*, **320**, 1163–1170.

295. Mills, J.J., Chari, R.S., Boyer, I.J. *et al.* (1995) Induction of apoptosis in liver tumors by the monoterpene perillyl alcohol. *Cancer Res.*, **55**, 979–983.

296. Reddy, B.S., Wang, C.X., Samaha, H. *et al.* (1997) Chemoprevention of colon carcinogenesis by dietary perillyl alcohol. *Cancer Res.*, **57**, 420–425.

297. Ariazi, E.A., Satomi, Y., Ellis, M.J. *et al.* (1999) Activation of the transforming growth factor β signaling pathway and induction of cytostasis and apoptosis in mammary carcinomas treated with the anticancer agent perillyl alcohol 1. *Cancer Res.*, **59**, 1917–1928.

298. Burke, Y.D., Ayoubi, A.S., Werner, S.R. *et al.* (2002) Effects of the isoprenoids perillyl alcohol and farnesol on apoptosis biomarkers in pancreatic cancer chemoprevention. *Anticancer Res.*, **22**, 3127–3134.

299. Clark, S.S., Perman, S.M., Sahin, M.B. *et al.* (2002) Antileukemia activity of perillyl alcohol (POH): uncoupling apoptosis from G0/G1 arrest suggests that the primary effect of POH on Bcr/Abl-transformed cells is to induce growth arrest. *Leuk. Res.*, **16**, 213–222.

300. Ren, Z., Elson, C.E. and Gould, M.N. (1997) Inhibition of type I and type II geranylgeranyl-protein transferases by the monoterpene perillyl alcohol in NIH3T3 cells. *Biochem. Pharmacol.*, **54**, 113–120.

301. Cates, C.A., Michael, R.L., Stayrook, K.R. *et al.* (1996) Prenylation of oncogenic human PTP (CAAX) protein tyrosine phosphatases. *Cancer Lett.*, **110**, 49–55.

302. Haluska, P., Dy, G.K. and Adjei, A.A. (2002) Farnesyl transferase inhibitors as anticancer agents. *Eur. J. Cancer*, **38**, 1685–1700.

303. Cerda, S.R., Wilkinson, J., Thorgeirsdottir, S. and Broitman, S.A. (1999) R-(+)-perillyl alcohol-induced cell cycle changes, altered actin cytoskeleton, and decreased ras and p34cdc2 expression in colonic adenocarcinoma SW480 cells. *J. Nutr. Biochem.*, **10**, 19–30.

304. Lluria-Prevatt, M., Morreale, J., Gregus, J. *et al.* (2002) Effects of perillyl alcohol on melanoma in the TPras mouse model 1. *Cancer Epidemiol. Biomarkers Prev.*, **11**, 573–579.

305. Holstein, S.A. and Hohl, R.J. (2003) Monoterpene regulation of Ras and Ras-related protein expression. *J. Lipid Res.*, **44**, 1209–1215.

306. Hohl, R.J. and Lewis, K. (1995) Differential effects of monoterpenes and lovastatin on RAS processing. *J. Biol. Chem.*, **270**, 17508–17512.

307. Satomi, Y., Miyamoto, S. and Gould, M.N. (1999) Induction of AP-1 activity by perillyl alcohol in breast cancer cells. *Carcinogenesis*, **20**, 1957–1961.

308. Wu, C.C., Chung, J.G., Tsai, S.J. *et al.* (2004) Differential effects of allyl sulfides from garlic essential oil on cell cycle regulation in human liver tumor cells. *Food Chem. Toxicol.*, **42**, 1937–1947.

309. Arora, A., Kalra, N. and Shukla, Y. (2006) Regulation of p21/ras protein expression by diallyl sulfide in DMBA induced neoplastic changes in mouse skin. *Cancer Lett.*, **242**, 28–36.

310. Zheng, G.Q., Kenney, P.M. and Lam, L.K.T. (1992) Myristicin: a potential cancer chemopreventive agent from parsley leaf oil. *J. Agric. Food Chem.*, **40**, 107–110.

311. Kurkin, V.A. (2003) Phenylpropanoids from medicinal plants: distribution, classification, structural analysis, and biological activity. *Chem. Nat. Comp.*, **39**, 123–153.

312. Pinto, J.T., Qiao, C., Xing, J. *et al.* (2000) Alterations of prostate biomarker expression and testosterone utilization in human LNCaP prostatic carcinoma cells by garlic-derived S-allylmercaptocysteine. *Prostate*, **45**, 304–314.

313. Wu, C.C., Sheen, L.Y., Chen, H.W. *et al.* (2001) Effects of organosulfur compounds from garlic oil on the antioxidation system in rat liver and red blood cells. *Food Chem. Toxicol.*, **39**, 563–569.

314. Knowles, L.M. and Milner, J.A. (2001) Possible mechanism by which allyl sulfides suppress neoplastic cell proliferation. *J. Nutr.*, **131**, 1061–1066.

315. Oommen, S., Anto, R.J., Srinivas, G. and Karunagaran, D. (2004) Allicin (from garlic) induces caspase-mediated apoptosis in cancer cells. *Eur. J. Pharmacol.*, **485**, 97–103.

316. Shukla, Y. and Kalra, N. (2007) Cancer chemoprevention with garlic and its constituents. *Cancer Lett.*, **247**, 167–181.

317. Chan, A.S.L., Pang, H., Yip, E.C.H. *et al.* (2005) Carvacrol and eugenol differentially stimulate intracellular Ca2 +mobilization and mitogen-activated protein kinases in Jurkat T-Cells and monocytic THP-1 cells. *Planta Med.*, **71**, 634–639.

318. Yoo, C.B., Han, K.T., Cho, K.S. *et al.* (2005) Eugenol isolated from the essential oil of *Eugenia caryophyllata* induces a reactive oxygen species-mediated apoptosis in HL-60 human promyelocytic leukemia cells. *Cancer Lett.*, **225**, 41–52.

319. Han, E.H., Hwang, Y.P., Jeong, T.C. *et al.* (2007) Eugenol inhibit 7, 12 dimethylbenz [a] anthracene-induced genotoxicity in MCF-7 cells: Bifunctional effects on CYP1 and NAD (P) H: quinone oxidoreductase. *FEBS Lett.*, **581**, 749–756.

320. Carrasco, A.H., Espinoza, C.L., Cardile, V. *et al.* (2008) Eugenol and its synthetic analogues inhibit cell growth of human cancer cells (Part I). *J. Braz. Chem. Soc.*, **19**, 543–548.

321. Tabanca, N., Ma, G., Pasco, D.S. *et al.* (2007) Effect of essential oils and isolated compounds from *Pimpinella* species on NF-kB: a target for antiinflammatory therapy. *Phytother. Res.*, **21**, 741–745.

322. Ka, H., Park, H.J., Jung, H.J. *et al.* (2003) Cinnamaldehyde induces apoptosis by ROS-mediated mitochondrial permeability transition in human promyelocytic leukemia HL-60 cells. *Cancer Lett.*, **196**, 143–152.

323. Imai, T., Yasuhara, K., Tamura, T. *et al.* (2002) Inhibitory effects of cinnamaldehyde on 4-(methylnitrosamino)-1-(3-pyridyl)-1-butanone-induced lung carcinogenesis in rasH 2 mice. *Cancer Lett.*, **175**, 9–16.

324. Wu, S.J., Ng, L.T. and Lin, C.C. (2005) Cinnamaldehyde-induced apoptosis in human PLC/PRF/5 cells through activation of the proapoptotic Bcl-2 family proteins and MAPK pathway. *Life Sci.*, **77**, 938–951.

325. Wu, S.J. and Ng, L.T. (2007) MAPK inhibitors and pifithrin-alpha block cinnamaldehyde-induced apoptosis in human PLC/PRF/5 cells. *Food Chem. Toxicol.*, **45**, 2446–2453.

326. Henry-Mowatt, J., Dive, C., Martinou, J.C. and James, D. (2004) Role of mitochondrial membrane permeabilization in apoptosis and cancer. *Oncogene*, **23**, 2850–2860.

327. Usta, J., Kreydiyyeh, S., Bajakian, K. and Nakkash-Chmaisse, H. (2002) In vitro effect of eugenol and cinnamaldehyde on membrane potential and respiratory chain complexes in isolated rat liver mitochondria. *Food Chem. Toxicol.*, **40**, 935–940.

328. Issa, A.Y., Volate, S.R. and Wargovich, M.J. (2006) The role of phytochemicals in inhibition of cancer and inflammation: new directions and perspectives. *J. Food Compost. Anal.*, **19**, 405–419.

329. Di Sotto, A., Evandri, M.G. and Mazzanti, G. (2008) Antimutagenic and mutagenic activities of some terpenes in the bacterial reverse mutation assay. *Mutat. Res.*, **653**, 130–133.

330. Sanyal, R., Darroudi, F., Parzefall, W. *et al.* (1997) Inhibition of the genotoxic effects of heterocyclic amines in human derived hepatoma cells by dietary bioantimutagens. *Mutagenesis*, **12**, 297–303.

331. Shaughnessy, D.T., Schaaper, R.M., Umbach, D.M. and DeMarini, D.M. (2006) Inhibition of spontaneous mutagenesis by vanillin and cinnamaldehyde in *Escherichia coli*: dependence on recombinational repair. *Mutat. Res.*, **602**, 54–64.

332. King, A.A., Shaughnessy, D.T., Mure, K. *et al.* (2007) Antimutagenicity of cinnamaldehyde and vanillin in human cells: global gene expression and possible role of DNA damage and repair. *Mutat. Res.*, **616**, 60–69.

333. Kerbel, R. and Folkman, J. (2002) Clinical translation of angiogenesis inhibitors. *Nat. Rev. Cancer*, **2**, 727–739.

334. Kesisis, G., Broxterman, H. and Giaccone, G. (2007) Angiogenesis inhibitors. Drug selectivity and target specificity. *Curr. Pharm. Des.*, **13**, 2795–2809.

335. Loutrari, H., Hatziapostolou, M., Skouridou, V. *et al.* (2004) Perillyl alcohol is an angiogenesis inhibitor. *J. Pharmacol. Exp. Ther.*, **311**, 568–575.

336. Lu, X.G., Zhan, L.B., Feng, B.A. *et al.* (2004) Inhibition of growth and metastasis of human gastric cancer implanted in nude mice by d-limonene. *World J. Gastroenterol.*, **10**, 2140–2144.

337. Loutrari, H., Magkouta, S., Pyriochou, A. *et al.* (2006) Mastic oil from *Pistacia lentiscus* var. *chia* inhibits growth and survival of human K562 leukemia cells and attenuates angiogenesis. *Nutr. Cancer*, **55**, 86–93.

338. Bailey, H.H., Attia, S., Love, R.R. *et al.* (2008) Phase II trial of daily oral perillyl alcohol (NSC 641066) in treatment-refractory metastatic breast cancer. *Cancer Chemother. Pharmacol.*, **62**, 149–157.

339. Meadows, S.M., Mulkerin, D., Berlin, J. *et al.* (2002) Phase II trial of perillyl alcohol in patients with metastatic colorectal cancer. *Int. J. Gastrointest. Cancer*, **32**, 125–128.

340. Bailey, H.H., Levy, D., Harris, L.S. *et al.* (2002) A phase II trial of daily perillyl alcohol in patients with advanced ovarian cancer: Eastern Cooperative Oncology Group Study E2E96. *Gynecol. Oncol.*, **85**, 464–468.

341. Liu, G., Oettel, K., Bailey, H. *et al.* (2003) Phase II trial of perillyl alcohol (NSC 641066) administered daily in patients with metastatic androgen independent prostate cancer. *Invest. New Drugs*, **21**, 367–372.

342. da Fonseca, C.O., Schwartsmann, G., Fischer, J. *et al.* (2008) Preliminary results from a phase I/II study of perillyl alcohol intranasal administration in adults with recurrent malignant gliomas. *Surg. Neurol.*, **70**, 259–266.

343. Dorai, T. and Aggarwal, B.B. (2004) Role of chemopreventive agents in cancer therapy. *Cancer Lett.*, **215**, 129–140.

344. Ghosh, D., Skinner, M. and Ferguson, L.R. (2006) The role of the Therapeutic Goods Administration and the Medicine and Medical Devices Safety Authority in evaluating complementary and alternative medicines in Australia and New Zealand. *Toxicology*, **221**, 88–94.

345. Forte, J.S. and Raman, A. (2000) Regulatory issues relating to herbal products. Part 1. Legislation in the European Union, North America, and Australia. *J. Med. Food*, **3**, 23–39.

346. Barnes, J. (2003) Quality, efficacy and safety of complementary medicines: fashions, facts and the future. Part I. Regulation and quality. *Br. J. Clin. Pharmacol.*, **55**, 226–233.

347. Beerling, J., Meakins, S. and Small, L. (2002) Eucalyptus oil products. Formulations and legislation, in *Eucalyptus: The Genus Eucalyptus* (ed. J.J.W. Coppen), Taylor & Francis, London, UK, pp. 345–364.

348. Margetts, J. (2005) Aroma chemicals V: natural aroma chemicals, in *Chemistry and Technology of Flavors and Fragrances* (ed. D.J. Rowe), Wiley-Blackwell, pp. 169–198.

349. Taylor, B. (2005) Other beverage ingredients, in *Chemistry and Technology of Soft Drinks and Fruit Juices* (ed. P.R. Ashurst), Wiley-Blackwell, pp. 90–128.

350. Ahmad, M.M. and Rehman, S. (2006) Sensory evaluation of citrus peel essential oils as flavouring agents in various food products. *J. Agric. Res.*, **44**, 325–333.

351. Schwab, W., Davidovich-Rikanati, R. and Lewinsohn, E. (2008) Biosynthesis of plant-derived flavor compounds. *Plant J.*, **54**, 712–732.

352. Serra, S., Fuganti, C. and Brenna, E. (2005) Biocatalytic preparation of natural flavours and fragrances. *Trends Biotechnol.*, **23**, 193–198.

353. Batish, D.R., Singh, H.P., Kohli, R.K. and Kaur, S. (2008) Eucalyptus essential oil as a natural pesticide. *For. Ecol. Manage.*, **256**, 2166–2174.

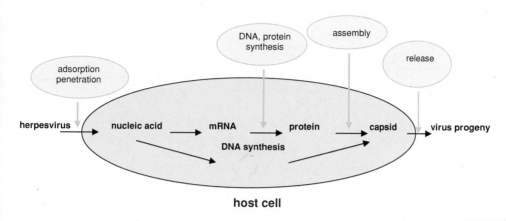

Figure 10.1 *Life cycle of HSV and possible antiviral targets.*

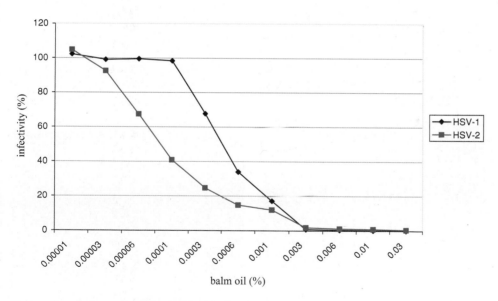

Figure 10.2 *Determination of IC$_{50}$ of lemon-balm oil on HSV-1 and HSV-2. Viruses were incubated for one hour at room temperature with increasing concentrations of lemon balm oil and immediately tested in a plaque-reduction assay. The percentage reduction was calculated relative to the amount of virus in the 1% ethanol-treated virus control. Experiments were repeated independently and data presented are the mean of three experiments. (Reprinted with permission from [50]. Copyright Elsevier.)*

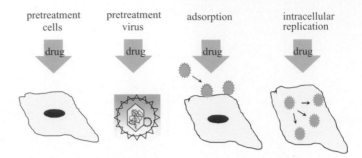

Figure 10.4 *Illustration of time-on-addition experiments during the HSV-replication cycle. Cells were pretreated with the essential oil prior to virus infection (pretreatment cells), viruses were pretreated prior to infection (pretreatment virus), the essential oil was added during the adsorption period (adsorption) or after penetration of the viruses into cells (intracellular replication). (Reprinted with permission from [50]. Copyright Elsevier.)*

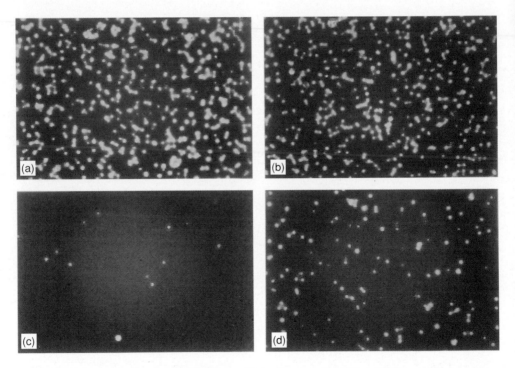

Figure 10.5 *Antiviral activity of* Melissa officinalis *essential oil on HSV-1 at different times during viral replication. (a) Untreated control. Lemon balm oil was added to cells prior to infection (b) and to virus prior to infection (c) and during the adsorption period (d). Infected cell cultures were incubated for three days, then fixed and stained with crystal violet to visualize plaques. (Reprinted with permission from [50]. Copyright Elsevier.)*

10

Antiviral Effects of Plant-Derived Essential Oils and Pure Oil Components

Paul Schnitzler[1], Akram Astani[1] and Jürgen Reichling[2]
[1]*Department of Infectious Diseases, Virology, University of Heidelberg, Heidelberg, Germany*
[2]*Department of Biology, Institute of Pharmacy and Molecular Biotechnology, University of Heidelberg, Heidelberg, Germany*

10.1 Introduction

Various kinds of essential oil have been used for the treatment of respiratory infections, asthma, dermatitis and gastrointestinal diseases [1]. Antibacterial and antifungal activities

---

of essential oils have been demonstrated against Gram-positive bacteria, Gram-negative bacteria and different fungi [2–4]. Furthermore, essential oils of different aromatic plant origins and various essential oil constituents have demonstrated antiviral properties against enveloped RNA and DNA viruses. Only a few non-enveloped viruses were included in these studies. In folk medicine, herpes labialis is traditionally treated with essential-oil preparations. Consequently, laboratories have focussed on testing essential oils against the enveloped herpes simplex viruses type 1 (HSV-1) and type 2 (HSV-2) [3, 5–9].

10.2   Characterization and Medicinal Use of Essential Oils

Essential oils are complex aromatic-smelling mixtures of various compounds with low molecular weights and diverse chemical structures. Predominant compounds are monoterpene hydrocarbons, sesquiterpene hydrocarbons, their corresponding oxidized products (e.g. alcohols, aldehydes, ethers and ketones), several phenylpropene derivatives, and miscellaneous volatile organic compounds (e.g. octanal, dodecanal, 2-undecanone). Essential oils are widely distributed in certain plant families, such as Alliaceae, Apiaceae, Asteraceae, Brassicaceae, Lamiaceae, Myrtaceae and Rutaceae. They are biosynthesized by the blossoms, leaves, fruits and roots of different aromatic plants and stored in special tissues such as glandular hairs, oil cells, oil receptacles and oil ducts. For medical and commercial use, essential oils are derived from plant material by hydro- or steam distillation. Essential oils are reported to reveal antibacterial, antifungal, antiviral, antiinflammatory, antirheumatic, antitussive, expectorant, immunomodulatory and sedative effects [3]. They may also act on cognition, memory and mood. The means of application depend on the pathophysiology, the desired outcome, safety and toxicity. For treatment of respiratory symptoms and nervous disorders, inhalation may be the best means of application, whereas topical application is the best way of treating skin diseases, including herpes labialis and herpes genitalis. A problem concerning the cutaneous use of essential oils is the degree of skin and mucous-membrane irritation. Recently, in view of a topic application of essential oils and to assess the possible irritation potential of these essential oils on skin and mucous membranes, their irritation threshold concentration was determined *in vitro* using a modified hen's egg test, chorioallontois membrane (HET-CAM) assay. In view of the results obtained by *in vitro* and *in vivo* experiments, it is commonly recommended by practitioners that essential oils are used only in diluted forms for external application [1, 2, 10].

10.3   Cytotoxicity of Essential Oils

Essential oils are screened for antiviral activity in cell cultures, such as the human tumour cell line HeLa and the monkey kidney cell lines Vero and RC-37. In order to ascertain that an observed reduction in virus infectivity is due to a direct effect of the essential oil on the virus and not to a toxic effect on the host cells, it is necessary to determine the cytotoxicity of essential oils. A procedure commonly used to evaluate cytotoxicity in cell cultures is the standard neutral red assay [6]. This assay determines the number of viable cells after their exposure to serial dilutions of a toxic substance by quantifying the uptake of neutral red dye by the cells in culture. The concentration of the drug which reduces the viable cell number by half is determined from a dose–response curve by optical density measurements and is defined as the 50% cytotoxic concentration (CC_{50}). The same method is used to determine the maximum noncytotoxic concentration of an essential oil. The ratio of CC_{50} to the effective antiviral concentration which inhibits viral infectivity by 50% (IC_{50}) is defined as the the the selectivity index and is a measure of the potential therapeutic applicability of the oil.

10.4   Antiviral Activities of Essential oils

10.4.1   *In Vitro* and *In Vivo* Studies of Antiherpesvirus Activity of Essential Oils

HSV-1 and HSV-2 are ubiquitous pathogens that cause a wide variety of infections, ranging from mild to severe, which might under certain circumstances develop to life-threatening diseases [11–13]. Herpesviruses represent a serious cause of morbidity, especially in immunocompromised patients [14]. HSV-1 predominantly causes epidermal lesions in and around the oral cavity [15]. Genital herpes is a chronic, persistent infection mainly caused by HSV-2 spreading efficiently and silently as a sexually transmitted disease (STD) through the population [16]. The hallmark of a herpes infection is the ability of the virus to establish a latent infection in the nervous system, to reactivate and to cause recrudescent lesions [17]. Herpetic infections proceed from a primary lytic infection in the periphery to a lifelong state of latency in sensory neurons, characterized by episodes of virus reactivation and lytic infection [12]. The latent virus is reactivated spontaneously or is induced to reactivate by a variety of stimuli.

Antiviral agents currently licensed for the treatment of herpesvirus infections include acyclovir and other inhibitors of the viral DNA polymerase, for example ganciclovir, foscarnet and cidofovir [18]. However, acyclovir-resistant herpesviruses have been increasingly isolated, particularly from immunocompromised hosts, such as patients with AIDS or malignancy and recipients of bone marrow or organ transplantation [19, 20]. The emergence of virus strains resistant to commonly used antiherpesvirus drugs is a growing problem. Given the need for the development of more antiviral drugs with diverse modes of action, the medicinal plants are regarded as promising sources of novel antivirals [8, 21, 22]. Thus the identification and development of novel antivirals with a mode of action different from the inhibition of DNA synthesis by nucleoside analogues, such as acyclovir, is highly desired (Figure 10.1). Such drugs could interfere with any of the steps required for HSV infection, such as virus attachment, penetration, intracellular replication and virus release from the infected cell [23]. Plant-derived antivirals with virucidal activity or which inhibit virus entry could prove valuable therapeutically, although their use would be limited mainly to topical application. However, antivirals that inhibit HSV replication after the virus has entered the cell might have far wider therapeutic implications.

Traditionally, medicinal plants have been used for the treatment of different kinds of ailments, including infectious diseases caused by bacteria, fungi and viruses, and their activity has been reported previously [24–31]. In many developing countries, traditional medicine is still used to cover major health-care needs and most drugs are derived from plants. [21, 32]. Plants have been utilized for the isolation of novel bioactive compounds as they synthesize a vast number of chemical compounds with complex structures. Natural products, either as pure compounds or as standardized plant extracts, provide unlimited

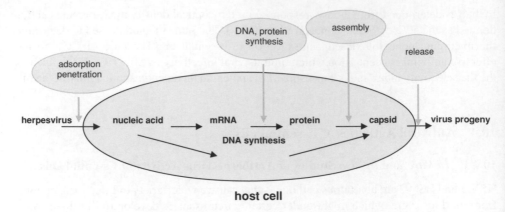

Figure 10.1   *Life cycle of HSV and possible antiviral targets. (See color figure).*

opportunities for new antiviral drugs, since their chemical diversity provides unmatched availability. Natural products are major sources of innovative therapeutic agents for various conditions, including infectious diseases. Viral infections remain an important worldwide problem, since many viruses have resisted prophylaxis or therapy longer than other microorganisms [33]. At the moment, only a few effective antiviral drugs are available for the treatment of viral diseases. There is a need to find new substances with not only intracellular but also extracellular antiviral properties. The methods commonly used for the evaluation of *in vitro* antiviral activities of synthetic and natural substances are based mainly on the inhibition of cytopathic effects, the reduction or inhibition of plaque formation, and reduction in the virus yield, but also on other viral functions in selected host-cell cultures. A large number of antiherpes screening experiments on medicinal plant extracts and plant-derived secondary metabolites (e.g. flavonoids, anthraquinones, naphthodianthrones, phenolics) have been reported [34–36], and antiherpes activity has been demonstrated for several essential oils of different plant origins (Table 10.1) and of various essential-oil constituents (Table 10.2) [5,8,22,39,41,47–49]. The potential antiviral effects against HSV of different essential oils were analysed in a plaque-reduction assay in six-well culture plates. Serial dilutions of essential oils were incubated with virus for one hour at room temperature and then adsorbed to cells. After adsorption, the remaining inoculum was removed and the infected cells were overlayed with medium containing 0.5% methylcellulose. By reference to the number of plaques observed in virus-infected controls, treated with 1% ethanol but without addition of essential oil, IC_{50} was determined from dose–response curves. A potential antiviral effect of balm oil against HSV-1 and HSV-2 was examined by plaque-reduction assays after incubation of the viruses with serial dilutions of lemon balm oil before infection of cells. Since the initial dilution of the essential oil was always performed in ethanol and all assays contained 1% ethanol final concentration, additional tubes containing virus mixed with 1% ethanol were used as controls. After adsorption of these pretreated viruses to cells, the remaining inoculum was removed and the infected cells were overlaid with medium containing methylcellulose. A clearly concentration-dependent activity was demonstrated for both types of herpesvirus in dose–response curves (Figure 10.2). Ethanol at a final concentration of 1% had no effect on virus titers. The IC_{50} of balm

Table 10.1 *Antiviral activity of different essential oils against HSV-1 and HSV-2.*

Virus	Source of oil/common name of oil	Inhibitory concentration IC_{50}[a]	Reference
HSV-1	*Aloysia gratissima*/whitebrush/common Beebrush	0.0065%	[37]
	Artemisia arborescen/great mugwort	0.24%	[38]
	Artemisia douglasiana/Californian mugwort	0.0083%	[37]
	Cinnamonum verum/cinnamon leaf	0.008%	[5]
	Cymbopogon citratus/lemongrass	0.1%	[39]
	Eucalyptus globulus/eucalyptus	0.009%	[6]
	Eupatorium patens/Little Joe	0.0125%	[37]
	Illicium verum/star anise	0.004%	[40]
	Juniperus oxycedrus/cade	0.02%	[41]
	Lavandula latifolia/lavender	1%	[37]
	Leptospermum scoparium/manuka	0.0001%	[42]
	Melissa officinalis/lemon balm	0.0004%	[50]
	Matricaria recutita/German chamomile	0.000 03%	[40]
	Melaleuca alternifolia/tea tree	0.0009%	[6]
	Mentha x piperita/peppermint	0.002%	[7]
	Origanum majorana/sweet majoram	1%	[37]
	Pinus mugo/dwarf pine	0.0007%	[40]
	Rosmarinus officinalis/rosemary	1%	[39]
	Santolina insularis/Santolina	0.0001%	[43]
	Tessaria absinthioides	0.0105%	[37]
HSV-2	*Artemisia arborescens*/great mugwort	0.41%	[38]
	Eucalyptus globulus/eucalyptus	0.008%	[6]
	Hyssopus officinalis/hyssop	0.0006%	[44]
	Illicium verum/star anise	0.003%	[40]
	Leptospermum scoparium/manuka	0.000 06%	[42]
	Melissa officinalis/lemon balm	0.000 08%	[50]
	Matricaria recutita/German chamomile	0.000 15%	[40]
	Melaleuca alternifolia/tea tree	0.0008%	[6]
	Mentha x piperita/peppermint	0.0008%	[7]
	Pinus mugo/dwarf pine	0.0007%	[40]
	Santalum album/sandalwood	0.0005%	[44]
	Santolina insularis/Santolina	0.0001%	[43]
	Thymus vulgaris/thyme	0.0007%	[44]
	Zingiber officinale/ginger	0.0001%	[44]

[a]IC_{50}: essential oil concentration (%) that reduces virus infectivity by 50%.

oil was determined at 0.0004 and 0.000 08% for HSV-1 and HSV-2, respectively, presented as a percentage of the virus control. Balm oil inhibited plaque formation of HSV-1 and HSV-2 in a dose-dependent manner [50]. At a concentration of 0.002% balm oil, the titres of HSV-1 and HSV-2 were reduced by 98.8 and 97.2%, respectively. Higher concentrations of the essential oil abolished viral infectivity nearly completely. Prior to application to the cell monolayer, the higher concentration of the essential oil was diluted to reach noncytotoxic levels.

Table 10.2   *Antiherpesvirus activity of typical pure constituents of essential oils against HSV-1 and HSV-2 measured by plaque-reduction assay.*

Compound	Essential oils[a]	Virus	IC$_{50}$[b]	Reference
Monoterpenes				
α-terpinene	tea tree	HSV-1	0.009 mg/ml	[9]
γ-terpinene	tea tree	HSV-1	0.008 mg/ml	[9]
α-pinene	dwarf mountain pine	HSV-1	0.005 mg/ml	[9]
β-pinene	dwarf mountain pine	HSV-1	0.004 mg/ml	[9]
limonene	common lime; Mexican lime	HSV-1	0.006 mg/ml	[9]
p-cymene	lemon; rosemary	HSV-1	0.016 mg/ml	[9]
terpinen-4-ol	tea tree	HSV-1	0.060 mg/ml	[9]
α-terpineol	tea tree; Caeput tree	HSV-1	0.022 mg/ml	[9]
thymol	thyme	HSV-1	0.030 mg/ml	[9]
citral	lemon balm	HSV-1	0.004 mg/ml	[9]
1,8-cineol	eucalyptus; rosemary	HSV-1	1.2 mg/ml	[9]
borneol	rosemary; common sage	HSV-2	16.0 mg/ml	[5]
1,8-cineol	eucalyptus; rosemary	HSV-2	6.9 mg/ml	[5]
citral	lemon balm	HSV-2	0.51 mg/ml	[5]
geraniol	lemon balm	HSV-2	7.0 mg/ml	[5]
limonene	lemon	HSV-2	3.61 mg/ml	[5]
linalool	lemon	HSV-2	1.38 mg/ml	[5]
menthol	peppermint	HSV-2	7.2 mg/ml	[5]
thymol	thyme	HSV-2	0.8 mg/ml	[5]
Sesquiterpenes				
bisabolol	German chamomile	HSV-1	n.d.[c]	Astani, unpubl.[d]
eudesmol	ginger	HSV-1	0.004 mg/ml	Astani, unpubl.
Phenylpropanoids				
curcumin	wild tumeric	HSV-2	0.32 mg/ml	[5]
eugenol	clove	HSV-2	1.5 mg/ml	[5]
eugenol	clove	HSV-2	16.2 µg/ml	[45]
safrol	ague tree; sassafras	HSV-2	3.1 mg/ml	[5]
eugenol	clove	HSV-1	25.6 µg/ml	[45]
Miscellaneous				
capryl aldehyde	fishwort; heartleaf	HSV-1	0.0004%	[57]
lauryl aldehyde	fishwort; heartleaf	HSV-1	0.0008%	[57]
methyl nonyl ketone	fishwort; heartleaf	HSV-1	0.009%	[57]

[a]The essential oils of which the compound is a major constituent.
[b]IC$_{50}$: essential oil concentration (%) that reduces virus infectivity by 50%.
[c]n.d.: not detected.
[d]unpubl.: unpublished results.

Essential oils from star anise (*Illicium verum*), dwarf pine (*Pinus mugo*), manuka (*Leptospermum scoparium*), chamomile (*Matricaria recutita*) and tea tree (*Melaleuca alternifolia*) have been investigated against acyclovir-sensitive and different acyclovir-resistant HSV-1 strains [40, 42, 44, 51, 52]. These essential oils were shown to strongly inhibit infectivity of drug-sensitive and drug-resistant HSV when viruses were incubated with essential oils prior to infection of cell cultures (Figure 10.3). This unique feature makes them promising plant-derived antivirals in the setting of resistance against synthetic drugs.

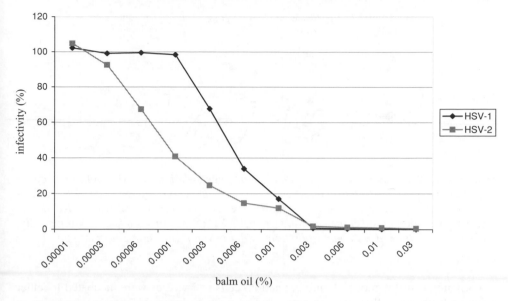

Figure 10.2 *Determination of IC$_{50}$ of lemon-balm oil on HSV-1 and HSV-2. Viruses were incubated for one hour at room temperature with increasing concentrations of lemon balm oil and immediately tested in a plaque-reduction assay. The percentage reduction was calculated relative to the amount of virus in the 1% ethanol-treated virus control. Experiments were repeated independently and data presented are the mean of three experiments. (Reprinted with permission from [50]. Copyright Elsevier.) (See color figure).*

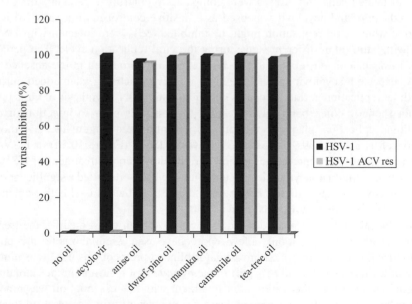

Figure 10.3 *Inhibition of acyclovir-sensitive and acyclovir-resistant HSV by acyclovir, anise oil, dwarf-pine oil, manuka oil, chamomile oil and tea-tree oil.*

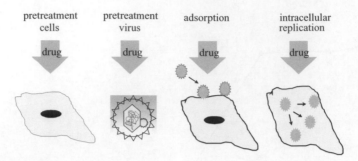

Figure 10.4 *Illustration of time-on-addition experiments during the HSV-replication cycle. Cells were pretreated with the essential oil prior to virus infection (pretreatment cells), viruses were pretreated prior to infection (pretreatment virus), the essential oil was added during the adsorption period (adsorption) or after penetration of the viruses into cells (intracellular replication). (Reprinted with permission from [50]. Copyright Elsevier.) (See color figure).*

For analysis of the mode of antiviral action, cells and viruses were incubated together during adsorption, cells were pretreated with extracts before viral infection, and viruses were incubated with extracts before cell infection or after penetration of the virus into the host cells (Figure 10.4) [50]. After three days of incubation the monolayers were fixed with 10% formalin and stained with 1% crystal violet, and plaques were counted. The extracts were used at their maximum noncytotoxic concentrations. In all experiments untreated virus-infected cells were used as controls. Untreated controls always contained 1% ethanol in order to exclude any effect of ethanol on cells or viruses. The number of plaques of treated cells and viruses were compared to controls to calculate the extent of plaque reduction, and acyclovir was used as a positive control in all assays. To identify the step at which viral replication might be inhibited, cells were infected with HSV and treated with balm oil at different steps during the viral-replication cycle. The percentage reduction was calculated relative to the amount of virus produced in the absence of the drug. As expected, acyclovir showed the highest antiviral activity when added during the intracellular replication period, since it is known to inhibit DNA synthesis of virus progeny (data not shown). When balm oil was added to host cells prior to infection or during intracellular replication, plaque formation was not influenced (Figure 10.5). A moderate reduction of 64.8% and 39.9% was detected when HSV-1 (Figure 10.5d) or HSV-2 was treated with balm oil during the adsorption period. However, pretreatment of HSV-1 and HSV-2 with maximum noncytotoxic concentrations of balm oil caused a significant decline in viral infectivity (Figure 10.5c). In summary, balm essential oil is effective against free HSV particles and exhibits virucidal activity [50].

Many clinical trials with plant-derived extracts have been performed for the treatment of herpes labialis [35, 53], but only a few reports have used essential oils for this purpose. A randomized, placebo-controlled, investigator-blinded protocol was used to evaluate the efficacy of tea-tree essential oil (6% gel) in the treatment of recurrent herpes labialis [27]. The median time to reepithelization after treatment with this essential oil was nine days, compared to 12.5 days after placebo, indicating some benefit from essential-oil treatment. Tea-tree oil might be a potentially useful cheaper alternative for other topical therapies,

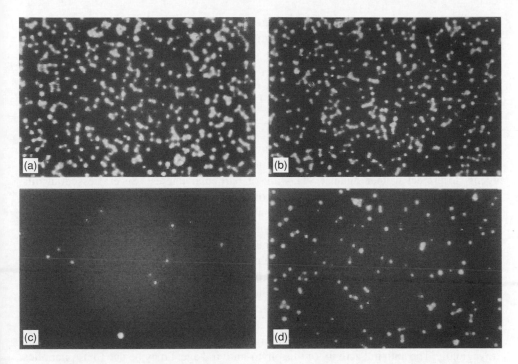

Figure 10.5 Antiviral activity of Melissa officinalis *essential oil on HSV-1 at different times during viral replication. (a) Untreated control. Lemon balm oil was added to cells prior to infection (b) and to virus prior to infection (c) and during the adsorption period (d). Infected cell cultures were incubated for three days, then fixed and stained with crystal violet to visualize plaques. (Reprinted with permission from [50]. Copyright Elsevier.) (See color figure).*

and poses little threat of inducing resistance as do the synthetic antiviral agents. Besides essential oils, many other herbal preparations with antiviral activity have been identified. Clinical trials have been performed and most describe benefits for the treated patients [35]. There remains a need for larger, stringently designed, randomized clinical trials with essential oils to provide conclusive evidence of their efficacy.

10.4.2 *In Vitro* and *In Vivo* Studies of Antiherpesvirus Activity of Pure Essential Oil Components

Essential oils are mixtures of about 50 or more volatile substances. In recent years, several studies have demonstrated that structurally diverse components of essential oils contribute to their overall antiherpesvirus activity [5, 8, 9, 46]. Like essential oils, the pure isolated essential-oil components exert their antiviral effect by a direct interaction with the cell-free virions. Intracellular viruses treated with the pure compounds are not affected. The tested compounds, including monoterpenes, sesquiterpenes and phenylpropanoids, exhibit variable antiviral activities against HSV-1 and HSV-2 (Table 10.2).

Essential oils, as well as single compounds of essential oils, could be used for topical treatments, particularly in combination with other drugs such as acyclovir. Recently it was reported [45] that acyclovir and eugenol, the main phenolic compound of clove oil, synergistically inhibited HSV-1 replication *in vitro*, which agrees with the fact that both compounds exert their antiviral effects on different viral targets – acylovir on DNA replication and eugenol on the virus envelope. In addition, pure eugenol was reported to inactivate HSV directly [49]. Furthermore, in a mouse model, eugenol delayed the development of herpesvirus-induced keratitis [45]. Isoborneol, a monoterpene alcohol and a component of several plant essential oils, showed effect against HSV-1 and specifically inhibited glycosylation of viral proteins [54]. The application of 1,8-cineole protected mice against infection with HSV-2 [5]. STDs are a worldwide health problem. The development of safe, inexpensive and effective topical microbicides remains a challenge. Intravaginal agents to protect women against STDs are of particular interest. Several essential oils, isolated oil components and some other secondary metabolites derived from herbal drugs were screened against HSV-2 *in vitro* using a plaque-reduction assay [5]. The most active substances, cineol, curcumin and eugenol, with IC_{50} values equal to or less than 7.0 mg/ml (Table 10.2) were tested for efficacy in a mouse model of genital HSV-2 infection. Cineol and eugenol were tested undiluted and curcumin in a concentration of 100 mg/ml, which is the limit of solubility. All compounds tested revealed significant protection against diseases caused by intravaginal HSV-2 challenge. The most effective compound was eugenol. In eugenol-treated female mice with HSV-2 infection, there was a significant reduction of virus titre (14 of 16 mice) in comparison to the control group (2 of 15 mice). None of the tested compounds caused apparent vaginal irritation in the concentration administered intravaginally. The antiviral activity of this phyto-antiviral agent may be due to its phenolics. Formerly, it could be demonstrated that phenolics are able to damage protein in envelopes of newly synthesized HSV virions [55].

10.4.3   *In Vitro* and *In Vivo* Studies of Antiviral Effects of Essential Oils against Other Viruses

The activity of essential oils from various aromatic plant origins has been explored against other enveloped DNA and RNA viruses besides HSV. Only a few non-enveloped viruses have been studied. The experimental methods used are similar to those described above. Non-enveloped adenoviruses (ADV) and enveloped mumps viruses (MV) were isolated from patients with respiratory tract infection. Essential oil from *Eucalyptus globulus* (eucalyptus), traditionally used to treat pharyngitis, bronchitis and sinusitis, revealed a mild antiviral activity against MV. Using the maximum noncytotoxic concentration of 0.25 µg/ml, plaque formation was reduced by about 40%. On the other hand, ADV was not affected by eucalyptus oil [56]. In a former experiment, an antiherpesvirus activity of eucalyptus oil was demonstrated [6]. Because HSV and MV are enveloped viruses, it has been shown that enveloped viruses respond sensitively to essential oils.

This hypothesis is sustained by Hayashi and coworkers [57], who investigated the virucidal potential of the essential oil derived from *Houttuynia cordata* against the enveloped viruses HSV-1, influenza virus (INF) and human immunodeficiency virus type 1 (HIV-1) as well as against the non-enveloped viruses poliovirus and coxsackievirus. *H. cordata* is a medicinal plant traditionally used in Japan and China for its antiinflammatory and antioedematous activities. The plant-derived essential oil was found to exhibit direct virucidal activity against HSV-1, INF (Table 10.3) and HIV-1, but not against poliovirus and

Table 10.3   Antiviral activity of essential oils extracted from different plant sources.

Virus	Source of oil/common name of oil	Inhibitory concentration	References
YFV[a]	*Lippia alba*/ginger grass, white lippia, bushy lippia	MIC:[b] 3.7 µg/ml	[58]
	Lippia origanoides/Mexican oregano	MIC: 3.7 µg/ml	
	Oreganum vulgare/oregano	MIC: 3.7 µg/ml	
	Artemisia vulgaris/mugwort	MIC: 11.1 µg/ml	
DEN-2[a]	*Artemisia douglasiana*/Californian mugwort	IC_{50}:[c] 60 ppm	[37]
	Eupatorium patens/Little Joe	IC_{50}: 150 ppm	[57]
	Heterothalamus alienus/romerillo	IC_{50}: 39.0 ppm	
INF[a]	*Houttuynia cordata*/fishwort, heartleaf	IC_{50}: 0.0048%	[57]
	Cynanchum stauntonii/Bai Qian	IC_{50}: 64 µg/ml	[62]
SARS-CoV[a]	*Laurus nobilis*/bay laurel	IC_{50}: 120 µg/ml	[41]
JUNV[a]	*Heterothalamus alienus*/romerillo	IC_{50}: 44.2 ppm	[59]
	Buddleja cordobensis/Palo Blanco	IC_{50}: 39.0 ppm	[59]
	Lippia junelliana	IC_{50}: 20 ppm	[37]
	Lippia turbinate/poleo	IC_{50}: 14 ppm	[37]
	Aloysia gratissima/whitebrush, common beebrush	IC_{50}: 52 ppm	[37]
	Heterotheca latifolia/camphorweed	IC_{50}: 90 ppm	[37]
	Tessaria absinthioides/bobo bird	IC_{50}: 63 ppm	[37]
NDV[a]	*Origanum vulgare*/oregano	0.025 %	[60]

[a]YFV = yellow fever virus; DEN-2 = dengue virus type 2; INF = influenza virus; SARS-CoV = SARS-associated coronarvirus; JUNV = Junin virus; NDV = Newcastle disease virus.
[b]MIC: essential oil concentration that reduces virus infectivity by more than 50%.
[c]IC50: essential oil concentration that reduces virus infectivity by 50%.

coxsackievirus. Three major components of the essential oil, methyl n-nonyl ketone, lauryl aldehyde and capryl aldehyde, also inactivated HSV-1 (Table 10.2), INF (IC_{50}: 0.0015 to 0.0062%) and HIV-1 (most active concentration: 0.0083%).

Quite recently, the essential oils of *Ridolfia segetum* and *Oenanthe crocata* have been assayed for inhibition of two enzyme-associated activities of the HIV-1 reverse transcriptase (RT): RNA-dependent DNA polymerase (RdDP) activity and ribonuclease H (RNase H) activity. *R. segetum* is used traditionally in the Mediterranean as a medicinal plant, for instance to regulate women's menstrual periods and to prevent respiratory tract infections. In biochemical assays, both essential oils inhibited HIV-1 RT RdDP activity in a dose-dependent manner, while they were inactive against RNase H activity [61]. INF was also inhibited by essential oil derived from the roots of *Cynanchum stauntonii*, with an IC_{50} value of 64 µg/ml. Interestingly, in *in vivo* experiments the root-derived essential oil prevented INF-induced death in mice in a dose-dependent manner [62].

Garcia and coworkers [37] explored essential oils of eight aromatic medicinal plants from Argentina for virucidal effects against the enveloped viruses HSV-1, JUNV (Junin virus) and DEN-2 (dengue virus type 2). Surprisingly, essential oils derived from different plant origins displayed high selective antiviral activities. For instance, essential oils of

Lippia junelliana and *Lippia turbinate* revealed a very potent antiviral activity against JUNV, with IC_{50} values of 14 and 20 ppm, respectively (Table 10.3), but both essential oils were more or less inactive against HSV-1 and DEN-2. In addition, plaque formation of JUNV was also reduced by *Aloysia gratissima* (IC_{50}: 52 ppm), *Heterotheca latifolia* (IC_{50}: 90 ppm) and *Tessaria absinthioides* (IC_{50}: 63 ppm). Plaque formation of DEN-2 was significantly reduced by the essential oils of *Artemisia douglasiana* (IC_{50}: 60 ppm) and *Eupatorium patens* (IC_{50}: 150 ppm). The observed virucidal effects proved to be time- and temperature-dependent.

In a similar work, the same scientific group [59] tested essential oils of seven other aromatic plants from Argentina against HSV-1, DEN-2 and JUNV. The highest antiviral action was observed with the essential oils of *Heterothalamus alienus* and *Buddleja cordobensis* against JUNV, with IC_{50} values of 44.2 and 39 ppm, respectively (Table 10.3). The inhibitory activity was exerted by a direct interaction of the virions with the oils. On the other hand, the attachment of the virions to the host cells was not impaired by the oils.

The essential oil of *Laurus nobilis* is used in folk medicine for the treatment of rheumatoid ailments. Loizzo and coworkers [41] tested the essential oil derived from plant leaves against SARS-associated coronavirus (SARS-CoV). The oil exerted an interesting antiviral activity, with an IC_{50} value of 120 µg/ml (Table 10.3) and a selectivity index (SI) of 4.6. Unfortunately, the relatively low SI seems to exclude the oil from therapeutical application.

Yellow fever (YF) is a viral haemorrhagic fever endemic in South America and Africa caused by the yellow fever virus (YFV). The virus is transmitted to humans by Aedes or Haemagogus mosquitoes. According to the WHO, at least 200 000 cases of YF are reported each year, including 30 000 deaths. Despite an existing vaccination, there are worldwide efforts to explore antiviral compounds against YFV. Meneses and coworkers [58] screened several essential oils derived from *Lippia alba*, *Lippia origanoides*, *Origanum vulgare* and *Artemisia vulgaris* for their antiviral properties against YFV *in vitro*. Preincubation of viruses with the selected essential oils for 24 hours at 4 °C before adsorption on host cells revealed a significant reduction of plaque-forming units. The antiviral activity was determined by means of minimum inhibitory concentration (MIC). MIC was defined as the essential oil concentration that inhibited virus replication by more than 50%. Based on this, the MIC for *L. alba*, *L. origanoides* and *O. vulgare* oils was 3.7 µg/ml, whereas for *A. vulgaris* it was determined at 11.1 µg/ml (Table 10.3). Interestingly, the CC_{50} : MIC ratios were within the range of 8.8 to 26.5. At 11.1 µg/ml the essential oil of *L. origanoides* exhibited a 100% reduction of virus plaque formation. The same result was obtained with *L. alba*, *O. vulgaris* and *A. vulgaris* oils at 100 µg/ml. The mode of antiviral action seems to be based on a direct virus inactivation.

10.5   Mode of Antiviral Action of Essential Oils

The best candidates for clinically useful antiviral drugs are substances which act on specific steps of viral biosynthesis. They inhibit specific processes in the viral-replication cycle, and thus little or no viral progeny is produced. These antiviral drugs should act at low concentrations and should not influence the host-cell machinery. They should prevent the spread of viruses, or ultimately cure infected cells. Unfortunately, viruses commonly develop resistance against such specific drugs. On the other hand, virucidal drugs denature viral structural proteins or glycopropteins, or disrupt the lipid envelope of enveloped viruses. Thus, the infectivity of virus particles is completely lost. As previously outlined, in the case of essential oils, cells and HSV were incubated with these volatile phyto-antiviral agents at different stages during the viral-infection cycle in order to determine the mode of antiviral action. Cells were pretreated with essential oils before viral infection, viruses were incubated with essential oils before infection, and cells and viruses were incubated together with essential oils during adsorption or after penetration of the virus into the host cells. The highest antiviral effect was observed when HSV was incubated with essential oils prior to addition to the cells, thus indicating a direct virucidal activity of essential oils against HSV [3, 63]. Most studies on the effect of essential oils on other viruses have shown similar results [37, 57–60]. Furthermore, it is generally agreed that virus infections develop when virions spread from infected cells to neighbouring uninfected cells. It is of great interest that essential oils may inhibit the cell-to-cell diffusion of HSV *in vitro* when added to cell cultures of already infected host cells [38, 43]. Most studies suggest that essential oils affect the viral envelope or mask viral components that are necessary for adsorption or entry into host cells. In particular, monoterpenes increase cytoplasmic membrane fluidity and permeability, and disturb the order of membrane-embedded proteins [2]. A recent electron-microscopic examination of the envelope of HSV-1 after treatment with oregano (*Origanum vulgare*) oil and clove (*Syzygium aromaticum*) oil supports the notion that the virus envelope could be the major target. It was demonstrated that these essential oils destroy the HSV-1 envelope [60]. Further experiments are necessary to corroborate this finding. Essential oils exert their antiviral activity at a concentration that is usually much lower than the cytotoxic concentration. This indicates that virion envelopes are more sensitive to essential oils than host-cell membranes.

10.6 Conclusions

Essential oils are mixtures of different compounds, for example monoterpenes, sesquiter-penes and phenylpropene derivatives, with a typical fragrance, and are derived from aromatic plants by steam or hydro-distillation. A number of studies have shown that essential oils and their structurally diverse chemical components have antiviral activities, particularly against HSV, but also against a number of other enveloped viruses, such as INF, JUNV, YFV and HIV, but not against non-enveloped viruses such as coxsackievirus. The antiviral effect on HSV of different essential oils, such as anise oil, dwarf-pine oil, chamomile oil and tea-tree oil, has been determined by measuring the reduction in viral infectivity by plaque assay after treatment with these oils. A reduction in infectivity of more than 99% has been observed. The greatest antiviral effect was observed when HSV was incubated with essential oils prior to addition to host cells, thus indicating a direct effect on cell-free virus. The same effect has been shown for most other viruses studied. Essential oils show antiherpetic activity in a mouse model for genital HSV-2 infection, and a beneficial effect of topical treatment of recurrent human herpes labialis with tea-tree oil (6% gel) has been shown in a randomized, placebo-controlled, investigator blinded clinical study. Thus, essential oils are promising candidates for topical treatment of recurrent herpes infections, even against drug-resistant virus strains. For self-medication, essential oils, diluted in olive oil or oil of almonds, can be applied to areas of herpes infection. Essential-oil formula should be applied three or four times per day locally to areas of recurrent outbreaks. In this context, essential oils are also effective for pain and inflammation control, prevent spreading of the infection and shorten the duration of typical symptoms of cold sores such as itching, tingling, burning and swelling. Furthermore, based on their antimicrobial activity, essential oils should also be able to prevent superinfection of virus blisters.

References

1. Lis-Balchin, M. (2006) *Aromatherapy Science: A Guide for Healthcare Professionals*, Pharmaceutical Press, London, UK, Chicago, IL, USA.
2. Reichling, J., Suschke, U., Schneele, J. and Geiss, H.K. (2006) Antibacterial activity and irritation potential of selected essential oil components: structure-activity relationship. *Nat. Prod. Comm.*, **1**, 1003–1012.
3. Reichling, J., Schnitzler, P., Suschke, U. and Saller, R. (2009) Essential oils of aromatic plants with antibacterial, antifungal, antiviral, and cytotoxic properties: an overview. *Forsch. Komplementmed.*, **16**, 79–90.
4. Reichling, J. (2007) Experimentelle Belege zur antimikrobiellen Wirkung von ausgewählten ätherischen Ölen, in *Aromatherapie – Wissenshaft: Klinik-Praxis* (eds M. Steflitsch and W. Steflitsch), Springer Verlag, Vienna, Austria, New York, NY, USA, pp. 181–189.
5. Bourne, K.Z., Bourne, N., Reising, S.F. and Stanberry, L.R. (1999) Plant products as topical microbicide candidates: assessment of in vitro and in vivo activity against herpes simplex type 2. *Antiviral Res.*, **42**, 219–226.
6. Schnitzler, P., Schön, K. and Reichling, J. (2001) Antiviral activity of Australian tea tree oil and eucalyptus oil against herpes simplex virus in cell culture. *Pharmazie*, **56**, 343–347.
7. Schuhmacher, A., Reichling, J. and Schnitzler, P. (2003) Virucidal effect of peppermint oil on the enveloped viruses herpes simplex virus type 1 and type 2 *in vitro*. *Phytomedicine*, **10**, 504–510.
8. Khan, M.T., Ather, A., Thompson, K.D. and Gambari, R. (2005) Extracts and molecules from medicinal plants against herpes simplex viruses. *Antiviral Res.*, **67**, 107–119.
9. Astani, A., Reichling, J. and Schnitzler, P. (2010) Comparative study on the antiviral activity of selected monoterpenes derived from essential oils. *Phytother. Res.*, **24**, 673–679.
10. Schäfer, U.F., Schneele, J., Schmitt, S. and Reichling, J. (2008) Efficacy, absorption, and safety of essential oils, in *Dermatologic, Cosmeceutic, and Cosmetic Development: Therapeutic and Novel Approaches* (eds H. A. Walters and M. S. Roberts), Informa Healthcare, New York, NY, USA, London, UK, pp. 401–418.
11. Corey, L. and Spear, P.G. (1986) Infections with herpes simplex virus. *N. Engl. J. Med.*, **314**, 686–691.
12. Whitley, R.J. and Roizman, B. (2001) Herpes simplex virus infections. *The Lancet*, **357**, 1513–1518.
13. Smith, J.S. and Robinson, N.J. (2002) Age-specific prevalence of infection with herpes simplex virus types 2 and 1: a global review. *J. Infect. Dis.*, **186** (suppl.), S3–S28.
14. Reusser, P. (1996) Herpesvirus resistance to antiviral drugs: a review of the mechanisms, clinical importance and therapeutic options. *J. Hosp. Infect.*, **3**, 235–248.
15. Gilbert, S.C. (2007) Management and prevention of recurrent herpes labialis in immunocompetent patients. *Herpes*, **14**, 56–64.
16. Freeman, E.E., Weiss, H.A., Glynn, J.R. *et al.* (2006) Herpes simplex virus 2 infection increases HIV acquisition in men and women: systematic review and meta-analysis of longitudinal studies. *AIDS*, **20**, 73–83.
17. Cunningham, A.L., Diefenbach, R.J., Miranda-Saksens, M. *et al.* (2006) The cycle of human herpes simplex virus infection: virus transport and immune control. *J. Infect. Dis.*, **194** (suppl.), S11–S18.
18. Balfour Jr, H.H. (1999) Antiviral drugs. *N. Engl. J. Med.*, **340**, 1255–1268.

19. Bacon, T.H., Levin, M.J., Leary, J.J. *et al.* (2003) Herpes simplex virus resistance to acyclovir and penciclovir after two decades of antiviral therapy. *Clin. Microbiol. Rev.*, **16**, 114–128.
20. Stranska, R., Schuurman, R., Nienhuis, E. *et al.* (2005) Survey of acyclovir-resistant herpes simplex virus in the Netherlands: prevalence and characterization. *J. Clin. Virol.*, **32**, 7–18.
21. Vlietinck, A.J. and Vanden Berghe, D.A. (1991) Can ethnopharmacology contribute to the development of antiviral drugs? *J. Ethnopharmacol.*, **32**, 141–153.
22. Jassim, S.A. and Naji, M.A. (2003) Novel antiviral agents: a medicinal plant perspective. *J. Appl. Microbiol.*, **95**, 412–427.
23. De Clercq, E. (2002) Strategies in the design of antiviral drugs. *Nature Rev.*, **1**, 13–25.
24. Hammer, K.A., Carson, C.F. and Riley, T.V. (1999) Antimicrobial activity of essential oils and other plant extracts. *J. Appl. Microbiol.*, **86**, 985–990.
25. Cowan, M.M. (1999) Plant products as antimicrobial agents. *Clin. Microbiol. Rev.*, **12**, 564–582.
26. Caelli, M., Porteous, J., Carson, C. *et al.* (2000) Tea tree oil as an alternative topical decolonization agent for methicillin-resistant *Staphylococcus aureus*. *J. Hosp. Infect.*, **46**, 236–237.
27. Carson, C.F., Ashton, L., Dry, L. *et al.* (2001) *Melaleuca alternifolia* (tea tree) oil gel (6%) for the treatment of current herpes labialis. *J. Antimicrob. Chemother.*, **48**, 450–451.
28. Carson, C.F., Hammer, K.A. and Riley, T.V. (2006) *Melaleuca alternifolia* (tea tree) oil: a review of antimicrobial and other medicinal properties. *Clin. Microbiol., Rev.*, **19**, 50–62.
29. Inouye, S., Takizawa, T. and Yamaguchi, H. (2001) Antibacterial activity of essential oils and their major constituents against respiratory tract pathogens by gaseous contact. *J. Antimicrob. Chemother.*, **47**, 565–573.
30. Cos, P., Vlietinck, A.J., Vanden Berghe, D. and Maes, L. (2006) Anti-infective potential of natural products: how to develop a stronger in vitro 'proof of concept'. *J. Ethnopharmacol.*, **106**, 290–302.
31. Furneri, P.M., Paolino, D., Saija, A. *et al.* (2006) In vitro antimycoplasmal activity of *Melaleuca alternifolia* essential oil. *J. Antimicrob. Chemother.*, **58**, 706–707.
32. Tolo, F.M., Rukunga, G.M., Muli, F.W. *et al.* (2006) Antiviral activity of the extracts of a Kenyan medicinal plant *Carissa edulis* against herpes simplex virus. *J. Ethnopharmacol.*, **104**, 92–99.
33. Morfin, F. and Thouvenot, D. (2003) Herpes simplex virus resistance to antiviral drugs. *J. Clin. Virol.*, **26**, 29–37.
34. Vermani, K. and Garg, S. (2002) Herbal medicines for sexually transmitted diseases and AIDS. *J. Ethnopharmacol.*, **80**, 49–66.
35. Martin, K.W. and Ernst, E. (2003) Antiviral agents from plants and herbs: a systematic review. *Antiviral Ther.*, **8**, 77–90.
36. Mukhtar, M., Arshad, M., Ahmad, M. *et al.* (2008) Antiviral potentials of medicinal plants. *Virus Res.*, **131**, 111–120.
37. Garcia, C.C., Talarico, L., Almeida, N. *et al.* (2003) Virucidal activity of essential oils from aromatic plants of San Luis, Argentina. *Phytother. Res.*, **17**, 1073–1075.
38. Saddi, M., Sanna, A., Cottiglia, F. *et al.* (2007) Antiherpes activity of *Artemisia arborescens* essential oil and inhibition of lateral diffusion in Vero cells. *Ann. Clin. Microbiol. Antimicrob.*, **6**, 1–10.
39. Minami, M., Kita, M., Nakaya, T. *et al.* (2003) The inhibitory effect of essential oils on herpes simplex virus type 1 replication in vitro. *Microbiol. Immunol.*, **47**, 681–684.
40. Koch, C., Reichling, J., Schneele, J. and Schnitzler, P. (2008) Inhibitory effect of essential oils against herpes simplex virus type 2. *Phytomedicine*, **15**, 71–78.
41. Loizzo, M.R., Saab, A.M., Tundis, R. *et al.* (2008) Phytochemical analysis and in vitro antiviral activities of the essential oils of seven Lebanon species. *Chem. Biodivers.*, **5**, 461–470.
42. Reichling, J., Koch, C., Stahl-Biskup, E. *et al.* (2005) Virucidal activity of a β-triketone-rich essential oil of *Leptospermum scoparium* (manuka oil) against HSV-1 and HSV-2 in cell culture. *Planta Med.*, **71**, 1123–1127.
43. De Logu, A., Loy, G., Pellerano, M.L. *et al.* (2000) Inactivation of HSV-1 and HSV-2 and prevention of cell-to-cell virus spread by *Santolina insularis* essential oil. *Antiviral Res.*, **48**, 177–185.

44. Koch, C., Reichling, J., Kehm, R. *et al.* (2008) Efficacy of anise oil, dwarf-pine oil and camomile oil against thymidine kinase positive and thymidine kinase negative herpesviruses. *J. Pharm. Pharmacol.*, **60**, 1545–1550.
45. Benencia, F. and Courrèges, M.C. (2000) In vitro and in vivo activity of eugenol on human herpesvirus. *Phytother. Res.*, **14**, 495–500.
46. Hayashi, K., Hayashi, T., Ujita, K. and Takaishi, Y. (1996) Characterization of antiviral activity of a sesquiterpene, triptofordin C-2. *J. Antimicrob. Chemother.*, **37**, 759–768.
47. Kalemba, D. and Kunicka, A. (2003) Antibacterial and antifungal properties of essential oils. *Curr. Med. Chem.*, **10**, 813–829.
48. Allahverdiyev, A., Duran, N., Ozguven, M. and Koltas, S. (2004) Antiviral activity of the volatile oils of *Melissa officinalis* L. against herpes simplex virus type-2. *Phytomedicine*, **11**, 657–661.
49. Tragoolpua, Y. and Jatisatieur, A. (2007) Anti-herpes simplex virus activities of *Eugenia caryophyllus* (Spreng.) Bullock & S.G. Harrison and essential oil, eugenol. *Phytother. Res.*, **21**, 1153–1158.
50. Schnitzler, P., Schuhmacher, A., Astani, A. and Reichling, J. (2008) *Melissa officinalis* oil affects infectivity of enveloped herpesviruses. *Phytomedicine*, **15**, 734–740.
51. Schnitzler, P., Koch, C. and Reichling, J. (2007) Susceptibility of drug-resistant clinical herpes simplex virus type 1 strains of essential oils of ginger, thyme, hyssop, and sandalwood. *Antimicrob. Agents Chemother.*, **51**, 1859–1862.
52. Koch, C., Reichling, J. and Schnitzler, P. (2008) Essential oils inhibit the replication of herpes simplex virus type 1 (HSV-1) and type 2 (HSV-2), in *Botanical Medicine in Clinical Practice* (eds V.R. Preedy and R.R. Watson), CAB International, pp. 192–197.
53. Saller, R., Büechi, S., Meyrat, R. and Schmidhauser, C. (2001) Combined herbal preparation for topical treatment of herpes labialis. *Forsch. Komplementmed.*, **8**, 373–382.
54. Armaka, M., Papanikolaou, E., Sivropoulou, A. and Arsenakis, M. (1999) Antiviral properties of isoborneol, a potent inhibitor of herpes simplex virus type 1. *Antiviral Res.*, **43**, 79–92.
55. Serkedjieva, J. and Manolova, N. (1992) Plant polyphenolics complex inhibits the reproduction of influenza and herpes simplex viruses. *Basic Life Sci.*, **59**, 705–715.
56. Cermelli, C., Fabio, A., Fabio, G. and Quaglio, P. (2008) Effect of eucalyptus essential oil on respiratory bacteria and viruses. *Curr. Microbiol.*, **56**, 89–92.
57. Hayashi, K., Kamiya, M. and Hayashi, T. (1995) Virucidal effects of the steam distillate from *Houttuynia cordata* and its components on HSV-1, influenza virus, and HIV. *Planta Med.*, **61**, 237–241.
58. Meneses, R., Ocazionez, R.E., Martinez, J.R. and Stashenko, E.E. (2009) Inhibitory effect of essential oils obtained from plants grown in Colombia on yellow fever virus replication in vitro. *Ann. Clin. Microbiol. Antimicrob.*, **8**, 8.
59. Duschatzky, C.B., Possetto, M.L., Talarico, L.B. *et al.* (2005) Evaluation of chemical and antiviral properties of essential oils from South American plants. *Antiviral Chem. Chemother.*, **16**, 247–251.
60. Siddiqui, Y.M., Ettayebi, M., Haddad, A.M. and Al-Ahdal, M.N. (1996) Effect of essential oils on the enveloped viruses: antiviral activity of oregano and glove oils on herpes simplex virus type 1 and Newcastle disease virus. *Med. Sci. Res.*, **24**, 185–186.
61. Bicchi, C., Rubiolo, P., Ballero, M. *et al.* (2009) HIV-1 inhibiting activity of the essential oil of *Ridolfia segetum* and *Oenanthe crocata*. *Planta Med*, **75**, 1331–1335.
62. Zai-Chang, Y., Bo-Chu, W., Xiao-Sheng, Y. and Qiang, W. (2005) Chemical composition of the volatile oil from *Cynanchum stauntonii* and its activities of anti-influenza virus. *Colloids Surf. B: Biointerfaces*, **43**, 198–202.
63. Schnitzler, P., Nolkemper, S., Stintzing, F.C. and Reichling, J. (2007) Comparative *in vitro* study on the anti-herpetic effect of aqueous and ethanolic extracts of *Salvia officinalis* grown at two different locations. *Phytomedicine*, **15**, 62–70.

11

Antibacterial and Antifungal Activities of Essential Oils

Katherine A. Hammer and Christine F. Carson

Discipline of Microbiology and Immunology, School of Biomedical, Biomolecular and Chemical Sciences, The University of Western Australia, Crawley, WA, Australia

Lipids and Essential Oils as Antimicrobial Agents   Halldor Thormar

11.1 Introduction

Essential oils are volatile, hydrophobic, typically fragrant plant extracts, which may also be referred to as volatile oils or plant oils. These oils are usually obtained by steam distillation and contain a range of oxygenated and nonoxygenated terpene hydrocarbons. Essential oils may be derived from any part of the plant, such as the foliage, bark, wood, fruit, seeds or rhizomes. It has been estimated that approximately 3000 essential oils are currently known and that about 10% of these are commercially important [1]. This chapter will focus on, although not be limited to, those oils that meet the broad criteria outlined above and for which a moderate amount of scientific data is available. This includes many of the traditional culinary herbs such as thyme, oregano, mint and rosemary, oils used in perfumery such as lavender, sandalwood and rosewood, and those used as topical antiseptics such as tea tree.

Great variation exists amongst antimicrobial essential oils in terms of both the diversity of plants from which they may be derived and the chemical composition of each oil. Despite this diversity, there are a number of generalizations that can be made about their antimicrobial activity. For example, most essential oils are inhibitory at concentrations well below 5% (v/v) and exhibit dose-dependent activity, with greater activity seen at higher oil concentrations. Essential oils tend to be bactericidal in action, meaning that organisms are inhibited and killed at approximately the same concentration [2]. In contrast, bacteriostatic agents inhibit growth but do not kill. Many essential oils also have a relatively rapid antimicrobial action, with significant cell death occurring at concentrations equivalent to or greater than the minimum bactericidal or fungicidal concentrations. Lastly, the majority of oils are broad-spectrum in activity, meaning that they are active against a wide range of bacteria and fungi.

11.2   Methods for Quantifying Antimicrobial Activity

Whilst an exhaustive review of all methods used for assessing the activity of essential oils is not warranted here, it is worthwhile mentioning those most commonly used. These include the agar-diffusion method and broth- or agar-dilution method.

The disc- or well-diffusion assay is arguably the easiest susceptibility assay to perform and is frequently used as a screening tool. In this assay, an agar petri dish is inoculated with the relevant test organism, then a small amount of oil is placed on to a paper disc or into a well that has been cut into the agar. The diffusion of the oil through the agar creates a zone adjacent to the well or disc where microbial growth is prevented due to the high concentration of oil. This area of inhibition is then quantified by measuring the zone diametre. There are several comprehensive studies that have screened a large number of essential oils and/or components using this assay [3–6] and these studies have provided a very useful indication of antimicrobial activity. The benefits of this assay are that it is not technically difficult and it uses a comparatively small quantity of essential oil. However, as discussed elsewhere [7–9], this method is not ideal for water-insoluble compounds such as essential oils, and methodological variations between laboratories may mean that data are not always directly comparable.

Agar- and broth-dilution methods are also commonly used for assessing activity, whereby a series of dilutions of the oil is performed in the relevant growth medium. After inoculation of the assay and incubation, the presence or absence of growth is recorded. These methods allow the minimum inhibitory concentration (MIC) of an oil to be determined. The MIC can be defined as the lowest amount of essential oil required to inhibit the visible growth of the test organism. When a broth dilution method is used, the sampling of each oil dilution to quantify the surviving organisms allows for the minimum bactericidal concentration (MBC) or minimum fungicidal concentration (MFC) to be determined. The MBC and MFC refer to the minimum concentration of oil required to kill 99.9% of the cells originally inoculated into the assay for bacteria and fungi, respectively. Like the disc-diffusion assay, dilution methods are also subject to interlaboratory variation, but because the assay outcome is quantified in terms of oil concentration rather than zone size, values obtained by different researchers are more easily compared. With these assays, variables such as the presence of a solubilizing agent [8, 9] or choice of test organism may also influence the result. Target organisms typically reflect the desired end-use of the oil, whether it be as a medicinal product for humans or other animals or as a food or crop preservative. As such, groups of organisms that may be targeted include foodborne pathogens, postharvest pathogens, food-spoilage fungi, veterinary pathogens and oral flora. However, since many studies investigate potential medicinal applications for essential oils, the most frequently selected test organisms are common human pathogens such as those causing infections of the skin and soft tissue, bloodstream or wounds, and nosocomial pathogens.

This chapter will focus on MICs derived by broth or agar dilution as a means of comparing the activity of different essential oils and components. Also, the MIC data are for oils in liquid rather than vapour phase, unless stated otherwise. Furthermore, where possible, data will be expressed as a percentage (vol/vol) to enhance the comparability of results.

11.3   Antibacterial and Antifungal Essential Oils by Plant Family

Important fundamental factors to bear in mind when discussing the activity of plant essential oils are the genus and species from which the oil was obtained and the composition of the plant oil. Although inferences about oil composition can be made from the plant species name, oil composition can be influenced by environmental factors, meaning that two identical plants grown under different conditions are unlikely to produce identical oils. In this chapter, plant scientific names, where possible, have been stated in conjunction with common names to limit confusion. Where the botanical source is not stated this means that the oil is from the typical plant source, for example peppermint oil from *Mentha* × *piperita*. The data are summarized in Table 11.1.

11.3.1   Apiaceae

Coriander (*Coriandrum sativum*), anise or aniseed (*Pimpinella anisum*), dill (*Anethum graveolens*) and fennel (*Foeniculum vulgare*) belong to the family Apiaceae. Many Apiaceae oils can be obtained from both seeds and leaf material and these oils may differ significantly in composition.

Coriander leaf oil contains predominantly aldehydes such as decanal and decenal and alcohols such as decen-1-ol and *n*-decanol [10]. MICs for this oil were 108 to 217 mg/ml (10.8 to 21.7%) for a selection of pathogenic bacteria including *Escherichia coli, Staphylococcus aureus* and *Klebsiella pneumoniae* and 163 mg/ml (16.3%) for *Candida albicans* [10]. In contrast, coriander seed oil is high in linalool [11] and has MICs in the range of 0.006 to 0.4% (v/v) [11] and 0.25 to 1% (v/v) [12] for a selection of pathogenic bacteria including *S. aureus, E. coli* and *Serratia marcescens*, but >2% for *Pseudomonas aeruginosa* [12] and 0.25% for *C. albicans* [12]. Anise seed oil MICs are 0.25 to 2% (v/v) for a selection of bacteria, 0.5% for *C. albicans* and >2% for *K. pneumoniae* and *P. aeruginosa* [12]. Similarly, for fennel seed oil organisms such as *E. coli, C. albicans* and *S. aureus* were susceptible with MICs of 0.25 to 1% (v/v) but MICs were >2% for several others [12]. Additional fennel oil MICs are 3 mg/ml (0.3%) for *E. coli* and *Staphylococcus epidermidis* and 0.8 mg/ml (0.08%) for *Saccharomyces cerevisiae* [13].

11.3.2   Lamiaceae

The Lamiaceae family includes many genera of well-known medicinal and culinary herbs such as mint (*Mentha*), rosemary (*Rosmarinus*), basil (*Ocimum*), sage (*Salvia*), marjoram and oregano (*Origanum*), thyme (*Thymus*) and lavender (*Lavandula*). Since these plants have a long history of use as culinary herbs, many have a wealth of research describing their antimicrobial activity.

Several oils from the mint genus, such as wild or corn mint (*Mentha arvensis*), peppermint (*Mentha* × *piperita*), spearmint (*Mentha spicata*) and pennyroyal (*Mentha pulegium*) have antimicrobial activity. *Mentha* oils generally contain high levels of the terpene alcohol

Table 11.1 *A selection of minimum inhibitory concentrations (% v/v) of essential oils.*

Common name of oil	Plant source	Organism	MIC [Reference]
Basil	*Ocimum basilicum*	*Escherichia coli*	0.25 [18]
			0.5 [12]
		Listeria monocytogenes	0.05 [18]
		Staphylococcus aureus	0.1 [18]
			2.0 [12]
		Candida albicans	0.5 [12]
Bay	*Pimenta racemosa*	*E. coli*	0.02 [54]
			0.05 [18]
			0.12 [12]
		S. aureus	0.05 [54]
			0.05 [18]
			0.25 [12]
		Salmonella spp.	0.05 [18]
			0.05 [54]
			0.25 [12]
		L. monocytogenes	0.02 [18]
		C. albicans	0.12 [218]
Clove	*Syzygium aromaticum*	*E. coli*	0.04 [18]
			0.125 [219]
			0.25 [12]
			0.43 [24]
		S. aureus	0.04 [18]
			0.125 [219]
			0.25 [12]
			0.6 [24]
		Pseudomonas aeruginosa	0.5 [219]
			>2 [12]
		C. albicans	0.12 [12]
			0.125 [219]
Eucalyptus	*Eucalyptus globulus*	*E. coli*	0.28 [13]
			>1 [18]
		L. monocytogenes	0.075 [18]
		Staphylococcus epidermidis	0.28 [13]
		S. aureus	0.1 [18]
Lavender	*Lavandula angustifolia*	*E. coli*	0.5 [12]
			0.62 [16]
		S. aureus	**0.5** [23]
			0.86 [16]
			1.0 [12]
		C. albicans	0.5 [12]
			1 [25]
Lemongrass	*Cymbopogon citratus*	*E. coli*	0.06 [12]
			0.2 [11]
		S. aureus	0.06 [12]
			0.1 [11]
		Pseudomonas putida	0.8 [50]
		C. albicans	0.12 [218]

(Continued)

Table 11.1   *A selection of minimum inhibitory concentrations (% v/v) of essential oils.*
(Continued)

Common name of oil	Plant source	Organism	MIC [Reference]
Oregano	*Origanum vulgare*	*E. coli*	0.05 [34]
			0.24 [24]
		S. aureus	0.12 [12]
			0.125 [32]
			0.24 [24]
		S. epidermidis	**0.125** [32]
		L. monocytogenes	0.02 [30]
		P. aeruginosa	2 [12]
		C. albicans	**0.04** [20]
			0.12 [218]
Peppermint	*Mentha × piperita*	*E. coli*	0.5 [12]
			0.57 [16]
			0.57 [13]
			>1 [18]
		S. aureus	0.04 [18]
			0.5 [23]
			1.0 [12]
			1.19 [16]
		C. albicans	0.24 [16]
			0.5 [12]
Rosemary	*Rosmarinus officinalis*	*E. coli*	0.25 [219]
			0.45 [16]
			1 [12]
			1.13 [13]
			>1 [18]
		S. aureus	0.04 [18]
			0.125 [219]
			0.62 [16]
			1 [12]
		C. albicans	0.25 [219]
			0.57 [16]
			1 [12]
Tea tree	*Melaleuca alternifolia*	*E. coli*	**0.25** [220]
			0.37 [16]
			0.56 [13]
			0.62 [24]
		S. aureus	**0.25** [23]
			0.5 [220]
			0.5 [48]
			0.86 [16]
			1.05 [24]
		Klebsiella pneumoniae	**0.25** [48]
		P. putida	0.8 [50]
			1 [221]

Table 11.1 *A selection of minimum inhibitory concentrations (% v/v) of essential oils.* (Continued)

Common name of oil	Plant source	Organism	MIC [Reference]
		P. aeruginosa	**4** [221]
		C. albicans	**0.25** [218]
			0.37 [16]
			0.5 [154]
Thyme	*Thymus vulgaris*	E. coli	0.05 [18]
			0.05 [16]
			0.15 [13]
			0.006 [11]
		S. aureus	0.02 [18]
			0.025 [11]
			0.13 [16]
			0.5 [23]
		P. putida	0.05 [50]
			0.2 [30]
		C. albicans	**0.16** [20]
		Campylobacter jejuni	0.04 [18]

Values in bold type are MIC$_{90}$s, which is the MIC occurring for 90% of the isolates tested and indicates that more than 10 isolates were tested. The majority of remaining values shown were for single test isolates.

menthol, which is likely to contribute to overall activity [14]. MICs for peppermint include 0.125 to 2% (v/v) [12], 5.7 mg/ml (0.57%) [13], 8 mg/ml (0.8%) [15] and 3.7 to 11.9 mg/ml (0.37–1.19%) [16] for a range of human pathogens including *S. aureus*, *E. coli* and *C. albicans* and MBCs of 6000 ppm (0.6%) and 1000 ppm (0.1%) for *Streptococcus mutans* and *Streptococcus pyogenes*, respectively [17]. *P. aeruginosa* was less susceptible with an MIC of >2% (v/v) [12]. Spearmint oil has similar activity to peppermint oil with MICs of 0.125 to 2% (v/v) [12] and 0.04 to 0.25% (v/v) [18] for a range of Gram-positive and -negative bacteria such as *E. coli*, *Klebsiella* and *Salmonella* sp., and *S. aureus*. Again, *P. aeruginosa* was less susceptible with an MIC of >2% (v/v) [12]. Peppermint oil also has antifungal activity, with MICs of 2.4 to 6.0 mg/ml (0.24 to 0.6%) [15,16], 0.4 mg/ml (0.04%) [13] and 0.5% (v/v) [12] for yeasts. Similarly, MICs for spearmint oil were 0.12% (v/v) for *C. albicans* [12] and 0.25 μl/ml (0.025%) for *Trichophyton rubrum* and *Trichosporon beigelii* [19].

Rosemary (*Rosmarinus officinalis*) oil generally contains moderate levels of 1,8-cineole and camphor [20]. MICs published for rosemary oil include 0.5 to 2% (v/v) [12], 6 to 8 mg/ml (0.6 to 0.8% w/v) [15], 2.8 to 11.3mg/ml (0.28 to 1.13% w/v) [13] and 0.125 to 0.5 mg/ml (0.012 to 0.05%) [21] for several medical pathogens including *E. coli*, *S. aureus* and *B. cereus*, and MBCs of 2000 ppm (0.2%) and 4000 ppm (0.4%) have been found for *Streptococcus mutans* and *Streptococcus pyogenes*, respectively [17]. However, no activity was seen against *Candida* spp. at concentrations up to 3200 μg/ml (0.32%) and MICs of >2% (v/v) were also found for organisms such as *Enterococcus faecalis*, *P. aeruginosa* and *S. Typhimurium* [12].

Lavender oil (English or French) is obtained from *Lavandula angustifolia* (syn *L. officinalis*). Additional oils have been obtained from *Lavandula stoechas*, *L. latifolia* (spike lavender), *L.* × *intermedia* (lavandin) and *L. dentata*. Linalyl acetate and linalool are major components of many lavender oils [11,22]. Lavender oil MICs of 0.25 to 2% (v/v) [12], 0.5 to 1% (v/v) [23], 6.2 to 32 mg/ml (0.62 to 3.2%) and 0.27 to 8.75% (v/v) [24] have been obtained for a range of bacterial pathogens including *Staphylococcus* spp., *Enterobacter* sp. and *E. coli*. Organisms with MICs in excess of 2% (v/v) [12] or 10% [24] have also been found. For fungi, MICs of 0.125 to 2% (v/v) were obtained for 50 *C. albicans* isolates [25], 4 to 5.7 mg/ml (0.4–0.57%) for *C. albicans* and *Cryptococcus neoformans* [16], 1 μl/ml (0.1% v/v) for *T. rubrum* and 2 μl/ml (0.2%) for *T. beigelii* [19] and 1000 to 2000 μg/ml (0.1 to 0.2%) for several yeast species [26].

Thyme (*Thymus vulgaris*) oil is one of the more antimicrobially active plant oils. *T. vulgaris* has several different chemotypes but the thymol chemotype, which has thymol and *p*-cymene as major components [27,28], is most commonly used for obtaining commercial thyme oil. MICs of thyme oil against bacteria include <0.2 to 0.5% [29], 4.0 to 4.7 mg/ml (0.40 to 0.47%) [15], 1.5 mg/ml (0.15%) [13], 0.12 to 0.5% (v/v) [12], 200 to 600 ppm (0.02 to 0.06%) [30] and 0.02 to 0.05% [18]. MICs for *P. fluorescens* and *P. putida* were 2000 ppm (0.2%) [30] and for *P. aeruginosa* and *S.* Typhimurium were >2.0% (v/v) [12]. MICs for yeasts include 0.4 to 0.7mg/ml (0.04 to 0.07%) for *S. cerevisiae* [13], 3.3 mg/ml (0.33%) [15], 0.08 to 0.32 μl/ml (0.008 to 0.032%) [2] and 0.12% (v/v) [12] for *C. albicans*, and 400 to 3200 μg/ml (0.04 to 0.32%) for a range of yeasts including *C. albicans*, *Candida dubliniensis* and *Candida glabrata* [20]. Thyme oil also achieved complete inhibition of *Aspergillus flavus* growth at 350 ppm (0.035%) in liquid culture [28]. Additional *Thymus* species that yield essential oils include *T. pulegioides*, *T. zygis* and *T. mastichina*. *T. pulegioides* MICs include 0.32 to 0.64 μl/ml (0.03 to 0.06%) for yeasts and 0.16 to 0.32 μl/ml (0.016–0.03%) for dermatophytes and *Aspergillus* [31]. For *T. zygis*, MICs were 0.08 to 0.32 μl/ml (0.008 to 0.032%) and for *T. mastichina* were 1.25 to 10 μl/ml (0.125 to 1% v/v) against *Candida* [2].

Oregano oil, obtained from *Origanum vulgare*, contains predominantly carvacrol [20,27]. MICs obtained for oregano oil include 0.12 to 0.25% (v/v) [12] and 200 to 400 ppm (0.02 to 0.04%) [30] for a range of bacterial pathogens including species of *Listeria*, *Enterobacter* and *Klebsiella*, 0.06 to 0.125% (v/v) for staphylococci [32], 68.2 μg/ml (0.007%) for *Mycobacterium* spp. [33], 0.5 μl/ml (0.05%) for *E. coli* [34] and 575 mg/l (0.057%) for *S. aureus* [35]. MICs for *Pseudomonas* spp. include 2000 ppm (0.2%) for *P. fluorescens* and *P. putida* [30] and 2% (v/v) [12] and 1648 mg/l (0.16%) [35] for *P. aeruginosa*. The activity of oregano oil is similar against both bacteria and fungi, with MICs of 0.12% (v/v) for *C. albicans* [12], 200 to 800 μg/ml (0.02 to 0.08%) for a range of *Candida* spp. [20], 1 μl/ml (0.1%) for *Malassezia furfur* and 0.25 μl/ml (0.025%) for *T. rubrum* and *T. beigelii* [19]. Oregano oil also inhibited the growth of several *Aspergillus* species at concentrations of 20 to 80 μl/ml (2 to 8%) [36]. Additional *Origanum* species, such as *O. majorana* (marjoram) [11, 12], *O. heracleoticum* [11] and *O. compactum* [11], also have antimicrobial activity. *O. compactum* oil, which has γ-terpinene, thymol and carvacrol as major components, has MICs of 0.12 to 0.34% (v/v) [24], 0.012 to 0.025% (v/v) [11] for a range of pathogens including species of *Listeria*, *Salmonella*, *Enterobacter* and *Citrobacter*, and 2.19% (v/v) [24] for *P. aeruginosa*.

The *Salvia* genus contains *Salvia officinalis*, the source of sage oil, and *S. sclarea*, from which clary sage is obtained. MICs for sage oil are 0.5 to 2% (v/v) for pathogenic bacteria [12], >2% for *P. aeruginosa* [12] and 0.5% for *C. albicans* [12]. No activity has been reported for yeasts up to 3200 μg/ml (0.32%) [20] and another report states MICs of 2000 to >2000 μg/ml (0.2 to >0.2%) for yeasts [26]. *S. fruticosa* (Greek sage) oil (major components cineole and camphor) had MICs of 2 μl/ml (0.2%) for *T. rubrum* and 4 μl/ml (0.4%) for *T. beigelii* [19]. Clary sage appears to be less active than sage, with no inhibition of pathogenic bacteria at concentrations up to 2% [12].

Basil oil, also known as sweet basil, is derived from *Ocimum basilicum*. Major components have been identified as linalool and methyl chavicol (estragole) [27,37,38]. However, as many as seven chemotypes have been identified [39] and antimicrobial data without accompanying essential-oil compositional analysis may be difficult to assess. MICs published for basil oil include 0.001 to 0.003% (v/v) [38], 0.05 to 0.25% (v/v) [18] and 0.5 to >2% (v/v) [12] for bacteria and 0.5% (v/v) for *C. albicans* [12]. The lower concentration of 3200 μg/ml (0.32%) was not inhibitory to yeasts [20].

Other antimicrobial essential oils from the Lamiaceae family include lemon balm (*Melissa officinalis*), catmint (*Nepeta cataria*) and savory (*Satureja hortensis*). Further genera yielding essential oils include *Calamintha, Hyssopus, Perilla, Sideritis* and *Thymbra*.

11.3.3 Lauraceae

This family contains several genera from which essential oils are obtained, such as *Cinnamomum, Aniba, Litsea* and *Laurus*. Cinnamon oil is one of the better-known essential oils from this plant family. True cinnamon is derived from *Cinnamomum verum*, also known as *C. zeylanicum*. Two distinct oils are obtainable from *C. verum*; bark oil, which is dominated by cinnamaldehyde [24,40], and leaf oil, which typically contains high levels of eugenol [40,41]. The oil derived from *C. aromaticum* (syn *C. cassia*) is referred to as cassia oil, Chinese cassia or Chinese cinnamon, and typically contains cinnamaldehyde as a major component [11,41]. Essential oil can also be obtained from *C. camphora* (camphor laurel). However, very little data has been published for this oil. MICs reported for cinnamon oil, where the bark or leaf source was not stated, include 0.03 to 0.05% (v/v) for *E. coli, C. jejuni, S. aureus, S. enteritidis* and *Listeria monocytogenes* [18]. MICs for cinnamon leaf oil were 0.31 to 1.25% (v/v) for a range of pathogens including *E. coli, Acinetobacter baumannii, Aeromonas hydrophila* and *S. aureus*. However, *P. aeruginosa* was not inhibited at 10% [24]. Another study reported MICs of 0.012 to 0.025% (v/v) for *L. monocytogenes* and *E. coli* [42]. For *Candida*, MICs of cinnamon leaf oil ranged from 800 to 1600 μg/ml (0.08 to 0.16% w/v) [20] or 0.31 to 0.63 μg/μl (0.03 to 0.06% w/v) [40]. Cinnamon bark oil MICs were 0.01 to 0.24% (v/v) [24] or 0.012 to 0.05% (v/v) [42] for pathogens including *S. aureus, E. coli* and *Salmonella* sp., 0.08 to 0.16 μg/μl (0.008 to 0.016% w/v) for *Candida* spp. [40], <0.04 to 0.16 μg/μl (<0.004 to 0.016%) for dermatophytes [40] and 0.05 to 1 μl/ml (0.005 to 0.1% v/v) for a range of filamentous fungi including *Alternaria, Aspergillus, Cladosporium* and *Penicillium* spp. [43]. These limited data suggest that bark oil is more antimicrobially active than leaf oil. MICs for cassia oil include 0.025 to 0.05% (v/v) for the four food-poisoning organisms *E. coli* O157 : H7, *S.* Typhimurium, *S. aureus* and *L. monocytogenes* [42], and 26.2 μg/ml (0.0026%

w/v) for strains of *Mycobacterium avium* ssp. *paratuberculosis* [33]. Oil from a further *Cinnamomum* species, *C. osmophloeum* (cinnamaldehyde type), had MICs of 250 to 500 µg/ml (0.025 to 0.05%) against several pathogens including *E. coli*, *Salmonella*, *Vibrio* and *Staphylococcus* spp., and *E. coli* [44].

Rosewood oil is obtained from *Aniba rosaeodora*, a tree native to the Amazon. Traditionally used in perfumery, the oil is linalool-rich [43]. Very few antimicrobial data are available for his oil, with MICs of 0.12 to 0.5% (v/v) reported for a range of bacteria and *C. albicans*, and >2% (v/v) for *P. aeruginosa* [12]. MICs ranging from 1.0 to 10.0 µl/ml (0.1 to 1%) have also been reported for a range of fungi including *Aspergillus*, *Cladosporium* and *Penicillium* spp. [43].

The oil obtained from *Laurus nobilis*, which is also known as bay laurel, contains 1,8-cineole, sabinene and α-terpinene as major components [43,45]. MICs range from 10 to 40 µl/ml (1 to 4%) for a selection of filamentous fungi [43] and the oil is also active at concentrations as low as 5 µl/ml (0.5%) against the foodborne pathogens *E. coli* O157, *S.* Typhimurium, *S. aureus* and *L. monocytogenes* [45].

11.3.4 Myrtaceae

The Myrtaceae family contains many commercially important essential oil-bearing genera, including *Melaleuca*, *Eucalyptus*, *Corymbia*, *Pimenta* and *Syzygium* (*Eugenia*), and to a lesser extent *Kunzea* and *Leptospermum*.

Although several *Melaleuca* species yield antimicrobial essential oils, the best known is *M. alternifolia*, from which tea-tree oil is obtained. This essential oil differs from many other oils in terms of the number of publications available describing both *in vitro* antimicrobial activity, *in vitro* or clinical efficacy and also toxicity, which have been reviewed elsewhere [46,47]. Major components of tea-tree oil include the terpene alcohol terpinen-4-ol, followed by γ- and α-terpinene. MICs for tea-tree oil include 0.39 to 1.87% (v/v) [24], 0.25 to 2% (v/v) [12], 0.06 to 3% (v/v) [48], 2.4 to 8.6 mg/ml (0.24 to 0.86%) [16] and 5.6 mg/ml (0.56%) [13] for a range of pathogenic bacteria such as species of *Staphylococcus*, *Enterococcus*, *Bacillus*, *Klebsiella*, *Salmonella* and *Enterobacter*. For a selection of anaerobic bacteria such as *Bacteroides*, *Prevotella* and *Fusobacterium*, MICs were 0.03 to 0.5% (v/v) [49]. MICs for *Pseudomonas* spp. include 0.8% (v/v) for *P. putida* [50] and 2 to 5% (v/v) [48] or >10% (v/v) [24] for *P. aeruginosa*. For fungi, MICs of 2.4 to 4 mg/ml (0.24 to 0.4%) [16], 2.8 mg/ml (0.28%) [13] and 0.06 to 0.5% (v/v) [51] have been published for a range of yeasts. MICs for dermatophytes include 0.004 to 0.06% (v/v) and 0.008 to 0.25% (v/v) for other filamentous fungi such as *Aspergillus* spp. [52]. Additional *Melaleuca* species from which antimicrobial essential oils are obtained include *M. linariifolia* [11], *M. cajuputi* (cajeput) [12,24] and *M. quinquenervia* (niaouli) [53].

Eucalyptus oil contains high levels of 1,8-cineole [24] and is obtained predominantly from *E. globulus*. Other species from which eucalyptus oils may be obtained are *E. polybractea* and *E. citriodora* (lemon-scented gum). MICs for eucalyptus oil are 1.25 to 8.75% (v/v) for organisms including *Streptococcus*, *Staphylococcus* and *Acinetobacter* species [24], 2.8 mg/ml (0.28%) for *E. coli* and *S. epidermidis* [13], and 0.7 mg/ml (0.07%) for *S. cerevisiae* [13]. However, *P. aeruginosa* was not inhibited at 10% [24]. MICs for *E. polybractea* oil are 0.5 to 2% for most bacterial pathogens but >2% for *P. aeruginosa* and *S.* Typhimurium [12].

The *Pimenta* genus yields two common essential oils: bay oil from *P. racemosa* and allspice from *P. dioica*. MICs for bay oil include 0.02 to 0.075% (v/v) for the foodborne pathogens *C. jejuni*, *S. enteritidis*, *E. coli*, *S. aureus* and *L. monocytogenes* [18], 0.12 to 1% (v/v) for a range of bacteria including *P. aeruginosa* [12], and 100 to >500 μg/ml (0.01 to >0.05%) for a range of bacteria including *Bacillus* spp., *E. coli*, *S. aureus* and *S.* Typhimurium [54]. MICs for allspice, containing the major components eugenol and myrcene [11], are 0.012 to 0.05% (v/v) for *E. coli* O157, *S.* Typhimurium, *S. aureus* and *L. monocytogenes* [11].

Clove oil is obtained from *Syzygium aromaticum* (syn *Eugenia caryophyllata* and *E. caryophyllus*) and contains predominantly eugenol [24]. MICs for clove oil include 0.03 to 0.04% (v/v) [18], 0.013 to 0.05% (v/v) [11], 0.25 to 0.62 % (v/v) [24], 0.12 to 0.5% (v/v) [12] for a range of pathogens including *S. aureus*, *E. coli* and *C. albicans*, and 0.31 mg/ml (0.031%) for *Propionibacterium acnes* [55]. However, no inhibition was seen at 2% (v/v) for *S.* Typhimurium and *P. aeruginosa* [12] and in another study *P. aeruginosa* was not inhibited at 10% (v/v) [24]. Clove oil also partially inhibited the growth of *A. flavus* at 500 ppm (0.05%) in liquid culture [28].

Several further myrtaceous oils notable for antimicrobial activity include Manuka (*Leptospermum scoparium*) [56], Kanuka (*Kunzea ericoides*) [56], myrtle (*Myrtus communis*) [57] and lemon-scented myrtle (*Backhousia citriodora*) [58].

11.3.5 Poaceae

This family of grasses (monocotyledons) contains the genus *Cymbopogon*, from which lemongrass (from *Cymbopogon citratus* or *C. flexuosus*), citronella (from *C. nardus* or *C. winterianus*) and palmarosa (*C. martinii*) oils are derived. The major components of lemongrass are neral and geranial, which are together known as citral [11, 24, 37], and the major components of palmarosa are geraniol and geranyl acetate [11, 24].

For lemongrass oil, MICs ranged from 0.12 to 1.25% (v/v) [24], 0.012 to 0.2% (v/v) [11], 0.03 to 0.25% (v/v) [12] or 0.16 to 0.34% (v/v) [59] for a range of Gram-negative and -positive bacteria including *S. aureus*, *E. coli*, species of *Klebsiella* and *Salmonella*, and *C. albicans*. *P. aeruginosa* was comparatively less susceptible, with an MIC of 1% (v/v) [12] in one study and no inhibition at 10% (v/v) [24] in another. Palmarosa oil was similar in activity to lemongrass, with MICs ranging from 0.13 to 0.62% (v/v) [24], 0.025 to 0.2% (v/v) [11] or 0.06 to 0.5% (v/v) [12] for the majority of test organisms. *P. aeruginosa* was again comparatively less susceptible, with no inhibition evident at 2% [12] or 10% [24]. For citronella (*C. nardus*) oil, MICs ranged from 0.12 to 1% (v/v) for susceptible organisms such as *Acinetobacter*, *Aeromonas* and *Staphylococcus* spp., whereas *P. aeruginosa*, *S.* Typhimurium and *S. marcescens* were not inhibited at 2% (v/v) [12]. However, another study found MICs in the range of 0.025 to 0.2% for *C. nardus* oil against *E. coli*, *S.* Typhimurium, *S. aureus* and *L. monocytogenes* [11]. Additional data for *Cymbopogon* oils includes MICs of 0.05 to 0.8% (v/v) for Indian lemongrass (*C. flexuosus*) oil and 0.05 to 0.4% (v/v) for citronella (*C. winterianus*) oil against *E. coli*, *S.* Typhimurium, *S. aureus* and *L. monocytogenes* [11]. As a generalization, *Cymbopogon* oils have high antifungal activity. MICs reported for lemongrass (*C. citratus*) oil against yeasts include ≤ 0.05% (v/v) for *S. cerevisiae* [60], 0.06% for *C. albicans* and 0.13 to 5μl/ml (0.013 to 0.5%) for several food-spoilage yeasts [61]. MICs of palmarosa and citronella were 0.06% (v/v) and 0.12%

(v/v), respectively, against *C. albicans* [12]. For filamentous fungi, MICs of lemongrass (*C. citratus*) oil include ≤0.05% (v/v) for *A. flavus* [62] and 0.062 to 0.31 µl/ml (0.006 to 0.03%) for several filamentous fungi [61].

The Poaceae family also contains the genus *Chrysopogon*, from which vetiver oil (*Chrysopogon zizanioides*, formerly known as *Vetiveria zizanioides*) is obtained. MICs for the Gram-positive bacteria *S. aureus* and *E. faecalis* were 0.06 to 0.12% (v/v), whilst the remaining Gram-negative test organisms were not inhibited at 2% (v/v) [12]. The MIC for *C. albicans* was 0.12% [12].

11.3.6   Rutaceae

Rutaceae genera yielding essential oils include *Citrus* and to a much lesser extent *Murraya* [63] and *Boronia* [64].

Citrus oils are derived from the fruit peel of mostly hybrid species including orange (*Citrus* × *sinensis*), lemon (*C.* × *limon*), mandarin (*C. reticulata*), tangerine (*C.* × *tangerina*), bergamot (*C.* × *aurantium* ssp. *bergamia*) and grapefruit (*C.* × *paradisi*). Citrus peel oils typically contain high levels of limonene [24, 65], although bergamot oil generally has a slightly lower limonene content (~50%) and is higher in linalool content (~15%) [66]. Citrus oils are primarily used as fragrances or solvents but have also been investigated as antimicrobial agents for the preservation of food and extension of food shelf-life. Of the citrus oils, a selection of common medical pathogens such as *S. aureus*, *E. coli*, *K. pneumoniae* and *E. faecalis* had MICs of 1 to >2% (v/v) oil for lime (*C. aurantifolia*), orange (*C.* × *aurantium*), bergamot, lemon and grapefruit [12]. Lower MICs of 0.25 and 0.5% (v/v) were reported against *C. albicans* and *E. coli* for petitgrain (*C. aurantium*) oil [12], which was obtained from leaf rather than peel. MICs of orange and lemon oil ranged from 0.25 to >4% (v/v) and for bergamot ranged from 0.125 to >4 % (v/v) against the foodborne pathogens *C. jejuni*, *E. coli* O157, *L. monocytogenes*, *B. cereus* and *S. aureus* [66]. A further study found MICs of orange oil of 0.62% (v/v) for *V. cholerae* and *A. hydrophila* but several other test organisms were not inhibited at 10% (v/v) [24]. Citrus oils also have antifungal activity, with MICs ranging from 0.25 to 2% (v/v) against *C. albicans* for lime, orange, bergamot, lemon and grapefruit [12]. Orange, lemon, mandarin and grapefruit oils all completely inhibited the outgrowth of *Aspergillus niger*, *A. flavus*, *Penicillium verrucosum* and *P. chrysogenum* at 0.94% [67]. Growth was inhibited in a dose-dependent manner at concentrations below this.

11.3.7   Other

There are several additional essential oils with antimicrobial properties that are not contained within the seven plant families already discussed, for example geranium oil from *Pelargonium graveolens* (Geraniaceae) [59, 68] and sandalwood oil from *Santalum* spp. (Santalaceae) [12].

11.3.8   Comparative Activities of Essential Oils

These data demonstrate that the majority of plant essential oils have broad-spectrum antimicrobial activity. Exceptions to this include sandalwood (*Santalum album*) and vetiver (*C. zizanioides*) oils, which are both highly active against Gram-positive bacteria and

C. albicans, but relatively inactive against Gram-negative bacteria [12]. In addition, Manuka oil shows selective activity against Gram-positive bacteria only [56]. It is also apparent from the data that some essential oils are relatively more active than others. In particular, thyme, oregano, cinnamon [11], *Cymbopogon*, bay [12] and possibly rosewood [12, 43] are amongst the most active. This corresponds with these oils containing mainly phenols (thyme, oregano, bay), aldehydes (cinnamon, *Cymbopogon*) or alcohols (rosewood) as major components, as discussed below.

11.4 Antimicrobial Essential-Oil Components

Essential-oil components are largely hydrocarbons, some of which may also be oxygenated. Many nonoxygenated monoterpenes share the chemical formula $C_{10}H_{16}$ and differ only by the position of the double bonds. This said, isomers, both structural and optical, are not uncommon amongst essential-oil components. Examples of isomerization for individual components are the +/− enantiomers of limonene, α-pinene and camphene. Individual isomers are not often tested for antimicrobial activity since they can be difficult to separate.

Grouping essential-oil components by chemical similarities, for example the presence of particular moieties such as alcohols, is one approach to evaluating both the overall, and relative, antimicrobial activity of each component or component type. Whilst the groups listed below have been used previously to discuss the relative activity of components [8], alternative classifications based on other chemical characteristics have also been used [69].

11.4.1 Aldehydes

Examples of aldehydes found in essential oils are cinnamaldehyde, benzaldehyde, neral and geranial. Cinnamaldehyde, usually *trans*-cinnamaldehyde, is a major component of cinnamon bark oil. MICs reported for cinnamaldehyde include 0.05% to 0.1% (v/v) [70], 250 to 1000 μg/ml (0.025 to 0.1%) [44] and 500 to 1000 mg/l (0.05 to 0.1%) [71] for a range of bacteria including *E. coli*, *S. aureus*, *B. subtilis* and *Streptococcus* spp. Additional MICs include 25.9 μg/ml (0.0026%) for *Mycobacterium* spp. [33] and <0.04 to 0.16 μg/μl (<0.004 to 0.016%) for *Candida* spp. and dermatophytes [40]. The isomers neral and geranial are together known as citral. Whilst little susceptibility data are available describing the activity of the individual isomers, MICs of citral were 0.03 to 0.06% (v/v) for *L. monocytogenes*, *Bacillus cereus* and *S. aureus* [66], 100 μg/ml (0.1%) for *V. vulnificus*, and 500 μg/ml (0.5%) for *E. coli*, *L. monocytogenes* and *S.* Typhimurium [72]. In this same study, MICs for the aldehyde citronellal were 250 μg/ml (0.025%) for *V. vulnificus* and 1000 μg/ml (0.1%) for *L. monocytogenes*, whereas *S.* Typhimurium and *E. coli* were not inhibited at 1000 μg/ml [72]. MICs of perillaldehyde were 250 μg/ml (0.025%) for *V. vulnificus*, 500 μg/ml (0.05%) for *E. coli* and *S.* Typhimurium, and 1000 μg/ml (0.1%) for *L. monocytogenes* [72]. Lastly, MICs for the aromatic cyclic terpene aldehyde vanillin ranged from 15 to 75 mmol/l (0.2 to 1.1%) for *E. coli*, *L. innocua* and *L. plantarum* [73].

11.4.2 Alcohols

The terpene alcohols include some of the most antimicrobially active essential-oil components. Their activity may be attributed to the alcohol moiety, which has intrinsic antimicrobial activity and enhances the solubility of these components in both aqueous laboratory media and microbial membranes. In terms of structure, terpene alcohols may be acyclic (linear) or cyclic and may be mono-, di- or sesquiterpenes.

Examples of linear monoterpene alcohols are linalool and geraniol. Linalool, the main component of lavender oil, is utilized extensively in perfumery due to its characteristic sweet, soft fragrance. Linalool has broad-spectrum antimicrobial activity, with published MICs including 0.06% (v/v) for *C. jejuni* [66], *C. albicans* [51] and *E. coli* [74], 0.125% (v/v) for *B. cereus* [66], *S. aureus* [66] and *Stenotrophomonas maltophilia* [75], and 0.25% (v/v) for *E. coli*, *L. monocytogenes* [66] and *S. aureus* [74]. In addition, MICs of 0.008 to 0.5% (v/v) have been found for *C. albicans* [25]. For geraniol, MICs include 500 to 1000 µg/ml (0.05 to 0.1% w/v) against *E. coli*, *S.* Typhimurium, *L. monocytogenes* and *Vibrio vulnificus* [72], 0.08 mg/ml (0.008%) for *B. subtilis*, and 0.7 to 1.4 mg/ml (0.07 to 0.14%) for *B. cereus*, *S. aureus* and *E. coli* [76]. Against fungi, MICs include ≤0.03% (v/v) [77] and 0.31 µg/µl (0.03%) for *Candida* spp. [40], <0.04 to 0.16 µg/µl (<0.004 to 0.016%) for the dermatophytes *Trichophyton* and *Microsporum* [40], and 160 ppm (0.016%) for other fungi including *Rhizoctonia*, *Colletotrichum*, *Alternaria*, *Fusarium*, *Botrytis* and *Aspergillus* spp. [78].

Cyclic alcohols include the monoterpenes menthol, terpinen-4-ol and α-terpineol, and the sesquiterpenes nerolidol and bisabolol. MICs of menthol, which is present in most mint oils, include 0.62 mg/ml (0.06%) for *S. aureus* [14], 3.08 mM (0.05%) [79], 2.5mg/ml (0.25%) [14] and 525 µg/ml (0.052%) [13] for *E. coli*. MICs of 0.5 to 2 g/l (0.05 to 0.2%) have also been published for a selection of bacteria, although *L. plantarum* was not inhibited at 3 g/l (0.3%) [80]. Terpinen-4-ol, a major component of tea-tree oil, has MICs of 0.25% (v/v) for *S. aureus* [74, 81], 0.06% (v/v) for *E. coli* [74] and 2% (v/v) for *P. aeruginosa* [82]. For yeasts, MICs range from 0.015 to 0.036 % (v/v) [83] or 0.12 to 0.25% (v/v) [51] and from 0.008 to 0.06% (v/v) for the dermatophytes *Trichophyton*, *Epidermophyton* and *Microsporon*, and for *Aspergillus* spp. and *Penicillium* sp. [51]. The component α-terpineol has similar activity to terpinen-4-ol, with MICs of 0.06% (v/v) and 0.25% (v/v) for *E. coli* and *S. aureus* respectively [74], 0.06 to 0.25% (v/v) for yeasts, and 0.008 to 0.03% (v/v) for the dermatophytes, *Aspergillus* spp. and *Penicillium* sp. [51]. For sesquiterpene alcohols, farnesol, nerolidol and bisabolol did not inhibit *E. coli* at 1% (v/v), whereas MICs for *S. aureus*, *S. pneumoniae* and *H. influenzae* ranged from 0.008 to 0.5% (v/v) [84].

11.4.3   Phenols

These components are by definition aromatic. The isomers thymol and carvacrol, and a third phenolic essential-oil component eugenol, are major components of thyme, oregano and clove oils, respectively, and have been extensively studied for their antimicrobial effects.

MICs reported for carvacrol include 13.3 to 20 mM (0.2 to 0.3%) [85], 250 to 500 µg/ml (0.025 to 0.05%) [72], 0.25 to 1.0 g/l (0.02 to 0.1%) [80] and 0.4 to 0.8% [70] for various pathogenic bacteria including *E. coli*, *S. aureus* and *B. subtilis*. However, *L. plantarum* was not inhibited at 3 g/l (0.3%) [80]. For fungi, MICs of 0.2% [86] and 0.08 to 0.16 µl/ml (0.008 to 0.016%) [2] have been reported for *C. albicans*, as well as 0.04 to 0.32 µl/ml (0.004 to 0.03% v/v) for a range of *Candida* yeasts, dermatophytes and *Aspergillus* spp. [31]. For thymol, MICs include 0.4 to 0.8% for *S. aureus*, *E. coli* and several streptococci [70], 0.31 mg/ml (0.03%) [14] and 500 µg/ml (0.05%) [87] for *S. aureus*, and 5 mg/ml

(0.5%) [14] and 500 μg/ml (0.05%) [87] for *E. coli*. For yeasts, MICs were 1.5 mM (0.02%) for *S. cerevisiae* [88], 0.16 to 0.32 μl/ml (0.016 to 0.032%) for *C. albicans* [2] and 0.08 to 0.32 μl/ml (0.008 to 0.03% v/v) for a range of yeasts, dermatophytes and *Aspergillus* spp. [31]. An MIC of 200 ppm (0.02%) has also been reported for fungi such as *Aspergillis* and *Fusarium* [78]. Eugenol has MICs of 0.4 to 0.8% (v/v) for mastitis-associated bacteria including *S. agalactiae*, *S. dysgalactiae*, *S. aureus*, *E. coli* and *Streptococcus uberis* [70], 0.05 to 0.1% (v/v) for *S. aureus* and *E. coli* [87], 500 to 1000 μg/ml (0.05 to 0.1%) for the foodborne pathogens *E. coli*, *S.* Typhimurium, *L. monocytogenes* and *V. vulnificus* [72], and 12.2 mM (0.2%) for *B. cereus* and *E. coli* [85]. However, no inhibition was seen against *S. aureus* at 195 mM (3.2%) [85], *L. monocytogenes* at 1000 μg/ml (0.1%) [72] or *L. plantarum* at 3 g/l (0.3%) [80]. Maximal sublethal concentrations, which are similar to MICs, were reported as 2.12 to 3.18 mg/l (0.0002 to 0.0003%) for several bacteria including *E. coli* and *Pseudomonas fluorescens* [89]. For fungi, MICs of 79 mM (1.3%) [85] and 0.05 to 0.2% (v/v) [86,90] have been published for *C. albicans*, in addition to 1.8 mM (0.03%) for *S. cerevisiae* [88].

11.4.4 Ketones

The most commonly encountered ketone essential-oil components include menthone, carvone, pulegone and piperitone, which are monocyclic, and camphor, fenchone, thujone and verbenone, which are bicyclic. Data for monocyclic components include pulegone MICs of 105.1 to 157.7 mM (1.6 to 2.4%) for bacteria and 26.3 mM (0.4%) for *C. albicans* [85]. Carvone MICs are 53.3 to 106.5 mM (0.8 to 1.6%) for bacteria and *C. albicans* [85] and 225 ppm (0.025%) for fungi [78]. For the bicyclic component camphor, MICs of 3 to 8 μl/ml (0.3 to 0.8%) for fungi and 2.5 μl/ml (0.25%) for *P. tolaasii* [91] and MBCs of 0.75 to 1.75 mg/ml (0.075 to 0.17%) have been reported for bacteria including *B. subtilis*, *E. coli* and *S. aureus*, and 2.15 mg/ml (0.21%) for *A. niger* [92]. Similarly, verbenone MICs are 1.25 to 1.85 mg/ml (0.12 to 0.18%) for organisms including *E. coli*, *B. subtilis*, *S. aureus* and *C. albicans*, and 2.25 mg/ml (0.22%) for *A. niger*. [92,93]. However, one study found no inhibition at 9 μl/ml (0.9%) [93].

11.4.5 Esters

Esters found in essential oils are oxygenated and may also contain a hydroxyl group. They may be acyclic, such as geranyl, linalyl and neryl acetate, or cyclic, such as methyl salicylate and benzyl acetate. MICs for linalyl acetate include 1.25 mg/ml (0.125%) for *S. aureus* and 5 mg/ml (0.5%) for *E. coli* [14], 19 300 ppm (1.93%) for *P. aeruginosa*, *E. coli*, *S. aureus* and *C. albicans* [69], 10 μl/ml (1%) for *P. tolaasii* [91] and 163mM (3.2%) for *B. cereus* [85]. No inhibition was seen at 163 mM (3.2%) for *S. aureus*, *E. coli* or *C. albicans* [85]. A further study found MICs of 1 to 4% (v/v) for *C. albicans* [25] and 7.5 to 10.5 μl/ml (0.75 to 1.05%) against fungi [91]. For geranyl acetate, MICs of 18 000 ppm (1.8%) for *P. aeruginosa*, *E. coli*, *S. aureus* and *C. albicans* [69] and 163 mM (3.2%) [85] for *B. cereus* have been published, but *S. aureus*, *E. coli* and *C. albicans* were not inhibited at 163 mM (3.2%) [85]. MICs for neryl acetate were 17 800 ppm (1.78%) for *P. aeruginosa*, *E. coli*, *S. aureus* and *C. albicans* [69], whereas no inhibition was seen for carvacryl acetate at 3 g/l (0.3%) [80] nor for bornyl acetate at 9 μl/ml (0.9%) [93].

11.4.6   Oxides and Epoxides

The major oxide found in essential oils is 1,8-cineole, which is also known as eucalyptol. It is the main component of eucalyptus oil and may also be present in high quantities in rosemary oil. MICs for 1,8-cineole include 0.25 to 0.5% (v/v) [74, 81], 207.5mM (3.2%) [85], 2.8 to 5.6 mg/ml (0.28 to 0.56%) [76] and 1.25 to 2 mg/ml (0.12 to 0.2%) [92] for a range of bacteria such as *E. coli*, *S. aureus* and *Bacillus* species. However, one study reported no inhibition at 207.5 mM (3.2%) for *B. cereus* and *E. coli* [85]. For yeasts, MICs of 0.5% (v/v) [51], 1 to >4% (v/v) [83], 5 to 20 μl/ml (0.5 to 2%) [2], 1.5 mg/ml (0.15%) [92] and >8% (v/v) [77] have been reported. MICs for a range of filamentous fungi include 0.06 to 8% (v/v) [51] and 2.25 mg/ml (0.22%) for *A. niger* [92]. MICs for the less common oxide component 1,4 cineole have been reported as 5900 to 17 500 ppm (0.59 to 1.75%) [69].

Epoxides found in essential oils include linalool oxide, limonene oxide and caryophyllene oxide. For *P. aeruginosa*, *E. coli*, *S. aureus* and *C. albicans*, MIC ranges for linalool oxide were 1900 to 18 500 ppm (0.19 to 1.85%) and for limonene oxide were 1900 to 18 200 ppm (0.19 to 1.82%) [69].

11.4.7   Methyl Ethers

This group of compounds includes methyl chavicol (estragole), safrole, asarone and anethole. For methyl eugenol, MICs were 1600 ppm (0.16%) for *C. albicans* and 10 300 to 20 300 ppm (1.03 to 2.03%) for bacteria [69]. For methyl carvacrol, the MIC for *S. cerevisiae* was 3 g/l (0.3%), whereas the five bacteria tested were not inhibited at this concentration [80].

11.4.8   Terpene Hydrocarbons

The terpene hydrocarbon group contains nonoxygenated compounds that are mostly monoterpenes or sesquiterpenes, which may be acyclic, monocyclic or bicyclic. The majority are unsaturated, meaning that they contain one or more double bond. The lack of alcohol moiety or other oxygenation generally means that these components are amongst the least water-soluble. The names of these components typically end with the suffix -ene.

The monocyclic terpene hydrocarbons are a large group of components that are found in a great number of different essential oils and include limonene, phellandrene, α- and γ-terpinene, terpinolene, thujene, *p*-cymene, cadinene and sabinene. Limonene is an alicyclic monoterpene that has activity similar to many citrus oils, since citrus-peel oils may contain more than 90% limonene. MICs for limonene include 176 to 235 mM (2.4 to 3.2%) for *S. aureus*, *B. cereus*, *E. coli* and *C. albicans* [85], 10 μl/ml (1%) for *P. tolaasii* [91], 0.5% (v/v) for *C. albicans* [51] and 5 to 9 μl/ml (0.5 to 0.9%) for fungi [91]. Lastly, MICs for both (+)-limonene and (−)-limonene ranged from 2 to 27 mg/ml (0.2 to 2.7%) [94] and were 16 500 ppm (1.65%) [69] for *S. aureus*, *P. aeruginosa* and *Cryptococcus neoformans*, indicating that these enantiomers do not differ greatly in activity. The aromatic monoterpene cymene has MICs of 1.25 to 10 μl/ml (0.125 to 1% v/v) for a range of yeasts and dermatophytes [31]. However, no activity was seen against *Aspergillus* spp. at 20 μl/ml (2%) [31], nor against *S. aureus*, *E. coli* and *C. albicans* at 8% (v/v) [74]. MICs for γ-terpinene were 4% (v/v) for *S. aureus* and >8% for *E. coli* and *C. albicans* [74], 1.25 to 10 μl/ml (0.125 to 1.0%

v/v) for a range of yeasts and dermatophytes, and 10 to >20 μl/ml (>2%) for *Aspergillus* spp. [31].

Acyclic unsaturated terpene hydrocarbons include ocimene and β-myrcene. The little data available for these two components suggests that they are not highly active [95].

Components such as α- and β-pinene and camphene are characterized by their bridged bicyclic structure. For α-pinene, MICs include 1.25 to 2.25 mg/ml (0.12 to 0.22%) for bacteria including *S. aureus, B. subtilis, P. aeruginosa* and *E. coli* [92], 1.5 mg/ml (0.15%) [92] and 0.12% (v/v) [51] for *C. albicans*, and 0.008 to 0.03% (v/v) for the dermatophytes *Aspergillus* spp. and *Penicillium* sp. [51]. For β-pinene, MICs were <0.016% (v/v) for *C. albicans* [51]. Finally, very little data are available for sesquiterpene hydrocarbons such as cadinene, caryophyllene and aromadendrene.

11.4.9   Generalizations about Component Activity

In general, the aldehydes, phenols and alcohols appear to be the most active component groups and the terpene hydrocarbons and methyl esters appear to be the least active. Similar rankings have been proposed previously [8,24]. The activity of the aldehydes may be due to the reactivity of the carbonyl group, which may react to form several antimicrobial moieties such as alcohols or acids. Activity amongst the phenols and alcohols is at least partly due to the hydroxyl group, which has intrinsic antimicrobial activity and also contributes to the relatively greater solubility of these components in biological membranes. Furthermore, the ability of components to release or accept protons has been postulated as an important factor in antimicrobial activity [80,96,97].

11.5   Factors Influencing Activity

11.5.1   Microorganism-Related Factors

Essential oils are generally broad-spectrum antimicrobial agents, with similar activity against Gram-positive and Gram-negative bacteria, anaerobic and aerobic organisms, yeasts and other fungi. Although it has been postulated that essential oils display differential activity towards Gram-positive and Gram-negative bacteria [21], current opinion seems to favour no general relationship between Gram reaction and activity. However, there are a few microorganisms and growth forms that are nonetheless worthy of further discussion.

Gram-negative bacteria belonging to the genus *Pseudomonas* are typically less susceptible to volatile oils than related Gram-negative or Gram-positive bacteria [12,24]. In particular, the species *P. aeruginosa* is generally less susceptible to a diverse range of antimicrobial compounds, including essential oils. This reduced susceptibility has been attributed to the outer membrane and associated properties such as drug efflux [82,98], which are discussed in more detail later in this chapter.

Bacterial spores, such as those formed by *Bacillus* and *Clostridium* spp., have been postulated as a growth form that may be less susceptible to essential oils, largely because spores are relatively resistant to a wide range of adverse environmental conditions such as temperature, pH, chemical exposure and desiccation. Essential oils from eucalyptus, cedar, grapefruit, orange and rosemary inactivated *B. cereus* spores after 18 hours at concentrations in the range of 100 to 400 ppm (0.01 to 0.04%) and after 48 hours for *C. botulinum* spores in the range of 300 to 500 ppm (0.03 to 0.05%) [99]. Spores and vegetative cells were approximately similar in susceptibility. *C. botulinum* spores were also inhibited by cinnamon, clove, thyme, pimenta and oregano oils at 200 ppm (0.02%) [100]. Treatment with tea-tree oil also resulted in a loss of *B. cereus* and *B. subtilis* spore viability [101]. Oregano oil at 2% (v/v) inhibited the germination and outgrowth of *C. perfringens* spores in experimental cooked ground beef and turkey experiments [102, 103]. The essential-oil component carvacrol at 2 mM (0.03%) elicited significant spore death and at 3mM (0.05%) resulted in a >4-fold reduction in *B. cereus* spore viability after 40 seconds' exposure. Interestingly, in this study spores were less susceptible to carvacrol than the vegetative cells [104]. Thymol and carvacrol at 0.6 mM (0.009%) also reduced the growth rate and extended the lag phase of germinating *B. megaterium* spores [105]. Cinnamaldehyde, a principal component of cinnamon oil, also decreases spore viability for *Alicyclobacillus acidoterrestris* at 200 ppm (0.02%) [106]. However, 3330 μg/ml (0.33%) of thyme oil, cinnamon oil, carvacrol or perillaldehyde showed negligible sporicidal activity against *B. cereus* after one hour, as did oregano oil at 667 μg/ml (0.07%) [107]. Filamentous fungi may also produce spores, in addition to a range of other growth forms including pseudohyphae, hyphae and fruiting bodies. Whilst very little data is available, one study suggests that *A. niger* conidia are comparatively less susceptible to tea-tree oil than germinated conidia [52].

As a generalization, antibiotic-resistant organisms do not have reduced susceptibility to essential oils since the majority of studies comparing the susceptibility of antibiotic-resistant and -sensitive microorganisms to plant essential oils have not found major differences. For example, methicillin-resistant *Staphylococcus aureus* (MRSA) did not differ from methicillin-susceptible *S. aureus* (MSSA) in susceptibility to tea-tree [24, 108], oregano [24, 32], thyme [24], clove [24], cinnamon leaf [24], lemongrass [24, 59], palmarosa [24], geranium [59] and lavender oils [24]. Also, MRSA and MSSA did not differ in susceptibility to the components carvacrol and thymol [32]. Similarly, vancomycin resistance enterococci (VRE) did not differ from susceptible enterococci to tea-tree, oregano, thyme, clove, cinnamon leaf, palmarosa, lavender [24] and lemongrass oils [24, 59]. For yeasts, *C. albicans* isolates resistant to fluconazole and/or itraconazole were similar to nonresistant isolates in susceptibility to tea-tree oil, terpinen-4-ol and 1,8-cineole in one study [83], and to rosemary, oregano, thyme, *Lippia graveolens*, basil, sage and cinnamon oils in another [20]. In all of these studies both resistant and susceptible isolates were studied. However, there are also a number of studies where only antibiotic-resistant isolates were studied. Comparison of the essential-oil susceptibility of these antibiotic-resistant organisms to previously published data for susceptible organisms also suggests that essential-oil susceptibility is not reduced for antibiotic-resistant organisms such as MRSA [38, 109], *Campylobacter* spp. [110], *P. aeruginosa* [38] and extended-spectrum β-lactamase-producing *E. coli* isolates [34]. The lack of difference seen in essential-oil susceptibility is most likely due to the mechanisms of antibiotic resistance and essential-oil activity being largely independent, where mechanisms of antibiotic resistance have little or no bearing on mechanisms of essential-oil activity.

11.5.2   Environmental Conditions

A range of physicochemical factors and other environmental conditions may influence the antimicrobial efficacy of essential oils. For example, activity appears to be enhanced at pH values lower, and possibly also higher, than neutral. MICs of oregano, thyme, lemon balm and marjoram oil were significantly decreased at lower pH values for *Listeria* spp. and to a lesser extent for *P. putida* [30]. Thyme oil was also more active against *S.* Typhimurium at lower pH levels [111]. Spruce essential oil (*Picea excelsa*) was more active against *L. monocytogenes* at lower pH [112]. Similarly, carvacrol was more antimicrobially active against *B. cereus* at lower pH values, but also at values higher than neutral [104]. This enhancement of activity at lower pH is though to occur because the hydrophobicity of plant oils is increased at lower pH values, which in turn enhances the ability of the oils to enter the microbial membranes [27]. With regard to temperature, essential oils appear to be less active at lower temperatures [75, 104, 113]. However, spruce essential oil was more active against *L. monocytogenes* at 13 °C compared to the standard temperature of 37 °C [112]. Less activity at lower temperatures is presumably due to the reduced solubility of essential oils and components or to alteration of membrane lipid composition in response to low temperature [104].

Several studies suggest that the presence of proteins and other organic matter may impair the activity of essential oils. The activity of spruce essential oil against *L. monocytogenes* was decreased in the presence of the milk protein caseinate [112] and the activity of tea-tree oil was reduced in the presence of bovine serum albumin, skim milk and baker's yeast in standard microdilution assays [114]. Tea-tree oil activity was also decreased when bovine albumin was present in suspension tests, although not against all test organisms [115]. Suspension tests indicated that eugenol was marginally less active under dirty conditions (3 g/l bovine albumin) but that thymol had similar activity in both clean and dirty conditions [116]. However, the presence of beef extract (6 and 12%) extended the lag growth phase for *L. monocytogenes* grown in the presence of oregano and thyme oil [27], indicating that in this instance the protein enhanced the activity of the essential oils. The most likely mechanism by which proteins may impair essential-oil activity is by interfering with interactions between the essential oil and cell surface. Similarly, oils or fats may interfere with essential-oil activity [27, 112, 117]. The addition of 5 and 10% sunflower oil to growth media containing 30 ppm oregano oil or 60 ppm thyme oil significantly reduced the lag growth phase of *L. monocytogenes*, indicating that the sunflower oil interfered with the activity of the essential oils [27]. Fat is proposed to shield cells by forming a protective coating around the outside of the cell or by sequestering the essential oil and thereby reducing its availability in solution [117].

Solubilizers such as surfactants are used in many *in vitro* antimicrobial assays but are also known to impair the antimicrobial actions of essential oils via encapsulation in surfactant micelles. Tween 80 has been shown to decrease the activity of thyme oil when used at 0.5ml/l (0.05%) [111] and tea-tree oil when present at 5% (v/v) [114]. Tween 20 (5% v/v) also decreases the activity of tea-tree oil [114]. These interactions between essential oil and surfactant have implications for both *in vitro* susceptibility assays and pharmaceutical preparations containing surfactants, given that activity may be compromised.

The majority of essential-oil studies examine antimicrobial activity in the liquid form. However, as stated at the beginning of this chapter, essential oils are volatile and as such exist in a gaseous or vapour phase in addition to a liquid phase. The antimicrobial activity of essential oils in vapour phase has been examined in numerous studies against both bacteria and fungi. The method is more commonly used with fungi as this relates directly to the potential use of essential oils as fumigants for stored food crops. Inhibition of bacterial and/or fungal growth by the vapour phase of essential oils has been demonstrated for geranium and lemongrass [59], orange and bergamot [118], spearmint and tea tree [119], cinnamon, oregano and thyme [120], and clove [121]. The components carvacrol, eugenol and methanol [80] are also active in vapour phase. A large study investigating inhibition of *T. mentagrophytes* by vapour phase demonstrated that 55 out of 72 oils inhibited growth to some extent [122]. In an earlier study by the same research group, oils of cinnamon bark, lemongrass, thyme, lavender and tea tree were also inhibitory in the vapour phase, and were inhibitory at significantly lower concentrations in vapour phase compared to liquid phase, which was assessed by agar dilution [123].

Other conditions or factors that may influence activity include modified atmosphere [124] or anaerobic conditions [111, 114] and the presence of additional cations, similar to hard water [114, 125].

11.6   Mechanisms of Action

11.6.1   Membrane and Membrane-Related Actions

It is generally accepted that the treatment of microorganisms with essential oils results in the impairment of membrane integrity and function. This over time may lead to the loss of cell homeostasis, the leakage of intracellular constituents and eventually cell death. This generalized model for the mechanism of action of essential oils is supported by data from numerous scientific publications [14, 35, 73, 81, 118, 126]. Also, these effects are generally seen in a time- and dose-dependent manner, with higher concentrations causing severe effects more rapidly and lower concentrations exerting either nonlethal effects or lethal activity only after a longer exposure time.

The initial interaction between an essential-oil component and a microbial cell is likely to be the passive diffusion of the component molecules through the cell wall of Gram-positive bacteria and fungi or the outer membrane of Gram-negative bacteria. Active transport via transmembrane pumps has not been demonstrated. Due to the lipophilic or hydrophobic nature of many essential-oil components, the components then preferentially partition into the cell membrane, resulting in the alteration of membrane properties. Evidence suggests that terpenes alter the physical properties of membranes as a result of their insertion between the fatty acyl chains of the lipid bilayer [127]. This disturbs the van der Waals interactions between acyl chains [128], disrupting lipid packing and decreasing lipid order [129]. The accumulation of terpene molecules in the bilayer results in an increase in lipid volume, which causes membranes to swell and increase in thickness [127]. The expansion of the membrane leads to a decrease in membrane integrity and eventually results in the loss of intracellular compounds. As such, changes in membrane fluidity may be amongst the first effects elicited by essential-oil treatment. The expansion of model membranes has been demonstrated after exposure to the essential-oil components carvacrol and cymene [97], and carvacrol treatment immediately increased membrane fluidity in *B. cereus* [128]. Carvacrol, thymol, γ-terpinene and *p*-cymene caused decreases in the lipid melting temperatures of model membranes, which is suggestive of increased membrane fluidity [130]. Similarly, the exposure of *C. albicans* to tea-tree oil and its components terpinen-4-ol, 1,8-cineole, terpinolene and α-terpinene resulted in increased membrane fluidity after 30 minutes [131]. The expansion and increased fluidity of the membrane may lead to breaches in membrane integrity that allow small intracellular components such as hydrogen, potassium and sodium to pass through the cell membrane. The loss of these ions is associated with decreased membrane potential, since an important component of membrane potential is the ion gradient between the cell interior and exterior.

The loss of potassium or sodium ions has historically been regarded as one of the earliest signs that exposure to an antimicrobial compound has resulted in membrane damage. This is because these comparatively small molecules can pass through a compromised membrane with relative ease. Oregano-oil treatment resulted in the loss of potassium ions from *S. aureus* and *P. aeruginosa* after as little as 16 minutes [35]. Similarly, *O. compactum* essential-oil treatment at the MIC resulted in the loss of potassium ions almost immediately for both *P. aeruginosa* (1% oil) and *S. aureus* (0.031% oil) [132]. Tea-tree oil (0.25%) treatment resulted in the loss of potassium ions within minutes for *E. coli*, but not *S. aureus* [133]. Treatment with *C. citratus* oil in vapour form resulted in the loss of potassium ions from both *A. flavus* [62] and *S. cerevisiae* [60]. Palmarosa oil (0.1%) resulted in significant leakage after 30 minutes from *S. cerevisiae* [134], as did the oil component geraniol at 0.06% (v/v). Interestingly, geranyl acetate treatment (0.02%, up to two hours) did not result in significant potassium loss. Farnesol, nerolidol and plaunotol cause the loss of potassium ions in *S. aureus* [135]. Thymol (0.1% w/v) and eugenol (0.1% v/v) treatment result in the loss of potassium ions from *E. coli* and *S. aureus* [87]. Carvacrol treatment results in the loss of potassium ions from *B. cereus* [126] and causes the release of carboxyfluoroscein from large unilamellar vesicles [130]. Similarly, thymol [130], menthol and to a lesser extent linalyl acetate [14] cause the release of carboxyfluoroscein from large unilamellar vesicles. Vanillin also increases potassium permeability in *E. coli*, *Lactobacillus plantarum* and *Listeria innocua* [73].

This initial interaction whereby membrane fluidity is increased and intracellular ions are lost leads to decreases in membrane potential and intracellular pH, both of which have been observed when *Enterococcus* spp. were treated with citrus oils [118] and *B. cereus* with carvacrol, where membrane potential was reduced at ≥ 0.01 mM and completely dissipated at 0.15 mM [97, 126]. Decreases in membrane potential or intracellular pH have also been seen for oregano [35], Spanish oregano, Chinese cinnamon and savory essential oils [42], *O. compactum* oil [132], thymol [136], carvacrol [96, 136] and cymene [97]. In several of these studies it was also shown that the treatment of cells with essential oil or component resulted in the depletion of internal ATP pools. This was observed for carvacrol [126] and citrus oils [118] but not vanillin [73]. Inhibition of respiration has also been demonstrated for tea-tree oil, where respiration was completely inhibited at 0.5% for *E. coli*, inhibited by approximately 90% at 0.75% oil for *C. albicans*, and inhibited by about 60% at 1% oil for *S. aureus* [133, 137]. Inhibition occurred after 5 to 10 minutes' exposure to tea-tree oil, indicating that this is an immediate effect [133]. However, respiration inhibition occurred at concentrations similar to or greater than the MIC, indicating that it may simply be a reflection of cell death. The essential-oil component vanillin inhibited respiration in *E. coli* and *Listeria innocua*, also at the MIC [73].

At higher essential-oil concentrations or after prolonged exposure, gross membrane damage may occur, which is commonly quantified by the loss of intracellular materials that absorb at wavelengths of 260 or 280 nm. Materials absorbing at 260 nm indicate DNA, whereas those absorbing at 280 nm indicate protein. The presence of high levels of these materials in cell-free preparations indicates that large molecules have been lost from the cell interior and that major membrane damage has occurred. The loss of macromolecules such as DNA or protein has been assessed in many different studies because it is relatively easy to measure in the laboratory. The essential oils Spanish oregano (*Corydothymus capitatus*), Chinese cinnamon and savory (*Satureja montana*) each resulted in the loss of significant amounts of 260 nm-absorbing materials from both *E. coli* O157 : H7 and *L. monocytogenes* at half the MIC [42]. Tea-tree oil treatment also results in the loss of 260 nm-absorbing material from *S. aureus* [81], *C. albicans* and *C. glabrata* [131]. With regard to essential-oil components, treatment with 0.25% terpinen-4-ol for 60 minutes resulted in significant loss from *S. aureus* [81] and from *C. albicans* treated with 0.5% (v/v) [131]. Interestingly, no significant leakage was seen from *S. aureus* after 30 minutes' treatment, indicating that at this concentration of terpinen-4-ol the effect is not immediate [81]. Treatment with 1,8-cineole also resulted in leakage from *C. albicans* [131] and *S. aureus* [81]. Carvacrol treatment at $\geq 0.05\%$ v/v for one hour resulted in significant leakage of 280 nm-absorbing materials in *C. albicans* [86]. Similarly, exposure to 12 mM thymol for 10 minutes resulted in significant leakage from *S. cerevisiae* [88]. Exposure to $\geq 0.05\%$ v/v of eugenol for one hour resulted in significant leakage of 280 nm-absorbing materials from *C. albicans* [86] and from *S. cerevisiae* after exposure to 12 mM eugenol for 10 minutes [88]. Treatment with 0.4 ml/l (0.04%) cinnamaldehyde did not result in significant protein leakage from *Bacillus cereus* over two hours, whereas this same treatment resulted in significant leakage from *S. aureus* [138]. Similarly, no significant potassium loss was seen after treatment with 0.02% (v/v) geranyl acetate [134]. The uptake of the fluorescent nucleic acid stain propidium iodide, which is excluded by intact biological membranes, has also been used to quantify membrane damage. Propidium iodide uptake has been observed after treatment with clove [55], oregano [139], thyme [31, 139] and tea-tree [133] oils, and to a lesser extent cinnamon oil [139]. Whilst the loss of low levels of small ions such as potassium may not necessarily be lethal for the cell, the loss of significant quantities of 260/280 nm-absorbing material almost certainly indicates cell death. In some instances significant quantities of 260/280 nm-absorbing materials have been detected after as little as 10 minutes [88], indicating that gross membrane damage and corresponding cell death can occur very rapidly. Interestingly, aldehyde components appear to differ in action from the remaining component classes, with most evidence indicating that they have lethal activity in the absence of extensive membrane damage [140–142].

Despite clearly causing major membrane damage, most essential oils do not appear to cause destruction of the cell wall of Gram-positive organisms and yeasts or the outer membrane of Gram-negative bacteria. This type of gross damage is also termed cell lysis and refers to the rupture or destruction of the cell wall such that the cell form or shape is no longer apparent. Cell lysis was not observed in *S. aureus* treated with tea-tree oil [81], *B. cereus* treated with cinnamaldehyde [138] or *P. aeruginosa* and *S. aureus* treated with *O. compactum* oil [132]. However, studies utilizing electron microscopy have revealed extensive cell-wall damage after exposure to several different essential oils [65, 88, 143–145], indicating that oils are capable of cell-wall damage, if not lysis.

11.6.2 Adaptation and Tolerance

When discussing the mechanism of action of essential oils, it is important to remember that microorganisms do not necessarily remain passive during exposure and in fact possess a range of responses that may help them to counter and potentially adapt to some of the deleterious effects. These responses and adaptations are most apparent after exposure to low, nonlethal concentrations of oil for extended time periods, as this allows the organisms to mount the appropriate response.

Microorganisms have a range of innate properties and functions that protect the cell from any number of insults. These include the barrier of a protective cell wall or outer membrane as well as energy-driven mechanisms such as efflux. These have both been demonstrated to have a role in protecting organisms from the actions of essential oils. For example, the treatment of *C. albicans* [131] and *P. aeruginosa* [82] with the ionophore carbonyl cyanide m-chlorophenylhydrazone to depolarize cell membranes resulted in organisms becoming significantly more susceptible to tea-tree oil or components than cells with normal cell membranes. Energy-dependent mechanisms that may protect cells from essential oils include active transport and efflux, which is used by many microorganisms to actively pump toxic substances out of the cell. It has been demonstrated that efflux is utilized by *P. aeruginosa* as a protective measure against tea-tree oil [146]. Similarly, *C. albicans* cells treated with diethylstilboestrol to inhibit the cell-membrane enzyme ATPase were significantly more susceptible to tea-tree oil than control cells without the inhibitor. These studies indicate that microorganisms possess several innate mechanisms which assist in preventing damage due to essential-oil exposure. Whether one or more of these mechanisms is induced, upregulated or overexpressed as an adaptive response to essential oil remains to be determined.

Modulation of membrane fluidity is one adaptive measure that has been demonstrated in response to essential oils [89, 128, 147]. Since essential-oil components increase membrane fluidity, it follows that microorganisms will compensate for this change to maintain homeostasis and optimal fluidity. Microorganisms adapted to grow in the presence of sublethal concentrations of essential oils or components generally have lower membrane fluidity and altered membrane lipid composition. For membrane composition, exposure of organisms to low levels of essential oils generally results in an overall shift towards shorter-chain fatty acids [89, 128, 147] and an increase in fatty acid saturation. Increased saturation has been shown for *Rhodococcus erythropolis* grown with limonene [148], *Yarrowia lipolytica* grown with citrus oil [147], *S. cerevisiae* grown with palmarosa (*Cymbopogon martinii*) oil [134] and for several foodborne or food-spoilage organisms treated with thymol, limonene, eugenol, carvacrol or cinnamaldehyde [89]. The end result of these changes in membrane composition is to decrease membrane fluidity, which has been demonstrated for *B. cereus* adapted to 0.4 mM carvacrol [128]. However, there are examples that do not fit these generalizations as changes in membrane composition appear to be both organism- and compound-dependent [89]. A further illustration of this is *S. cerevisiae* treated with *Cymbopogon citratus* oil, which resulted in decreased levels of saturated fatty acids and increased unsaturated fatty acids [60]. Also, changes in composition are assumed to occur via homeoviscous adaptation but have also been postulated to occur, at least in *Y. lipolytica*, as a result of the inhibition of the enzymes that are responsible for the elongation of fatty acids [147]. Finally, it is relevant to note that altered membrane composition has not been demonstrated to correspond with reduced susceptibility to essential oils or components.

Very few additional adaptive measures have been demonstrated for microorganisms exposed to low levels of essential oils. The accumulation of intracellular solutes, a common response to many adverse conditions or challenges, did not occur in *C. albicans* or

S. cerevisiae exposed to tea-tree oil [131]. Two publications have described changes in the antimicrobial susceptibility profiles of bacteria cultured with subinhibitory tea-tree oil [149, 150], indicating that microbes are altered by incubation with the essential oil. However, these alterations have not been characterized and further investigation is required. Finally, changes in the transcription of genes after exposure to terpenes have been investigated to shed light on the responses of microorganisms to essential oils or components at the genetic level [151, 152]. In *S. cerevisiae* treated with a nonlethal concentration of α-terpinene (0.02%), the expression of 435 genes was increased, while the expression of 358 was decreased [151]. Genes with increased expression belonged primarily to ergosterol and phospholipid biosynthesis, cell-wall organization, transport and detoxification [151]. Similarly, exposure of *C. albicans* and *C. parapsilosis* to farnesol resulted in both the up- and downregulation of genes [152]. Upregulated genes were primarily associated with sterol metabolism and the expression of two oxidoreductases was also significantly increased [152]. These observations are consistent with the mechanisms discussed above and also demonstrate that microorganisms are capable of responding to essential-oil components in ways not previously illustrated.

11.6.2.1   Resistance

Only a small number of studies have investigated whether resistance to essential oils or components can, or is likely to, develop *in vitro*. One of the earliest studies investigating this subject reported that resistance to tea-tree oil was induced in one strain of *S. aureus* [153]. However, this is not supported by more recent studies, indicating that essential oil resistance is either unlikely or extremely uncommon. Resistance to tea-tree oil was not induced in *C. albicans* [154] and the exposure of *S. aureus* [155, 156], *S. epidermidis* and *E. faecalis* [155] to tea-tree oil also did not yield resistant mutants. Attempts to generate resistance to eugenol and cinnamaldehyde in five *Helicobacter pylori* strains were also unsuccessful [157].

There are several different mechanisms by which microorganisms may become resistant to one or more antimicrobial agents. These include alteration of the drug target site, inactivation of the antimicrobial agent, reduced permeability and the upregulation of efflux, and may be the result of genetic mutations within the organism or the acquisition of external genetic material. This said, it is difficult to envisage a single genetic mutation or series of mutations that would render microbial membranes impervious to essential oils yet allow normal cell functioning. Any potential mutations to membrane properties that confer resistance to essential oils are likely to also negatively impact general microbial growth and survival. This may be more unlikely given that some essential oils and components such as basil and linalool [158, 159] and sage, 1,8-cineole and camphor [160] demonstrate antimutagenic effects, meaning that spontaneous mutagenesis is decreased in the presence of these oils.

Whilst true resistance to essential oils is not apparent, organisms with reduced susceptibility or tolerance to essential oils or components have been isolated. Most changes in susceptibility were relatively minor and were generally no greater than one doubling dilution. Tolerance to tea-tree oil and the components terpinen-4-ol, α-terpineol and linalool was also demonstrated in strains of *E. coli* exhibiting the multiple antibiotic resistance (MAR) phenotype and in *S. aureus* strains showing the salicylate-inducible MAR phenotype [161]. Five *E. coli* mutants displaying reduced susceptibility to thymol have been generated by transposon mutagenesis [79]. For some mutants, reduced susceptibility was also observed for menthol and carvacrol [79]. Examination of transposon insertion sites for these mutants indicated that the disruption of membrane proteins may have been responsible for the reduced susceptibility. Isolates of *V. parahaemolyticus* with reduced susceptibility to basil and sage oils also showed reduced susceptibility to heat and hydrogen peroxide [162]. Interestingly, heat-adapted *V. parahaemolyticus* showed reduced susceptibility to sage and basil oils, indicating that the response to heat stress is cross-protective towards essential oils [162]. Lastly, MRSA isolates selected on the basis of reduced susceptibility to household cleaners in some instances also showed reduced susceptibility to pine oil, tea-tree oil or α-terpineol [163]. Disabling of the *S. aureus* sigB operon, which encodes for the alternative transcription factor SigB and is intimately involved in the general stress response of this organism, resulted in increased susceptibility to these compounds, in turn supporting a role for sigB, and possibly the general stress response, in the protection of *S. aureus* cells against harm caused by essential oils [163].

11.6.3 Other Antimicrobial Effects

11.6.3.1 Toxins, Enzymes and Virulence Factors

Microorganisms produce an array of factors that enhance their ability to infect a host and cause disease. These may include the production of extracellular enzymes, toxins or adhesins. The possibility that essential oils may modulate the expression or production of these virulence factors has been investigated as a potential means by which essential oils may reduce the pathogenicity of microbes in the absence of cell death.

Essential oils of bay, cinnamon leaf, clove stem or thyme (0.01% v/v), as well as nutmeg oil (0.0025% v/v) [113], inhibited the production of the extracellular protein listeriolysin O by *L. monocytogenes* [113]. Decreased lysin production was not due to reduced overall growth or protein production, nor to denaturation of the lysin by the essential oil, indicating a specific inhibition of listeriolysin O production in the presence of essential oil. Clove oil also reduced the production of the extracellular enzyme phospholipase C in the absence of decreased growth [113]. The same essential oils, although at different concentrations (0.01% v/v bay, clove, cinnamon and thyme oil and 1% nutmeg oil), also reduced the production of α-toxin by *S. aureus* [164]. The production of enterotoxins A and B was significantly inhibited by cinnamon and clove oils and enterotoxin A production was additionally inhibited by bay oil [164]. Carvacrol (0.02 to 0.06 mg/ml) reduced the production of enterotoxin by *B. cereus* after 24-hour incubation in the absence of a reduction in levels of total protein [165]. Several monoterpenes inhibited swarming motility and decreased haemolytic activity in *Proteus mirabilis* [166]. There is also evidence that *O. gratissimum* essential oil reduces the production of extracellular proteases by *Shigella* but there was no indication of whether overall growth was also affected [167]. The essential-oil component farnesol (250 mM) significantly inhibited the production of the virulence factor pyocyanin by *P. aeruginosa* [168]. Decreased production was possibly a result of quorum-sensing inhibition, since farnesol was also demonstrated to inhibit quorum sensing, which is known to regulate pyocyanin production [168]. Carvacrol inhibited the synthesis of flagellin in *E. coli*, meaning that cells became nonmotile [169]. It has been postulated that loss of motility renders cells less virulent as they become less invasive. However, this same study also demonstrated that carvacrol induced the production of heat-shock protein 60 (HSP60), which is a proposed virulence factor. The implications of these findings are at this stage unclear [169].

The production of aflatoxin B$_1$ by *A. flavus* was inhibited by *C. citratus* essential oil in vapour phase [62]. However, this treatment also significantly reduced biomass, indicating that reduced toxin production may be a function of reduced growth. Additional essential oils shown to inhibit aflatoxin production by *Aspergillus* spp. include cinnamon and clove oils [170], *P. graveolens* oil [68], *Satureja hortensis* oil [171] and thyme oils [172].

A major virulence factor of *C. albicans* is its ability to switch from a single-celled morphology to a hyphal phase, also known as germ-tube formation. Essential oils or components that inhibit the formation of germ tubes include lavender oil [25], tea-tree oil [173,174], *Thymbra capitata* oil [175] and three *Thymus* essential oils [2]. The components carvacrol, thymol, cineole, cymene [2], farnesol [176], linalool and linalyl acetate [25] also inhibit germ-tube formation. Farnesol inhibits pseudohyphae production by *Candida dubliniensis* [176] at subinhibitory concentrations, suggesting that inhibition is specific to pseudohyphae formation rather than being a result of generalized growth inhibition.

11.6.3.2   Biofilms and Quorum Sensing

The majority of tests with bacteria are performed with cells that are assumed to be in a single-celled, free-floating state, which is also known as a planktonic state. However, current thinking suggests that many microorganisms are much more likely to exist in communities attached to surfaces known as biofilms. Biofilms are typically difficult to eradicate and substantially less susceptible to antimicrobial agents than their planktonic counterparts. Several studies suggest that essential oils may be active against biofilms [177] and some studies also indicate that this activity is specific to biofilm formation and is not due to a generalized inhibition of growth [17]. Rosemary and peppermint oils were inhibitory at 200 to 800 ppm (0.02 to 0.08%) against *Streptococcus mutans* biofilms. In this study, biofilm formation was corrected for any reduction in growth caused by the oils [17]. Oregano oil, thymol and carvacrol inhibited biofilm production by *S. aureus* and *S. epidermidis* [178]. *C. albicans* biofilm formation was inhibited by eucalyptus, peppermint and clove oil [179] and also by thymol, carvacrol, geraniol, 1,8-cineole, citral, eugenol, farnesol, linalool, menthol and α-terpinene [77]. Thymol, carvacrol and geraniol were the most inhibitory, which correlated with their high activity in standard MIC assays. *Satureja thymbra* oil (1.0%) [180], oregano oil, thymol and carvacrol have also been demonstrated to eradicate existing biofilm [178].

A critical factor in the development of biofilm is a specific type of signalling between cells known as quorum sensing. The production and secretion of various signal molecules by microorganisms is important for the switching on and off of various properties such as virulence factors and biofilm production. The sesquiterpene farnesol has been demonstrated to reduce the production of *Pseudomonas* quinolone signal (PQS), which is involved in quorum sensing in *P. aeruginosa* [168]. Farnesol also appears to regulate morphogenesis in *Candida dubliniensis* and reduces the production of hyphae and pseudohyphae [176]. Cinnamaldehyde interferes with quorum sensing in *Vibrio harveyi* [181]. The ability of an organism to adhere to any given surface is also important for biofilm formation and is influenced to a degree by cell-surface hydrophobicity. Carveol and carvone both decrease cell-surface hydrophobicity in *R. erythropolis* [182] and since decreased hydrophobicity correlates with less biofilm, this may be one mechanism by which biofilm is inhibited.

Thymol has also been shown to inhibit the adhesion of *S. aureus* and *E. coli* to human cells *in vitro* [183], suggesting that the cell surface may be altered such that adhesion is reduced. These data indicate that the inhibition of biofilm formation by essential-oil components may be mediated by interference with quorum sensing and decreased cell-surface hydrophobicity.

11.6.3.3 Synergy between Essential Oils or with Other Compounds

Many studies have investigated synergy between essential oils or components and conventional antibiotics. Not surprisingly, outcomes vary according to the test organism, antibiotic and essential oil.

Synergism has been observed between cinnamaldehyde and clindamycin against *C. difficile* [184] and between oregano oil and sarafloxacin, levofloxacin, maquindox, florfenicol and doxycycline against *E. coli* [34]. Preliminary data also suggest that nerolidol and bisabolol are synergistic with a range of antibiotics against *S. aureus* [185]. *P. graveolens* essential oil and norfloxacin were synergistic in action against *B. cereus* and two *S. aureus* strains, but not *B. subtilis* or *E. coli* [76]. The combination of norfloxacin with citronellol or geraniol produced similar effects [76]. A combination of antagonistic, synergistic and additive interactions were seen between ciprofloxacin and tea-tree, thyme, peppermint and rosemary oils against *K. pneumoniae*, depending on the concentrations of each agent. Promising synergy was observed between rosemary oil and ciprofloxacin [15]. Largely indifferent interactions have been observed between tea-tree oil and lysostaphin, mupirocin and vancomycin against MRSA, although antagonism was observed for vancomycin with one of the two test organisms [109]. Further antagonistic relationships have been demonstrated between ciprofloxacin and tea-tree, thyme, peppermint and rosemary oils against *S. aureus* [15]. Combination of these same oils with amphotericin B resulted in mostly antagonistic activity against *C. albicans* [15]. Overall, there do not appear to be any clear trends for any particular essential oils or antibiotics. Whether the combination of two antimicrobial agents results in an antagonistic, indifferent or synergistic effect depends on the mechanisms exerted by each agent against the test organism, the characteristics of the test organism and any chemical interactions between the two antimicrobial agents. Furthermore, the specific ratio of oil to antibiotic appears to influence the outcome. Given that very little is known about any of these factors it is difficult to postulate mechanisms by which synergy or antagonism may be occurring. Nonantibiotic agents shown to act synergistically with essential oils include the permeabilizer EDTA [116, 186], organic acids [186, 187], sodium chloride [81] and nisin, a bacteriocin produced by *Lactococcus lactis* [188]. However, one study found no synergy between nisin and carvacrol, thymol, eugenol or cinnamic acid against *Enterobacter sakazakii* [188].

The possibility that one or more essential-oil component interacts synergistically within an essential oil has long been proposed as a potential factor contributing to the overall antimicrobial activity of an oil. Combinations that have displayed synergy include cinnamaldehyde with thymol or carvacrol against *S.* Typhimurium [189], cineole combined with thymol, *p*-cymene or carvacrol against *Candida* spp. [2], and nerolidol combined with eugenol and thymol against *S. aureus* [116]. Interestingly, nerolidol was not synergistic with eugenol or thymol against *E. coli* or *P. aeruginosa* [116], just as carvacrol combined with thymol or cymene was indifferent against *C. albicans* [2]. Similar to the combinations of essential oils and antibiotics described above, the type of interaction occurring between essential-oil components appears to depend on the organism, components examined and component ratios. This is supported by two studies investigating terpene combinations, which found that interactions could be synergistic, indifferent or antagonistic, depending on concentrations and ratios of components [94, 190].

11.6.3.4   Other

There are several additional antimicrobial effects which do not readily fall into any of the modes of action already discussed. For example, spore-forming bacteria such as *B. cereus* [71, 138] and *C. botulinum* [99] have been observed to form long chains of cells in the presence of essentials oils or components. Inhibition of cell separation in *Bacillus cereus* by cinnamaldehyde [71,138] occurs when the cinnamaldehyde binds to FtsZ, a cell-division regulator, and disturbs Z-ring formation [71]. Cinnamaldehyde did not inhibit cell division in *S. aureus* [138]. Treatment with peppermint oil and the component menthol resulted in the elimination of plasmids from *E. coli* K12 [13]. Eucalyptus and rosemary oils also resulted in some plasmid loss, although to a much lesser extent [13].

11.7   Clinical Efficacy of Essential Oils and Components

A limited number of studies have assessed the clinical efficacy of essential oils or components in treating infectious diseases. Furthermore, the majority of these studies focussed on Australian tea-tree oil, with very few studies published for other plant essential oils.

11.7.1   Animal Studies

Studies assessing the effectiveness of essential oils and components in treating infections in animals are summarized in Table 11.2. The majority of studies investigated fungal diseases such as mucosal or systemic *Candida* infections and dermatophyte infections. In most studies, essential oils were applied topically in a solution or cream. However, one study used inhalation as the route of administration, another used gavage and in two studies the essential oil was injected into an abscess or cavity.

Of the 14 studies summarized in Table 11.2, only one did not report any clinical improvement [190]. This study investigated the treatment of vaginal candidiasis with several different essential oils, using a mouse model. Given that all of the remaining studies reported favourable outcomes it may be speculated that the lack of response was due to inadequate dosing, whereby the volume applied was either too small or not applied for long enough, or both. Tea-tree oil has also shown promise in the treatment of pruritis in dogs [192, 193], which is likely to be a function of the antiinflammatory activity of the oil. It is of note that the studies summarized in Table 11.2 cover a broad range of both bacterial and fungal infections occurring in a variety of domestic and laboratory animals and that despite the heterogeneity of the studies, most showed a positive clinical outcome. This indicates that essential oils have potential as therapeutic agents in animals, with the potential for use in humans.

Table 11.2 *Summary of studies evaluating essential oils for the treatment of infectious diseases in animals.*

Disease or condition	Animal	Aetiological agent	Treatment groups (number of evaluable animals)[a]	Treatment application	Outcomes	Adverse events	References
Case series							
Prostate abscess	Dog	Various bacteria	Tea-tree oil (6)	Pus removed via syringe and tea-tree oil injected into abscess cavity	Clinical symptoms resolved three to six weeks after treatment. No relapses evident after one year	None stated	[222]
Comparative trials							
Otitis media	Rat (male Sprague-Dawley)	a) *Streptococcus pneumoniae* b) *Haemophilus influenzae*	1) *Ocimum basilicum* oil 2) Carvacrol (4%) and thymol (4%) mixture 3) Carvacrol (4%), thymol (4%) and salicylate (2%) mixture 4) Control (olive oil)	Two drops of treatment on cotton wool placed in the external ear canal twice daily for five days	Treatment with basil oil or component mix resulted in 56 to 81% cure (*H. influenzae*) and 6 to 75% cure (*S. pneumoniae*). Cure of 5.6 to 6% in placebo group	None stated	[223]
Necrotizing fasciitis (air-pouch model)	Mouse	*Streptococcus pyogenes*	0.2, 1 or 2% tea-tree oil solution versus no treatment	0.5 ml solution injected immediately after infection into air pouch	Untreated mice died within two days. No survival after treatment with 0.2%, 70% survival with 1%, 100% survival with 2%	None stated	[224]

(Continued)

Table 11.2 (Continued)

Disease or condition	Animal	Aetiological agent	Treatment groups (number of evaluable animals)[a]	Treatment application	Outcomes	Adverse events	References
Ringworm	Horse	*Trichophyton equinum*	1) Tea-tree oil, 25% in sweet almond oil (30) 2) Enilconazole 2% solution (30)	1) Tea-tree oil applied twice daily for 15 days 2) Enilconazole rinse applied every three days for 12 days (total of four applications)	No significant differences between treatments. Excellent clinical scores for 24 and 21 horses in the tea-tree oil and control groups respectively and good clinical scores for the remainder	None stated	[225]
Dermatophyte infection (skin)	Rat (male Wistar)	*Trichophyton rubrum*	1) *Origanum vulgare* ssp. *hirtum* oil, 1% in petroleum jelly 2) *O. vulgare* ssp. *hirtum* oil, 1% in lotion 3) petroleum jelly control 4) lotion control	Ointment applied once daily until infection cleared	Infections treated with essential oils cleared after 17 days' treatment whereas both controls remained infected	None stated	[19]

Dermatophyte infection (skin)	Rat (male Wistar)	a) *Trichophyton tonsurans* b) *Trichophyton mentagrophytes* c) *T. rubrum*	1) *Thymus vulgaris* oil, 0.1% (5) 2) Thymol, 0.1% (5) 3) Bifonazole (5) 4) Untreated (5)	Ointment containing *T. vulgaris* oil or thymol applied once daily until infection cleared	Infections treated with thyme oil or thymol ointment were cleared after 24 to 37 days' and 14 days' treatment, respectively. Bifonazole-treated infections were cleared after 14 to 15 days and untreated rats remained infected	None stated	[226]
Dermatophyte infection (skin)	Guinea pig	1) *Epidermophyton floccosum* 2) *Microsporum gypseum*	1) *Ocimum gratissimum* oil ointment 2) *Trachyspermum ammi* oil ointment 3) Untreated control	Ointment applied twice daily	*E. floccosum* infections cured after 11 days (*O. gratissimum*) and 13 days (*T. ammi*). *M. gypseum* infections cured after 9 days (*O. gratissimum*) and 11 days (*T. ammi*). Controls 100% infected at 15 days	None stated	[227]

Table 11.2 (Continued)

Disease or condition	Animal	Aetiological agent	Treatment groups (number of evaluable animals)[a]	Treatment application	Outcomes	Adverse events	References
Oral candidiasis	Rat (male Wistar)	Candida albicans	1) Carvacrol, 0.2% (7) 2) Eugenol, 0.4% (7) 3) Nystatin, 50 IU/ml (7) 4) Untreated (7)	0.5 ml applied twice daily for eight days	Treatments 1, 2 and 3 significantly reduced yeast counts compared to untreated control	None stated	[86]
Vaginal candidiasis	Rat (female Wistar)	C. albicans	1) Tea-tree oil, 1% (5) 2) Tea-tree oil, 2.5% (5) 3) Tea-tree oil, 5% (5) 4) Terpinen-4-ol, 1% (5) 5) Fluconazole, 100 µg (5) 6) Untreated (5)	0.1 ml tea-tree oil or terpinen-4-ol solution applied 1, 24 and 48 hours after infection	Significantly fewer viable organisms recovered from rats receiving treatments 2 to 5 than from controls at all time points	None stated	[83]
Vaginal candidiasis	Rat (female Wistar)	C. albicans	1) Tea-tree oil, 1% (5) 2) Tea-tree oil, 2.5% (5) 3) Tea-tree oil, 5% (5) 4) Fluconazole, 100 µg (5) 5) Untreated (5)	0.1 ml tea-tree oil solution applied 1, 24 and 48 hours after infection	Treatments 1 to 4 did not differ significantly and all cleared the infection by three weeks. Controls remained infected	None stated	[154]
Vaginal candidiasis	Rat (female Wistar)	C. albicans	1) Carvacrol, 0.2% (9) 2) Eugenol, 0.4% (9) 3) Nystatin, 54 mg/l (9) 4) Untreated (9)	0.5ml applied intravaginally twice daily for seven days	Treatments 1, 2 and 3 significantly reduced yeast counts compared to untreated control	None stated	[90]

Vaginal candidiasis	Mouse (BALB/c)	C. albicans	1) Geranium oil, 1 or 5% 2) Tea-tree oil, 1 or 5% 3) Lemongrass oil, 0.2 or 1% 4) Oregano oil, 0.2 or 1% 5) Clove oil, 0.2 or 1% 6) Clotrimazole, 1 mg/10 μl	10 μl applied intravaginally once daily for three continuous days	No reduction in yeast numbers after any essential-oil treatment. Significant reduction with clotrimazole	None stated	[191]
Systemic candidiasis	Mouse (BALB/c)	C. albicans	1) Oregano oil, 8.7 to 52 mg/kg body wt (6) 2) Carvacrol, 8.7 to 17.3 mg/kg (6) 3) Vehicle alone 4) Amphotericin B, 1 mg/kg	0.1 ml administered by gavage for 8 or 30 days	After 30 days, all mice receiving 17.3 to 52 mg/kg alive and no C. albicans evident. All mice receiving AmB survived, whilst all mice receiving vehicle died by day 10	Lethal effects at doses ≥650 mg/kg	[228]
Aspergillosis	Mouse (BALB/c)	Aspergillus fumigatus	1) Uninfected untreated controls (5) 2) Uninfected treated controls (5) 3) Infected untreated controls (11) 4) Infected early treatment (11) 5) Infected late treatment (12)	Treated for one hour on three consecutive days by inhalation of Leptospermum petersonii oil volatiles commencing one day (early) or five days (late treatment) after inoculation	No viable A. fumigatus detected in mice treated with L. petersonii volatiles	None stated	[229]

a Numbers of animals are omitted where not clearly stated in the publication.

11.7.2   Human Studies

Studies assessing the effectiveness of essential oils for treating bacterial and fungal infections in humans are summarized in Table 11.3. These are several additional publications describing the effectiveness of essential oils for treating nonbacterial and fungal infections such as viral infections [194, 195], *Demodex* mite infestation [196] and head lice [197].

Essential oils and derivatives have been used to treat a variety of bacterial and fungal infections in humans. Of the studies listed in Table 11.3, all but one describe the efficacy of tea-tree oil or a tea-tree oil-derived extract, reflecting the popularity of this essential oil. Most studies indicated a degree of clinical efficacy, with the exception of the oral-hygiene study [198] and one of the onychomycosis studies, which showed no improvement after treatment with 5% tea-tree oil [199]. This was likely due to inadequate dosing since another onychomycosis study which used neat (100%) oil for six months found resolution of infection in 60% of patients [200]. This study demonstrated that the nail infection can be cleared if significantly higher essential-oil concentrations are used and highlights the issue of appropriate dosing. Treatment with too low a concentration may be ineffective whereas treatment with too high a concentration can cause adverse reactions, as discussed in more detail below. Optimal dosing must be determined for each essential oil and is likely to be influenced by the formulation of the product vehicle or method of delivery.

11.8   Toxicity of Essential Oils

Despite their image as 'natural', essential oils are far from nontoxic. The majority of essential oils, in high enough doses, will cause toxic effects. The toxicity of essential oils can be evaluated in laboratory assays such as cytotoxicity tests, where the effects of oils on animal cells are examined, or *in vivo*, where laboratory animals are exposed to essential oil and the toxic effects are recorded. Some information may also be extracted from instances where humans have shown toxic effects after essential-oil exposure. Furthermore, toxic effects may occur after ingestion or from dermal exposure. This section will provide a brief overview of the toxic effects elicited by essential oils.

11.8.1   *In Vitro* Toxicity Assays

Determining the cytotoxicity of plant essential oils against human or animal cell lines *in vitro* is a reasonably straightforward and simple way of investigating the toxicity of essential oils. As such, many essential oils, including lavender [201], *Origanum* [202], *N. cataria* (catnip) and *M. officinalis* (lemon balm) [203], have been investigated with tests demonstrating that essential oils show toxicity at very low concentrations. It is postulated that one of the primary mechanisms of cytotoxicity is membrane damage, similar to that seen in bacteria and yeasts.

11.8.2   Animal Studies

A widely accepted measure of a compound's toxicity is its LD_{50}, which is the lethal dose for 50% of the test animals. The LD_{50} values for some popular essential oils are shown in Table 11.4; acute oral toxicity values range from 1.4 g/kg for basil oil to >5 g/kg

Table 11.3 Summary of studies evaluating essential oils for the treatment of infectious diseases in humans.

Study type and condition	Aetiological agent	Treatment groups (number of evaluable patients)	Treatment application	Outcomes	Adverse events	References
Case reports or series						
Bacterial vaginosis	Mixed anaerobic bacteria	Tea-tree oil pessary with 200 mg oil per pessary (1)	Once daily for five days	Normal vaginal flora restored	None stated	[230]
Oral candidiasis (HIV patients)	C. albicans	Melaleuca oral solution, 15 ml (12)	Four times daily for two to four weeks	Clinical response in 67% of patients	None	[231]
Tuberculosis	Mycobacterium tuberculosis	Tea-tree oil-derived oil blend, aerosolized (2)	Once daily for 5 to 10 days	Clinical improvement and negative cultures at days 4 and 5. Conventional therapy then commenced	None	[232]
Comparative trials						
Acne	Propionibacterium acnes	1) Tea-tree oil gel, 5% (58) 2) Benzoyl peroxide, 5% (61)	Three months	Both treatments significantly reduced inflamed and noninflamed lesions	Yes; both groups	[233]

(Continued)

Table 11.3 (Continued)

Study type and condition	Aetiological agent	Treatment groups (number of evaluable patients)	Treatment application	Outcomes	Adverse events	References
Acne	*P. acnes*	1–16) *Ocimum gratissimum* oil, 0.5, 1, 2 or 5% in polysorbate 80, cetomacrogol petrolatum or alcohol base (7 patients per group) 17) Benzoyl peroxide, 10% (7)	Twice daily for four weeks	50% reduction in lesion count achieved within three days for 2% and 5% *Ocimum* oil in 50% alcohol and 5% *Ocimum* oil in cetomacrogol. 50% reduction after five days with benzoyl peroxide	Yes; *Ocimum* groups	[212]
MRSA colonization	Methicillin-resistant *S. aureus*	1) Tea-tree oil, 4% ointment, 5% bodywash (15) 2) Mupirocin, 2% ointment, triclosan bodywash (15)	Minimum of three days	No significant differences between treatments	Yes; tea-tree oil group	[234]
MRSA colonization	Methicillin-resistant *S. aureus*	1) Tea-tree oil (10%) cream and bodywash (5%) (110) 2) Mupirocin nasal ointment, triclosan bodywash, silver sulfadiazine cream (114)	Once daily for five days	Clearance rates of 41% for tea-tree oil group and 49% for mupirocin group	None	[235]

Dandruff	_Malassezia_ spp.	1) Tea-tree oil, 5% shampoo 2) Placebo shampoo	Once daily for four weeks	Significant improvement in whole-scalp lesion score of 41.2% for tea-tree oil group and 11.2% for placebo	Yes; both groups [236]
Oral hygiene	Mixed oral bacteria	1) Mouthrinse with melaleuca oil, 0.67%, manuka oil, 0.33%, calendula flower extract, 1%, green-tea extract, 0.5%, ethanol, 12.8% in water (9) 2) Placebo with ethanol, 12.8% in water (8)	Twice daily for 12 weeks	No significant differences between groups for gingival or plaque index, or relative abundance of _Actinobacillus actinomycetem-comitans_ or _Tanerella forsythensis_	Yes; both groups [198]
Oral candidiasis	_C. albicans_	1) Melaleuca oral solution (12) 2) Melaleuca solution, alcohol-free (13)	Four times daily for two to four weeks	Clinical response for 58% and 54% of patients receiving treatments 1 and 2, respectively	Yes; both groups [237]

(Continued)

Table 11.3 (Continued)

Study type and condition	Aetiological agent	Treatment groups (number of evaluable patients)	Treatment application	Outcomes	Adverse events	References
Onychomycosis	*Trichophyton* spp.	1) Tea-tree oil, neat (64) 2) Clotrimazole, 1% (53)	Twice daily for six months	Full or partial resolution for 60% of tea-tree oil and 61% of clotrimazole patients	Yes; both groups	[200]
Onychomycosis	*Trichophyton* spp.	1) Tea-tree oil, 5% and butenafine, 2% cream 2) Tea-tree oil, 5% cream	Thrice daily for eight weeks	80% cure in butenafine + tea-tree group, 0% cure in tea-tree oil group	Yes; both groups	[199]
Tinea pedis	Dermatophytes	1) Tea-tree oil, 10% in sorbolene 2) Tolnaftate, 1% 3) Placebo (sorbolene)	Twice daily for four weeks	Clinical improvement and mycological cure rates of 46%, 22% and 9% for tolnaftate, tea-tree oil and placebo, respectively	None	[238]
Tinea pedis	Dermatophytes	1) Tea-tree oil, 25% 2) Tea-tree oil, 50% 3) Placebo	Twice daily for four weeks	48% cure with 25% tea-tree oil, 50% cure with 50% tea-tree oil and 13% cure for placebo	Yes; tea-tree oil groups	[239]

for *L. cubeba* and lavender. Essential oils may also be toxic if applied in high doses to the skin, as demonstrated by the dermal toxicity LD_{50} values, which ranged from 4.8 to >5 g/kg. As a generalization, dermal LD_{50} values are much higher than oral values, simply because when applied to the skin the oil must penetrate through the skin and then into the bloodstream before reaching vital organs, which in itself has a diluting effect, whereas ingested oil can pass into the bloodstream relatively quickly. In addition to scientific studies where essential-oil toxicity is systematically investigated in a controlled environment such as those described above, several case reports from individual animals showing essential-oil toxicity have also been published [204].

11.8.3 Toxic Effects in Humans

A spectrum of toxic effects may be observed in humans following essential-oil exposure. At one extreme, poisoning has occurred after ingestion of essential oils including tea-tree [205], lavandin [206], eucalyptus [207] and clove [208]. Poisonings largely occurred in children after accidental ingestion, with the major clinical symptom being central nervous system depression. Poisoning, however, is a relatively rare occurrence; the less severe adverse events of irritancy and allergy are more common. Although the ingestion of essential oils is generally not recommended, studies evaluating the clinical effectiveness of ingested capsules of essential oils or components have been published [209–211].

Irritant reactions to essential oils may manifest as redness, itchiness, a burning sensation or blistering. Reactions are generally dose-dependent, with less irritancy occurring at lower oil concentrations. For example, an evaluation of the irritancy of *Ocimum gratissimum* oil found that neat and 5% oil caused irritation but 2% did not [212]. These irritant thresholds vary between individual essential oils. By definition, essential-oil allergy develops after repeated exposure and will resolve after removal of the allergen. A factor in the development of essential-oil allergy is the use of aged or oxidized oil, whereby exposure of the oil to air has led to the formation of oxidation products which have higher allergenic potential [213]. Allergy to essential oils including lavender [214], tea-tree [215], cinnamon [216] and bergamot [217] has been reported in the literature.

11.9 Conclusions

Essential oils are bactericidal and fungicidal agents that are typically active *in vitro* at concentrations of 5% or less. Oils rich in aldehydes, phenols or alcohols generally show the greatest antimicrobial activity, since these component groups tend to be the most antimicrobially active. Lethal action is directly related to the solubility of oil components in microbial membranes and disruption of associated functions. More subtle mechanisms such as changes in metabolic or biosynthetic processes or enzyme function have not been investigated to any great extent. However, a range of effects occurring at nonlethal concentrations such as inhibition of virulence, adhesion and biofilm, and the up- and downregulation of numerous genes have been shown. On the basis of these actions, end-use applications such as in food and crop preservation and in medicinal agents for humans and domesticated animals have been investigated for a selection of oils. Essential oils have demonstrated clinical efficacy for the treatment of otitis media, dermatophyte infections and oral and

Table 11.4 Toxicity data for a selection of essential oils.

Essential oil	Common name	Acute oral toxicity		Acute dermal toxicity		References
		Animal	LD_{50}	Animal	LD_{50}	
Coriandrum sativum	Coriander	Rat	4.13 g/kg	Rat	5 g/kg	Hart (1971), cited in [240]
Lavandula	Lavender	Rat	>5 g/kg	Rat	5 g/kg	Moreno (1973), cited in [241]
Litsea cubeba		Rat	>5 g/kg	Rabbit	4.8 g/kg	Moreno (1976), cited in [242]
Origanum majorana	Marjoram	Rat	2.34 g/kg	Rat	5 g/kg	Wohl (1974), cited in [243]
Melaleuca alternifolia	Tea tree	Rat	1.9 g/kg	Rabbit	5 g/kg	Moreno (1982), cited in [244]
Ocimum basilicum	Basil	Rat	1.4 ml/kg	Rat	5 ml/kg	Levenstein (1972), cited in [245]
Pimenta racemosa	Bay	Rat	1.8 g/kg	Rabbit	5 ml/kg	Owen (1971), cited in [246]

vaginal candidiasis in animals, in addition to oral and vaginal infections, acne, MRSA colonization, dandruff, nail infections and tinea in humans. There is therefore broad scope for utilizing essential oils as antimicrobial agents in an array of settings, providing that critical issues such as effective delivery systems and potential toxicity to humans and the environment are addressed.

References

1. Bakkali, F., Averbeck, S., Averbeck, D. and Waomar, M. (2008) Biological effects of essential oils: a review. *Food Chem. Toxicol.*, **46**, 446–475.
2. Pina-Vaz, C., Rodrigues, A.G., Pinto, E. *et al.* (2004) Antifungal activity of *Thymus* oils and their major compounds. *J. Eur. Acad. Dermatol. Venereol.*, **18**, 73–78.
3. Yousef, R.T. and Tawil, G.G. (1980) Anti-microbial activity of volatile oils. *Pharmazie*, **35**, 698–701.
4. Morris, J.A., Khettry, A. and Seitz, E.W. (1979) Anti-microbial activity of aroma chemicals and essential oils. *J. Am. Oil Chem. Soc.*, **56**, 595–603.
5. Deans, S.G. and Ritchie, G. (1987) Antibacterial properties of plant essential oils. *Int. J. Food Microbiol.*, **5**, 165–180.
6. Dorman, H.J.D. and Deans, S.G. (2000) Antimicrobial agents from plants: antibacterial activity of plant volatile oils. *J. Appl. Microbiol.*, **88**, 308–316.
7. Janssen, A.M., Scheffer, J.J.C. and Baerheim Svendsen, A. (1987) Antimicrobial activity of essential oils: a 1976–1986 literature review. Aspects of the test methods. *Planta Med.*, **53**, 396–398.
8. Kalemba, D. and Kunicka, A. (2003) Antibacterial and antifungal properties of essential oils. *Curr. Med. Chem.*, **10**, 813–829.
9. Lahlou, M. (2004) Methods to study the phytochemistry and bioactivity of essential oils. *Phytother. Res.*, **18**, 435–448.
10. Matasyoh, J.C., Maiyo, Z.C., Ngure, R.M. and Chepkorir, R. (2009) Chemical composition and antimicrobial activity of the essential oil of *Coriandrum sativum*. *Food Chem.*, **113**, 526–529.
11. Oussalah, M., Caillet, S., Saucier, L. and Lacroix, M. (2007) Inhibitory effects of selected plant essential oils on the growth of four pathogenic bacteria: *E. coli* O157 : H7, *Salmonella* Typhimurium, *Staphylococcus aureus* and *Listeria monocytogenes*. *Food Cont.*, **18**, 414–420.
12. Hammer, K.A., Carson, C.F. and Riley, T.V. (1999) Antimicrobial activity of essential oils and other plant extracts. *J. Appl. Microbiol.*, **86**, 985–990.
13. Schelz, Z., Molnar, J. and Hohmann, J. (2006) Antimicrobial and antiplasmid activities of essential oils. *Fitoterapia*, **77**, 279–285.
14. Trombetta, D., Castelli, F., Sarpietro, M.G. *et al.* (2005) Mechanisms of antibacterial action of three monoterpenes. *Antimicrob. Agents Chemother.*, **49**, 2474–2478.
15. vanVuuren, S.F., Suliman, S. and Viljoen, A.M. (2009) The antimicrobial activity of four commercial essential oils in combination with conventional antimicrobials. *Lett. Appl. Microbiol.*, **48**, 440–446.
16. vanVuuren, S.F. and Viljoen, A.M. (2006) A comparative investigation of the antimicrobial properties of indigenous South African aromatic plants with popular commercially available essential oils. *J. Ess. Oil Res.*, **18**, 66–71.
17. Rasooli, I., Shayegh, S., Taghizadeh, M. and Astaneh, S.D.A. (2008) Phytotherapeutic prevention of dental biofilm formation. *Phytother. Res.*, **22**, 1162–1167.
18. Smith-Palmer, A., Stewart, J. and Fyfe, L. (1998) Antimicrobial properties of plant essential oils and essences against five important food-borne pathogens. *Lett. Appl. Microbiol.*, **26**, 118–122.
19. Adam, K., Sivropoulou, A., Kokkini, S. *et al.* (1998) Antifungal activities of *Origanum vulgare* subsp. *hirtum, Mentha spicata, Lavandula angustifolia*, and *Salvia fruticosa* essential oils against human pathogenic fungi. *J. Agricult. Food Chem.*, **46**, 1739–1745.

20. Pozzatti, P., Scheid, L.A., Spader, T.B. *et al.* (2008) In vitro activity of essential oils extracted from plants used as spices against fluconazole-resistant and fluconazole-susceptible *Candida* spp. *Can. J. Microbiol.*, **54**, 950–956.
21. Genena, A.K., Hense, H., Smania, A. and deSouza, S.M. (2008) Rosemary (*Rosmarinus officinalis*): a study of the composition, antioxidant and antimicrobial activities of extracts obtained with supercritical carbon dioxide. *Ciencia Tecnol. Aliment.*, **28**, 463–469.
22. Evandri, M., Battinelli, L., Daniele, C. *et al.* (2005) The antimutagenic activity of *Lavandula angustifolia* (lavender) essential oil in the bacterial reverse mutation assay. *Food Chem. Toxicol.*, **43**, 1381–1387.
23. Nelson, R.R.S. (1997) In-vitro activities of five plant essential oils against methicillin-resistant *Staphylococcus aureus* and vancomycin-resistant *Enterococcus faecium*. *J. Antimicrob. Chemother.*, **40**, 305–306.
24. Mayaud, L., Carricajo, A., Zhiri, A. and Aubert, G. (2008) Comparison of bacteriostatic and bactericidal activity of 13 essential oils against strains with varying sensitivity to antibiotics. *Lett. Appl. Microbiol.*, **47**, 167–173.
25. D'Auria, F.D., Tecca, M., Strippoli, V. *et al.* (2005) Antifungal activity of *Lavandula angustifolia* essential oil against *Candida albicans* yeast and mycelial form. *Med. Mycol.*, **43**, 391–396.
26. Araujo, C., Sousa, M.J., Ferreira, M.F. and Leao, C. (2003) Activity of essential oils from Mediterranean Lamiaceae species against food spoilage yeasts. *J. Food Protect.*, **66**, 625–632.
27. Gutierrez, J., Barry-Ryan, C. and Bourke, R. (2008) The antimicrobial efficacy of plant essential oil combinations and interactions with food ingredients. *Int. J. Food Microbiol.*, **124**, 91–97.
28. Omidbeygi, M., Barzegar, M., Hamidi, Z. and Naghdibadi, H. (2007) Antifungal activity of thyme, summer savory and clove essential oils against *Aspergillus flavus* in liquid medium and tomato paste. *Food Cont.*, **18**, 1518–1523.
29. Rota, M.C., Herrera, A., Martinez, R.M. *et al.* (2008) Antimicrobial activity and chemical composition of *Thymus vulgaris*, *Thymus zygis* and *Thymus hyemalis* essential oils. *Food Cont.*, **19**, 681–687.
30. Gutierrez, J., Barry-Ryan, C. and Bourke, P. (2009) Antimicrobial activity of plant essential oils using food model media: efficacy, synergistic potential and interactions with food components. *Food Microbiol.*, **26**, 142–150.
31. Pinto, E., Pina-Vaz, C., Salgueiro, L. *et al.* (2006) Antifungal activity of the essential oil of *Thymus pulegioides* on *Candida*, *Aspergillus* and dermatophyte species. *J. Med. Microbiol.*, **55**, 1367–1373.
32. Nostro, A., Blanco, A.R., Cannatelli, M.A. *et al.* (2004) Susceptibility of methicillin-resistant staphylococci to oregano essential oil, carvacrol and thymol. *FEMS Microbiol. Lett.*, **230**, 191–195.
33. Wong, S.Y.Y., Grant, I.R., Friedman, M. *et al.* (2008) Antibacterial activities of naturally occurring compounds against *Mycobacterium avium* subsp. *paratuberculosis*. *Appl. Environ. Microbiol.*, **74**, 5986–5990.
34. Si, H.B., Hu, J.Q., Liu, Z.C. and Zeng, Z.L. (2008) Antibacterial effect of oregano essential oil alone and in combination with antibiotics against extended-spectrum beta-lactamase-producing *Escherichia coli*. *Fems Immunol. Med. Microbiol.*, **53**, 190–194.
35. Lambert, R.J.W., Skandamis, P.N., Coote, P.J. and Nychas, G.J.E. (2001) A study of the minimum inhibitory concentration and mode of action of oregano essential oil, thymol and carvacrol. *J. Appl. Microbiol.*, **91**, 453–462.
36. Carmo, E.S., Lima, E.D. and de Souza, E.L. (2008) The potential of *Origanum vulgare* L. (Lamiaceae) essential oil in inhibiting the growth of some food-related *Aspergillus* species. *Braz. J. Microbiol.*, **39**, 362–367.
37. Baratta, M.T., Dorman, H.J.D., Deans, S.G. *et al.* (1998) Antimicrobial and antioxidant properties of some commercial essential oils. *Flavour Fragrance J.*, **13**, 235–244.
38. Opalchenova, G. and Obreshkova, D. (2003) Comparative studies on the activity of basil – an essential oil from *Ocimum basilicum* L. – against multidrug resistant clinical isolates of the genera *Staphylococcus*, *Enterococcus* and *Pseudomonas* by using different test methods. *J. Microbiol. Meth.*, **54**, 105–110.

39. Zheljazkov, V.D., Callahan, A. and Cantrell, C.L. (2008) Yield and oil composition of 38 basil (*Ocimum basilicum* L.) accessions grown in Mississippi. *Agricult. Food Chem.*, **56**, 241–245.

40. bin Jantan, I., Moharam, B.A.K., Santhanam, J. and Jamal, J.A. (2008) Correlation between chemical composition and antifungal activity of the essential oils of eight *Cinnamomum* species. *Pharmaceut. Biol.*, **46**, 406–412.

41. Wang, R., Wang, R.J. and Yang, B. (2009) Extraction of essential oils from five cinnamon leaves and identification of their volatile compound compositions. *Innov. Food Sci. Emerg. Technol.*, **10**, 289–292.

42. Oussalah, M., Caillet, S. and Lacroix, M. (2006) Mechanism of action of Spanish oregano, Chinese cinnamon, and savory essential oils against cell membranes and walls of *Escherichia coli* O157 : H7 and *Listeria monocytogenes*. *J. Food Protect.*, **69**, 1046–1055.

43. Simic, A., Sokovic, M.D., Ristic, M. *et al.* (2004) The chemical composition of some Lauraceae essential oils and their antifungal activities. *Phytother. Res.*, **18**, 713–717.

44. Chang, S.T., Chen, P.F. and Chang, S.C. (2001) Antibacterial activity of leaf essential oils and their constituents from *Cinnamomum osmophloeum*. *J. Ethnopharmacol.,*, **77**, 123–127.

45. Dadalioglu, I. and Evrendilek, G.A. (2004) Chemical compositions and antibacterial effects of essential oils of Turkish oregano (*Origanum minutiflorum*), bay laurel (*Laurus nobilis*), Spanish lavender (*Lavandula stoechas* L.), and fennel (*Foeniculum vulgare*) on common foodborne pathogens. *J. Agricult. Food Chem.*, **52**, 8255–8260.

46. Carson, C.F., Hammer, K.A. and Riley, T.V. (2006) *Melaleuca alternifolia* (tea tree) oil: a review of antimicrobial and other medicinal properties. *Clin. Microbiol. Rev.*, **19**, 50–62.

47. Hammer, K.A., Carson, C.F., Riley, T.V. and Nielsen, J.B. (2006) A review of the toxicity of *Melaleuca alternifolia* (tea tree) oil. *Food Chem. Toxicol.*, **44**, 616–625.

48. Hammer, K.A., Carson, C.F. and Riley, T.V. (1996) Susceptibility of transient and commensal skin flora to the essential oil of *Melaleuca alternifolia* (tea tree oil). *Am. J. Infect. Cont.*, **24**, 186–189.

49. Hammer, K.A., Carson, C.F. and Riley, T.V. (1999) In vitro susceptibilities of lactobacilli and organisms associated with bacterial vaginosis to *Melaleuca alternifolia* (tea tree) oil. *Antimicrob. Agents Chemother.*, **43**, 196.

50. Oussalah, M., Caillet, S., Saucier, L. and Lacroix, M. (2006) Antimicrobial effects of selected plant essential oils on the growth of a *Pseudomonas putida* strain isolated from meat. *Meat Sci.*, **73**, 236–244.

51. Hammer, K.A., Carson, C.F. and Riley, T.V. (2003) Antifungal activity of the components of *Melaleuca alternifolia* (tea tree) oil. *J. Appl. Microbiol.*, **95**, 853–860.

52. Hammer, K.A., Carson, C.F. and Riley, T.V. (2002) In vitro activity of *Melaleuca alternifolia* (tea tree) oil against dermatophytes and other filamentous fungi. *J. Antimicrob. Chemother.*, **50**, 195–199.

53. Lee, Y.S., Kim, J., Shin, S.C. *et al.* (2008) Antifungal activity of Myrtaceae essential oils and their components against three phytopathogenic fungi. *Flavour Fragrance J.*, **23**, 23–28.

54. Saenz, M.T., Tornos, M.P., Alvarez, A. *et al.* (2004) Antibacterial activity of essential oils of *Pimenta racemosa* var. *terebinthina* and *Pimenta racemosa* var. *grisea*. *Fitoterapia*, **75**, 599–602.

55. Fu, Y.J., Chen, L.Y., Zu, Y.G. *et al.* (2009) The antibacterial activity of clove essential oil against *Propionibacterium acnes* and its mechanism of action. *Arch. Dermatol.*, **145**, 86–88.

56. Porter, N.G. and Wilkins, A.L. (1999) Chemical, physical and antimicrobial properties of essential oils of *Leptospermum scoparium* and *Kunzea ericoides*. *Phytochem.*, **50**, 407–415.

57. Omidbaigi, R., Yahyazadeh, M., Zare, R. and Taheri, H. (2007) The in vitro action of essential oils on *Aspergillus flavus*. *J. Ess. Oil Bearing Plants*, **10**, 46–52.

58. Wilkinson, J.M., Hipwell, M., Ryan, T. and Cavanagh, H.M.A. (2003) Bioactivity of *Backhousia citriodora*: antibacterial and antifungal activity. *J. Agricult. Food Chem.*, **51**, 76–81.

59. Doran, A.L., Morden, W.E., Dunn, K. and Edwards-Jones, V. (2009) Vapour-phase activities of essential oils against antibiotic sensitive and resistant bacteria including MRSA. *Lett. Appl. Microbiol.*, **48**, 387–392.

60. Helal, G.A., Sarhan, M.M., Abu Shahla, A.N.K. and Abou El-Khair, E.K. (2006) Effect of *Cymbopogon citratus* L. essential oil on growth and morphogenesis of *Saccharomyces cerevisiae* ML2-strain. *J. Basic Microbiol.*, **46**, 375–386.
61. Irkin, R. and Korukluoglu, M. (2009) Effectiveness of *Cymbopogon citratus* L. essential oil to inhibit the growth of some filamentous fungi and yeasts. *J. Med. Food*, **12**, 193–197.
62. Helal, G.A., Sarhan, M.M., Abu Shahla, A.N.K. and Abou El-Khair, E.K. (2007) Effects of *Cymbopogon citratus* L. essential oil on the growth, morphogenesis and aflatoxin production of *Aspergillus flavus* ML2-strain. *J. Basic Microbiol.*, **47**, 5–15.
63. Chowdhury, J.U., Bhuiyan, N.I. and Yusuf, M. (2008) Chemical composition of the leaf essential oils of *Murraya koenigii* (L.) Spreng and *Murraya paniculata* (L.) Jack. *Bangladesh J. Pharmacol.*, **3**, 59–63.
64. Plummer, J.A., Wann, J.M. and Spadek, Z.E. (1999) Intraspecific variation in oil components of *Boronia megastigma* Nees. (Rutaceae) flowers. *Ann. Bot.*, **83**, 253–262.
65. Sharma, N. and Tripathi, A. (2008) Effects of *Citrus sinensis* (L.) Osbeck epicarp essential oil on growth and morphogenesis of *Aspergillus niger* (L.) Van Tieghem. *Microbiol. Res.*, **163**, 337–344.
66. Fisher, K. and Phillips, C.A. (2006) The effect of lemon, orange and bergamot essential oils and their components on the survival of *Campylobacter jejuni*, *Escherichia coli* O157, *Listeria monocytogenes*, *Bacillus cereus* and *Staphylococcus aureus* in vitro and in food systems. *J. Appl. Microbiol.*, **101**, 1232–1240.
67. Viuda-Martos, M., Ruiz-Navajas, Y., Fernandez-Lopez, J. and Perez-Alvarez, J. (2008) Antifungal activity of lemon (*Citrus lemon* L.), mandarin (*Citrus reticulata* L.), grapefruit (*Citrus paradisi* L.) and orange (*Citrus sinensis* L.) essential oils. *Food Cont.*, **19**, 1130–1138.
68. Singh, P., Srivastava, B., Kumar, A. *et al.* (2008) Assessment of *Pelargonium graveolens* oil as plant-based antimicrobial and aflatoxin suppressor in food preservation. *J. Sci. Food Agricult.*, **88**, 2421–2425.
69. Griffin, S.G., Wyllie, S.G., Markham, J.L. and Leach, D.N. (1999) The role of structure and molecular properties of terpenoids in determining their antimicrobial activity. *Flavour Fragrance J.*, **14**, 322–332.
70. Baskaran, S.A., Kazmer, G.W., Hinckley, L. *et al.* (2009) Antibacterial effect of plant-derived antimicrobials on major bacterial mastitis pathogens in vitro. *J. Dairy Sci.*, **92**, 1423–1429.
71. Domadia, P., Swarup, S., Bhunia, A. *et al.* (2007) Inhibition of bacterial cell division protein FtsZ by cinnamaldehyde. *Biochem. Pharmacol.*, **74**, 831–840.
72. Kim, J.M., Marshall, M.R. and Wei, C. (1995) Antibacterial activity of some essential oil components against 5 foodborne pathogens. *J. Agricult. Food Chem.*, **43**, 2839–2845.
73. Fitzgerald, D.J., Stratford, M., Gasson, M.J. *et al.* (2004) Mode of antimicrobial action of vanillin against *Escherichia coli*, *Lactobacillus plantarum* and *Listeria innocua*. *J. Appl. Microbiol.*, **97**, 104–113.
74. Carson, C.F. and Riley, T.V. (1995) Antimicrobial activity of the major components of the essential oil of *Melaleuca alternifolia*. *J. Appl. Bacteriol.*, **78**, 264–269.
75. Driffield, K.L., Mooney, L. and Kerr, K.G. (2006) Temperature-dependent changes in susceptibility of *Stenotrophomonas maltophilia* to the essential oils of sweet basil (*Ocimum basilicum*) and black pepper (*Piper nigrum*). *Pharmaceut. Biol.*, **44**, 113–115.
76. Rosato, A., Vitali, C., De Laurentis, N. *et al.* (2007) Antibacterial effect of some essential oils administered alone or in combination with Norfloxacin. *Phytomedicine*, **14**, 727–732.
77. Dalleau, S., Cateau, E., Berges, T. *et al.* (2008) In vitro activity of terpenes against *Candida* biofilms. *Int. J. Antimicrob. Agents*, **31**, 572–576.
78. Sridhar, S.R., Rajagopal, R.V., Rajavel, R. *et al.* (2003) Antifungal activity of some essential oils. *J. Agricult. Food Chem.*, **51**, 7596–7599.
79. Shapira, R. and Mimran, E. (2007) Isolation and characterization of *Escherichia coli* mutants exhibiting altered response to thymol. *Microb. Drug Resist.-Mech. Epidemiol. Dis.*, **13**, 157–165.
80. Ben Arfa, A., Combes, S., Preziosi-Belloy, L. *et al.* (2006) Antimicrobial activity of carvacrol related to its chemical structure. *Lett. Appl. Microbiol.*, **43**, 149–154.

81. Carson, C.F., Mee, B.J. and Riley, T.V. (2002) Mechanism of action of *Melaleuca alternifolia* (tea tree) oil on *Staphylococcus aureus* determined by time-kill, lysis, leakage, and salt tolerance assays and electron microscopy. *Antimicrob. Agents Chemother.*, **46**, 1914–1920.

82. Longbottom, C.J., Carson, C.F., Hammer, K.A. *et al.* (2004) Tolerance of *Pseudomonas aeruginosa* to *Melaleuca alternifolia* (tea tree) oil is associated with the outer membrane and energy-dependent cellular processes. *J. Antimicrob. Chemother.*, **54**, 386–392.

83. Mondello, F., De Bernardis, F., Girolamo, A. *et al.* (2006) In vivo activity of terpinen-4-ol, the main bioactive component of *Melaleuca alternifolia* Cheel (tea tree) oil against azole-susceptible and -resistant human pathogenic *Candida species*. *BMC Infect. Dis.*, **6**, 158.

84. Reichling, J., Suschke, U., Schneele, J. and Geiss, H.K. (2006) Antibacterial activity and irritation potential of selected essential oil components: structure–activity relationship. *Nat. Prod. Comm.*, **1**, 1003–1012.

85. van Zyl, R.L., Seatlholo, S.T. and vanVuuren, S.F. (2006) The biological activities of 20 nature identical essential oil constituents. *J. Ess. Oil Res.*, **18**, 129–133.

86. Chami, N., Bennis, S., Chami, F. *et al.* (2005) Study of anticandidal activity of carvacrol and eugenol in vitro and in vivo. *Oral Microbiol. Immunol.*, **20**, 106–111.

87. Walsh, S.E., Maillard, J.Y., Russell, A.D. *et al.* (2003) Activity and mechanisms of action of selected biocidal agents on Gram-positive and -negative bacteria. *J. Appl. Microbiol.*, **94**, 240–247.

88. Bennis, S., Chami, F., Chami, N. *et al.* (2004) Surface alteration of *Saccharomyces cerevisiae* induced by thymol and eugenol. *Lett. Appl. Microbiol.*, **38**, 454–458.

89. Di Pasqua, R., Betts, G., Hoskins, N. *et al.* (2007) Membrane toxicity of antimicrobial compounds from essential oils. *J. Agricult. Food Chem.*, **55**, 4863–4870.

90. Chami, F., Chami, N., Bennis, S. *et al.* (2004) Evaluation of carvacrol and eugenol as prophylaxis and treatment of vaginal candidiasis in an immunosuppressed rat model. *J. Antimicrob. Chemother.*, **54**, 909–914.

91. Sokovic, M. and van Griensven, L. (2006) Antimicrobial activity of essential oils and their components against the three major pathogens of the cultivated button mushroom, *Agaricus bisporus*. *Eur. J. Plant Pathol.*, **116**, 211–224.

92. Santoyo, S., Cavero, S., Jaime, L. *et al.* (2005) Chemical composition and antimicrobial activity of *Rosmarinus officinalis* l. essential oil obtained via supercritical fluid extraction. *J. Food Protect.*, **68**, 790–795.

93. Angioni, A., Barra, A., Cereti, E. *et al.* (2004) Chemical composition, plant genetic differences, antimicrobial and antifungal activity investigation of the essential oil of *Rosmarinus officinalis* L. *J. Agricult. Food Chem.*, **52**, 3530–3535.

94. VanVuuren, S.F. and Viljoen, A.M. (2007) Antimicrobial activity of limonene enantiomers and 1,8-cineole alone and in combination. *Flavour Fragrance J.*, **22**, 540–544.

95. Onawunmi, G.O., Yisak, W.-A. and Ogunlana, E.O. (1984) Antibacterial constituents in the essential oil of *Cymbopogon citratus* (DC.) Stapf. *J. Ethnopharmacol.*, **12**, 279–286.

96. Veldhuizen, E.J.A., Tjeersma-Van Bokhoven, J.L.M., Zweijtzer, C. *et al.* (2006) Structural requirements for the antimicrobial activity of carvacrol. *J. Agricult. Food Chem.*, **54**, 1874–1879.

97. Ultee, A., Bennik, M.H.J. and Moezelaar, R. (2002) The phenolic hydroxyl group of carvacrol is essential for action against the food-borne pathogen *Bacillus cereus*. *Appl. Environ. Microbiol.*, **68**, 1561–1568.

98. Griffin, S.G., Wyllie, S.G. and Markham, J.L. (2001) Role of the outer membrane of *Escherichia coli* AG100 and *Pseudomonas aeruginosa* NCTC 6749 and resistance/susceptibility to monoterpenes of similar chemical structure. *J. Ess. Oil Res.*, **13**, 380–386.

99. Chaibi, A., Ababouch, L.H., Belasri, K. *et al.* (1997) Inhibition of germination and vegetative growth of *Bacillus cereus* T and *Clostridium botulinum* 62A spores by essential oils. *Food Microbiol.*, **14**, 161–174.

100. Ismaiel, A. and Pierson, M.D. (1990) Inhibition of growth and germination of *C. botulinurn* 33A, 40B, and 1623E by essential oil of spices. *J. Food Sci.*, **55**, 1676–1678.

101. Messager, S., Hammer, K.A., Carson, C.F. and Riley, T.V. (2006) Sporicidal activity of tea tree oil. *Aust. Infect. Cont.*, **11**, 112–121.

102. Juneja, V.K., Thippareddi, H. and Friedman, M. (2006) Control of *Clostridium perfringens* in cooked ground beef by carvacrol, cinnamaldehyde, thymol, or oregano oil during chilling. *J. Food Protect.*, **69**, 1546–1551.
103. Juneja, V.K. and Friedman, M. (2007) Carvacrol, cinnamaldehyde, oregano oil, and thymol inhibit *Clostridium perfringens* spore germination and outgrowth in ground turkey during chilling. *J. Food Protect.*, **70**, 218–222.
104. Ultee, A., Gorris, L.G.M. and Smid, E.J. (1998) Bactericidal activity of carvacrol towards the food-borne pathogen *Bacillus cereus*. *J. Appl. Microbiol.*, **85**, 211–218.
105. Periago, P.M., Conesa, R., Delgado, B. *et al.* (2006) *Bacillus megaterium* spore germination and growth inhibition by a treatment combining heat with natural antimicrobials. *Food Technol. Biotechnol.*, **44**, 17–23.
106. Bevilacqua, A., Corbo, M.R. and Sinigaglia, M. (2008) Combined effects of low pH and cinnamaldehyde on the inhibition of *Alicyclobacillus acidoterrestris* spores in a laboratory medium. *J. Food Proc. Preserv.*, **32**, 839–852.
107. Friedman, M., Buick, R. and Elliott, C.T. (2004) Antibacterial activities of naturally occurring compounds against antibiotic-resistant *Bacillus cereus* vegetative cells and spores, *Escherichia coli*, and *Staphylococcus aureus*. *J. Food Protect.*, **67**, 1774–1778.
108. Carson, C.F., Cookson, B.D., Farrelly, H.D. and Riley, T.V. (1995) Susceptibility of methicillin-resistant *Staphylococcus aureus* to the essential oil of *Melaleuca alternifolia*. *J. Antimicrob. Chemother.*, **35**, 421–424.
109. LaPlante, K.L. (2007) In vitro activity of lysostaphin, mupirocin, and tea tree oil against clinical methicillin-resistant *Staphylococcus aureus*. *Diagn. Microbiol. Infect. Dis.*, **57**, 413–418.
110. Aslim, B. and Yucel, N. (2008) In vitro antimicrobial activity of essential oil from endemic *Origanum minutiflorum* on ciprofloxacin-resistant *Campylobacter* spp. *Food Chem.*, **107**, 602–606.
111. Juven, B.J., Kanner, J., Schved, F. and Weisslowicz, H. (1994) Factors that interact with the antibacterial action of thyme essential oil and its active constituents. *J. Appl. Bacteriol.*, **76**, 626–631.
112. Canillac, N. and Mourey, A. (2004) Effects of several environmental factors on the anti-*Listeria monocytogenes* activity of an essential oil of *Picea excelsa*. *Int. J. Food Microbiol.*, **92**, 95–103.
113. Smith-Palmer, A., Stewart, J. and Fyfe, L. (2002) Inhibition of listeriolysin O and phosphatidylcholine-specific production in *Listeria monocytogenes* by subinhibitory concentrations of plant essential oils. *J. Med. Microbiol.*, **51**, 567–574.
114. Hammer, K.A., Carson, C.F. and Riley, T.V. (1999) Influence of organic matter, cations and surfactants on the antimicrobial activity of *Melaleuca alternifolia* (tea tree) oil in vitro. *J. Appl. Microbiol.*, **86**, 446–452.
115. Messager, S., Hammer, K.A., Carson, C.F. and Riley, T.V. (2005) Assessment of the antibacterial activity of tea tree oil using the European EN 1276 and EN 12054 standard suspension tests. *J. Hosp. Infect.*, **59**, 113–125.
116. Walsh, S.E., Maillard, J.Y., Russell, A.D. *et al.* (2003) Development of bacterial resistance to several biocides and effects on antibiotic susceptibility. *J. Hosp. Infect.*, **55**, 98–107.
117. Smith-Palmer, A., Stewart, J. and Fyfe, L. (2001) The potential application of plant essential oils as natural food preservatives in soft cheese. *Food Microbiol.*, **18**, 463–470.
118. Fisher, K. and Phillips, C. (2009) The mechanism of action of a citrus oil blend against *Enterococcus faecium* and *Enterococcus faecalis*. *J. Appl. Microbiol.*, **106**, 1343–1349.
119. Szczerbanik, M., Jobling, J., Morris, S. and Holford, P. (2007) Essential oil vapours control some common postharvest fungal pathogens. *Aust. J. Exp. Agricult.*, **47**, 103–109.
120. Lopez, P., Sanchez, C., Batlle, R. and Nerin, C. (2007) Vapor-phase activities of cinnamon, thyme, and oregano essential oils and key constituents against foodborne microorganisms. *J. Agricult. Food Chem.*, **55**, 4348–4356.
121. Lopez, P., Sanchez, C., Batlle, R. and Nerin, C. (2005) Solid- and vapor-phase antimicrobial activities of six essential oils: susceptibility of selected foodborne bacterial and fungal strains. *J. Agricult. Food Chem.*, **53**, 6939–6946.
122. Inouye, S., Uchida, K. and Abe, S. (2006) Vapor activity of 72 essential oils against a *Trichophyton mentagrophytes*. *J. Infect. Chemother.*, **12**, 210–216.

123. Inouye, S., Uchida, K. and Yamaguchi, H. (2001) In-vitro and in-vivo anti-*Trichophyton* activity of essential oils by vapour contact. *Mycoses*, **44**, 99–107.
124. Matan, N., Rimkeeree, H., Mawson, A.J. *et al.* (2006) Antimicrobial activity of cinnamon and clove oils under modified atmosphere conditions. *Int. J. Food Microbiol.*, **107**, 180–185.
125. Tomlinson, S. and Palombo, E.A. (2005) Characterisation of antibacterial Australian medicinal plant extracts by investigation of the mechanism of action and the effect of interfering substances. *J. Basic Microbiol.*, **45**, 363–370.
126. Ultee, A., Kets, E.P.W. and Smid, E.J. (1999) Mechanisms of action of carvacrol on the food-borne pathogen *Bacillus cereus*. *Appl. Environ. Microbiol.*, **65**, 4606–4610.
127. Sikkema, J., de Bont, J.AM. and Poolman, B. (1994) Interactions of cyclic hydrocarbons with biological membranes. *J. Biol. Chem.*, **269**, 8022–8028.
128. Ultee, A., Kets, E.P.W., Alberda, M. *et al.* (2000) Adaptation of the food-borne pathogen *Bacillus cereus* to carvacrol. *Arch. Microbiol.*, **174**, 233–238.
129. Weber, F.J. and de Bont, J.A.M. (1996) Adaptation mechanisms of microorganisms to the toxic effects of organic solvents on membranes. *Biochim. Biophys. Acta*, **1286**, 225–245.
130. Cristani, M., D'Arrigo, M., Mandalari, G. *et al.* (2007) Interaction of four monoterpenes contained in essential oils with model membranes: implications for their antibacterial activity. *J. Agricult. Food Chem.*, **55**, 6300–6308.
131. Hammer, K.A., Carson, C.F. and Riley, T.V. (2004) Antifungal effects of *Melaleuca alternifolia* (tea tree) oil and its components on *Candida albicans*, *Candida glabrata* and *Saccharomyces cerevisiae*. *J. Antimicrob. Chemother.*, **53**, 1081–1085.
132. Bouhdid, S., Skali, S.N., Idaomar, M. *et al.* (2008) Antibacterial and antioxidant activities of *Origanum compactum* essential oil. *Afr. J. Biotechnol.*, **7**, 1563–1570.
133. Cox, S.D., Mann, C.M., Markham, J.L. *et al.* (2000) The mode of antimicrobial action of the essential oil of *Melaleuca alternifolia* (tea tree oil). *J. Appl. Microbiol.*, **88**, 170–175.
134. Prashar, A., Hili, P., Veness, R.G. and Evans, C.S. (2003) Antimicrobial action of palmarosa oil (*Cymbopogon martinii*) on *Saccharomyces cerevisiae*. *Phytochem.*, **63**, 569–575.
135. Inoue, Y., Shiraishi, A., Hada, T. *et al.* (2004) The antibacterial effects of terpene alcohols on *Staphylococcus aureus* and their mode of action. *FEMS Microbiol. Lett.*, **237**, 325–331.
136. Xu, J., Zhou, F., Ji, B.P. *et al.* (2008) The antibacterial mechanism of carvacrol and thymol against *Escherichia coli*. *Lett. Appl. Microbiol.*, **47**, 174–179.
137. Cox, S.D., Gustafson, J.E., Mann, C.M. *et al.* (1998) Tea tree oil causes K+ leakage and inhibits respiration in *Escherichia coli*. *Lett. Appl. Microbiol.*, **26**, 355–358.
138. Kwon, J.A., Yu, C.B. and Park, H.D. (2003) Bacteriocidal effects and inhibition of cell separation of cinnamic aldehyde on *Bacillus cereus*. *Lett. Appl. Microbiol.*, **37**, 61–65.
139. Paparella, A., Taccogna, L., Aguzzi, I. *et al.* (2008) Flow cytometric assessment of the antimicrobial activity of essential oils against *Listeria monocytogenes*. *Food Cont.*, **19**, 1174–1182.
140. Gill, A.O. and Holley, R.A. (2004) Mechanisms of bactericidal action of cinnamaldehyde against *Listeria monocytogenes* and of eugenol against *L. monocytogenes* and *Lactobacillus sakei*. *Appl. Environ. Microbiol.*, **70**, 5750–5755.
141. Gill, A.O. and Holley, R.A. (2006) Disruption of *Escherichia coli*, *Listeria monocytogenes* and *Lactobacillus sakei* cellular membranes by plant oil aromatics. *Int. J. Food Microbiol.*, **108**, 1–9.
142. Gill, A.O. and Holley, R.A. (2006) Inhibition of membrane bound ATPases of *Escherichia coli* and *Listeria monocytogenes* by plant oil aromatics. *Int. J. Food Microbiol.*, **111**, 170–174.
143. Park, M.J., Gwak, K.S., Yang, I. *et al.* (2007) Antifungal activities of the essential oils in *Syzygium aromaticum* (L.) Merr. Et Perry and *Leptospermum petersonii* Bailey and their constituents against various dermatophytes. *J. Microbiol.*, **45**, 460–465.
144. Chami, F., Chami, N., Bennis, S. *et al.* (2005) Oregano and clove essential oils induce surface alteration of *Saccharomyces cerevisiae*. *Phytother. Res.*, **19**, 405–408.
145. Rhayour, K., Bouchikhi, T., Tantaoui-Elaraki, A. *et al.* (2003) The mechanism of bactericidal action of oregano and clove essential oils and of their phenolic major components on *Escherichia coli* and *Bacillus subtilis*. *J. Ess. Oil Res.*, **15**, 356–362.
146. Papadopoulos, C.J., Carson, C.F., Chang, B.J. and Riley, T.V. (2008) Role of the MexAB-OprM efflux pump of *Pseudomonas aeruginosa* in tolerance to tea tree (*Melaleuca altenifolia*) oil

and its monoterpene components terpinen-4-ol, 1,8-cineole, and alpha-terpineol. *Appl. Environ. Microbiol.*, **74**, 1932–1935.

147. Papanikolaou, S., Gortzi, O., Margeli, E. *et al.* (2008) Effect of *Citrus* essential oil addition upon growth and cellular lipids of *Yarrowia lipolytica* yeast. *Eur. J. Lipid Sci.Technol.*, **110**, 997–1006.

148. de Carvalho, C., Parreno-Marchante, B., Neumann, G. *et al.* (2005) Adaptation of *Rhodococcus erythropolis* DCL14 to growth on n-alkanes, alcohols and terpenes. *Appl. Microbiol. Biotechnol.*, **67**, 383–388.

149. McMahon, M.A.S., Blair, I.S., Moore, J.E. and McDowell, D.A. (2007) Habituation to sub-lethal concentrations of tea tree oil (*Melaleuca alternifolia*) is associated with reduced susceptibility to antibiotics in human pathogens. *J. Antimicrob. Chemother.*, **59**, 125–127.

150. McMahon, M.A.S., Tunney, M.M., Moore, J.E. *et al.* (2008) Changes in antibiotic susceptibility in staphylococci habituated to sub-lethal concentrations of tea tree oil (*Melaleuca alternifolia*). *Lett. Appl. Microbiol.*, **47**, 263–268.

151. Parveen, M., Hasan, M.K., Takahashi, J. *et al.* (2004) Response of *Saccharomyces cerevisiae* to a monoterpene: evaluation of antifungal potential by DNA microarray analysis. *J. Antimicrob. Chemother.*, **54**, 46–55.

152. Rossignol, T., Logue, M.E., Reynolds, K. *et al.* (2007) Transcriptional response of *Candida parapsilosis* following exposure to farnesol. *Antimicrob. Agents Chemother.*, **51**, 2304–2312.

153. Nelson, R.R.S. (2000) Selection of resistance to the essential oil of *Melaleuca alternifolia* in *Staphylococcus aureus*. *J. Antimicrob. Chemother.*, **45**, 549–550.

154. Mondello, F., De Bernardis, F., Girolamo, A. *et al.* (2003) In vitro and in vivo activity of tea tree oil against azole-susceptible and -resistant human pathogenic yeasts. *J. Antimicrob. Chemother.*, **51**, 1223–1229.

155. Ferrini, A.M., Mannoni, V., Aureli, P. *et al.* (2006) *Melaleuca alternifolia* essential oil posesses potent anti-Staphylococcal activity extended to strains resistant to antibiotics. *Int. J. Immunopathol. Pharmacol.*, **19**, 539–544.

156. Hammer, K.A., Carson, C.E. and Riley, T.V. (2008) Frequencies of resistance to *Melaleuca alternifolia* (tea tree) oil and rifampicin in *Staphylococcus aureus*, *Staphylococcus epidermidis* and *Enterococcus faecalis*. *Int.J. Antimicrob. Agents*, **32**, 170–173.

157. Ali, S.M., Khan, A.A., Ahmed, I. *et al.* (2005) Antimicrobial activities of eugenol and cinnamaldehyde against the human gastric pathogen *Helicobacter pylori*. *Ann Clin Microbiol Antimicrob.*, **4**, 20.

158. Beric, T., Nikolic, B., Stanojevic, J. *et al.* (2008) Protective effect of basil (*Ocimum basilicum* L.) against oxidative DNA damage and mutagenesis. *Food Chem. Toxicol.*, **46**, 724–732.

159. Stanojevic, J., Beric, T., Opacic, B. *et al.* (2008) The effect of essential oil of basil (*Ocimum basilicum* L.) on UV-induced mutagenesis in *Escherichia coli* and *Saccharomyces cerevisiae*. *Arch. Biol. Sci.*, **60**, 93–102.

160. Vukovic-Gacic, B., Nikcevic, S., Beric-Bjedov, T. *et al.* (2006) Antimutagenic effect of essential oil of sage (*Salvia officinalis* L.) and its monoterpenes against UV-induced mutations in *Escherichia coli* and *Saccharomyces cerevisiae*. *Food Chem. Toxicol.*, **44**, 1730–1738.

161. Gustafson, J.E., Cox, S.D., Liew, Y.C. *et al.* (2001) The bacterial multiple antibiotic resistant (Mar) phenotype leads to increased tolerance to tea tree oil. *Pathology*, **33**, 211–215.

162. Koga, T., Hirota, N. and Takumi, K. (1999) Bactericidal activities of essential oils of basil and sage against a range of bacteria and the effect of these essential oils on *Vibrio parahaemolyticus*. *Microbiol. Res.*, **154**, 267–273.

163. Davis, A., O'Leary, J., Muthaiyan, A. *et al.* (2005) Characterization of *Staphylococcus aureus* mutants expressing reduced susceptibility to common house-cleaners. *J. Appl. Microbiol.*, **98**, 364–372.

164. Smith-Palmer, A., Stewart, J. and Fyfe, L. (2004) Influence of subinhibitory concentrations of plant essential oils on the production of enterotoxins A and B and alpha-toxin by *Staphylococcus aureus*. *J. Med. Microbiol.*, **53**, 1023–1027.

165. Ultee, A. and Smid, E.J. (2001) Influence of carvacrol on growth and toxin production by *Bacillus cereus*. *Int. J. Food Microbiol.*, **64**, 373–378.

166. Echeverrigaray, S., Michelim, L., Delamare, A.P.L. *et al.* (2008) The effect of monoterpenes on swarming differentiation and haemolysin activity in *Proteus mirabilis*. *Molecules*, **13**, 3107–3116.
167. Iwalokun, B.A., Gbenle, G.O., Adewole, T.A. *et al.* (2003) Effects of *Ocimum gratissimum* L. essential oil at subinhibitory concentrations on virulent and multidrug-resistant *Shigella* strains from Lagos, Nigeria. *APMIS*, **111**, 477–482.
168. Cugini, C., Calfee, M.W., Farrow, J.M. *et al.* (2007) Farnesol, a common sesquiterpene, inhibits PQS production in *Pseudomonas aeruginosa*. *Mol. Microbiol.*, **65**, 896–906.
169. Burt, S.A., Van Der Zee, R., Koets, A.P. *et al.* (2007) Carvacrol induces heat shock protein 60 and inhibits synthesis of flagellin in *Escherichia coli* O157 : H7. *Appl. Environ. Microbiol.*, **73**, 4484–4490.
170. Bullerman, L.B., Lieu, F.Y. and Seier, S.A. (1977) Inhibition of growth and aflatoxin production by cinnamon and clove oils, cinnamic aldehyde and eugenol. *J. Food Sci.*, **42**, 1107–1109.
171. Razzaghi-Abyaneh, M., Shams-Ghahfarokhi, M., Yoshinari, T. *et al.* (2008) Inhibitory effects of *Satureja hortensis* L. essential oil on growth and aflatoxin production by *Aspergillus parasiticus*. *Int. J. Food Microbiol.*, **123**, 228–233.
172. Rasooli, I. and Abyaneh, M.R. (2004) Inhibitory effects of Thyme oils on growth and aflatoxin production by *Aspergillus parasiticus*. *Food Cont.*, **15**, 479–483.
173. Hammer, K.A., Carson, C.F. and Riley, T.V. (2000) *Melaleuca alternifolia* (tea tree) oil inhibits germ tube formation by *Candida albicans*. *Med. Mycol.*, **38**, 355–362.
174. D'Auria, F.D., Laino, L., Strippoli, V. *et al.* (2001) In vitro activity of tea tree oil against *Candida albicans* mycelial conversion and other pathogenic fungi. *J. Chemother.*, **13**, 377–383.
175. Salgueiro, L.R., Pinto, E., Goncalves, M.J. *et al.* (2004) Chemical composition and antifungal activity of the essential oil of *Thymbra capitata*. *Planta Med.*, **70**, 572–575.
176. Henriques, M., Martins, M., Azeredo, J. and Oliveira, R. (2007) Effect of farnesol on *Candida dubliniensis* morphogenesis. *Lett. Appl. Microbiol.*, **44**, 199–205.
177. Quave, C.L., Plano, L.R.W., Pantuso, T. and Bennett, B.C. (2008) Effects of extracts from Italian medicinal plants on planktonic growth, biofilm formation and adherence of methicillin-resistant *Staphylococcus aureus*. *J. Ethnopharmacol.*, **118**, 418–428.
178. Nostro, A., Roccaro, A.S., Bisignano, G. *et al.* (2007) Effects of oregano, carvacrol and thymol on *Staphylococcus aureus* and *Staphylococcus epidermidis* biofilms. *J. Med. Microbiol.*, **56**, 519–523.
179. Agarwal, V., Lal, P. and Pruthi, V. (2008) Prevention of *Candida albicans* biofilm by plant oils. *Mycopathol.*, **165**, 13–19.
180. Chorianopoulos, N.G., Giaouris, E.D., Skandamis, P.N. *et al.* (2008) Disinfectant test against monoculture and mixed-culture biofilms composed of technological, spoilage and pathogenic bacteria: bactericidal effect of essential oil and hydrosol of *Satureja thymbra* and comparison with standard acid-base sanitizers. *J. Appl. Microbiol.*, **104**, 1586–1596.
181. Niu, C., Afre, S. and Gilbert, E.S. (2006) Subinhibitory concentrations of cinnamaldehyde interfere with quorum sensing. *Lett. Appl. Microbiol.*, **43**, 489–494.
182. deCarvalho, C. and de Fonseca, M.M.R. (2007) Preventing biofilm formation: promoting cell separation with terpenes. *FEMS Microbiol. Ecol.*, **61**, 406–413.
183. Dal Sasso, M., Culici, M., Braga, P.C. *et al.* (2006) Thymol: inhibitory activity on *Escherichia coli* and *Staphylococcus aureus* adhesion to human vaginal cells. *J. Ess. Oil Res.*, **18**, 455–461.
184. Shahverdi, A.R., Monsef-Esfahani, H.R., Tavasoli, F. *et al.* (2007) Trans-cinnamaldehyde from *Cinnamomum zeylanicum* bark essential oil reduces the clindamycin resistance of *Clostridium difficile* in vitro. *J. Food Sci.*, **72**, S55–S58.
185. Brehm-Stecher, B.F. and Johnson, E.A. (2003) Sensitization of *Staphylococcus aureus* and *Escherichia coli* to antibiotics by the sesquiterpenoid nerolidol, farnesol, bisabolol, and apritone. *Antimicrob. Agents Chemother.*, **47**, 3357–3360.
186. Zhou, F., Ji, B.P., Zhang, H. *et al.* (2007) Synergistic effect of thymol and carvacrol combined with chelators and organic acids against *Salmonella typhimurium*. *J. Food Protect.*, **70**, 1704–1709.

187. Dimitrijevic, S.I., Mihajlovski, K.R., Antonovic, D.G. *et al.* (2007) A study of the synergistic antilisterial effects of a sub-lethal dose of lactic acid and essential oils from *Thymus vulgaris* L., *Rosmarinus officinalis* L. and *Origanum vulgare* L. *Food Chem.*, **104**, 774–782.
188. Lee, S.Y. and Jin, H.H. (2008) Inhibitory activity of natural antimicrobial compounds alone or in combination with nisin against *Enterobacter sakazakii*. *Lett. Appl. Microbiol.*, **47**, 315–321.
189. Zhou, F., Ji, B.P., Zhang, H. *et al.* (2007) The antibacterial effect of cinnamaldehyde, thymol, carvacrol and their combinations against the foodborne pathogen *Salmonella typhimurium*. *J. Food Safety*, **27**, 124–133.
190. Cox, S.D., Mann, C.M. and Markham, J.L. (2001) Interactions between components of the essential oil of *Melaleuca alternifolia*. *J. Appl. Microbiol.*, **91**, 492–497.
191. Maruyama, N., Takizawa, T., Ishibashi, H. *et al.* (2008) Protective activity of geranium oil and its component, geraniol, in combination with vaginal washing against vaginal candidiasis in mice. *Biol. Pharm. Bull.*, **31**, 1501–1506.
192. Fitzi, J., Furst-Jucker, J., Wegener, T. *et al.* (2002) Phytotherapy of chronic dermatitis and pruritus of dogs with a topical preparation containing tea tree oil (Bogaskin (R)). *Schweiz. Arch. Tierheilk.*, **144**, 223–231.
193. Reichling, J., Fitzi, J., Hellman, K. *et al.* (2004) Topical tea tree oil effective in canine localised pruritic dermatitis: a multi-centre randomised double-blind controlled clinical trial in the veterinary practice. *Dtsch. Tierärzt. Wochenschr.*, **111**, 408–414.
194. Burke, B., Baillie, J. and Olson, R. (2004) Essential oil of Australian lemon myrtle (*Backhousia citriodora*) in the treatment of molluscum contagiosum in children. *Biomed. Pharmacother.*, **58**, 245–247.
195. Carson, C.F., Ashton, L., Dry, L. *et al.* (2001) *Melaleuca alternifolia* (tea tree) oil gel (6%) for the treatment of recurrent herpes labialis. *J. Antimicrob. Chemother.*, **48**, 450–451.
196. Gao, Y.Y., Di Pascuale, M.A., Elizondo, A. and Tseng, S.C. (2007) Clinical treatment of ocular demodecosis by lid scrub with tea tree oil. *Cornea*, **26**, 136–143.
197. McCage, C.M., Ward, S.M., Paling, C.A. *et al.* (2002) Development of a paw paw herbal shampoo for the removal of head lice. *Phytomedicine*, **9**, 743–748.
198. Lauten, J., Boyd, L., Hanson, M. *et al.* (2005) A clinical study: Melaleuca, Manuka, Calendula and green tea mouth rinse. *Phytother. Res.*, **19**, 951–957.
199. Syed, T.A., Qureshi, Z.A., Ali, S.M. *et al.* (1999) Treatment of toenail onychomycosis with 2% butenafine and 5% *Melaleuca alternifolia* (tea tree) oil in cream. *Trop. Med. Internat. Health*, **4**, 284–287.
200. Buck, D.S., Nidorf, D.M. and Addino, J.G. (1994) Comparison of two topical preparations for the treatment of onychomycosis: *Melaleuca alternifolia* (tea tree) oil and clotrimazole. *J. Fam. Pract.*, **38**, 601–605.
201. Prashar, A., Locke, I.C. and Evans, C.S. (2004) Cytotoxicity of lavender oil and its major components to human skin cells. *Cell Prolif.*, **37**, 221–229.
202. Sivropoulou, A., Papanikolaou, E., Nikolaou, C. *et al.* (1996) Antimicrobial and cytotoxic activities of origanum essential oils. *J. Agr. Food Chem.*, **44**, 1202–1205.
203. Suschke, U., Sporer, F., Schneele, J. *et al.* (2007) Antibacterial and cytotoxic activity of *Nepeta cataria* L., *N. cataria* var. *citriodora* (Beck.) Balb. and *Melissa officinalis* L. essential oils. *Nat. Prod. Comm.*, **2**, 1277–1286.
204. Villar, D., Knight, M., Hansen, S. and Buck, W. (1994) Toxicity of melaleuca oil and related essential oils applied topically on dogs and cats. *Vet. Hum. Toxicol.*, **36**, 139–142.
205. Morris, M.C., Donoghue, A., Markowitz, J.A. and Osterhoudt, K.C. (2003) Ingestion of tea tree oil (Melaleuca oil) by a 4-year-old boy. *Pediatr. Emergency Care*, **19**, 169–171.
206. Landelle, C., Francony, G., Sam-Lai, N.F. *et al.* (2008) Poisoning by lavandin extract in a 18-month-old boy. *Clin. Toxicol.*, **46**, 279–281.
207. Flaman, Z., Pellechia-Clarke, S., Bailey, B. and McGuigan, M. (2001) Unintentional exposure of young children to camphor and eucalyptus oils. *Paediatr. Child Health*, **6**, 80–83.
208. Janes, S.E., Price, C.S. and Thomas, D. (2005) Essential oil poisoning: N-acetylcysteine for eugenol-induced hepatic failure and analysis of a national database. *Eur. J. Pediatr.*, **164**, 520–522.

209. Juergens, U., Dethlefsen, U., Steinkamp, G. *et al.* (2003) Anti-inflammatory activity of 1.8-cineol (eucalyptol) in bronchial asthma: a double-blind placebo-controlled trial. *Respir. Med.*, **97**, 250–256.
210. Bradley, B., Brown, S., Chu, S. and Lea, R. (2009) Effects of orally administered lavender essential oil on responses to anxiety-provoking film clips. *Hum. Psychopharmacol.*, **24**, 319–330.
211. Worth, H., Schacher, C. and Dethlefsen, U. (2009) Concomitant therapy with cineole (eucalyptole) reduces exacerbations in COPD: a placebo-controlled double-blind trial.. *Respir. Res.*, **10**.
212. Orafidiya, L., Agbani, E., Oyedele, A. *et al.* (2002) Preliminary clinical tests on topical preparations of *Ocimum gratissimum* Linn leaf essential oil for the treatment of acne vulgaris. *Clin. Drug Invest.*, **22**, 313–319.
213. Brared Christensson, J., Forsstrom, P., Wennberg, A.M. *et al.* (2009) Air oxidation increases skin irritation from fragrance terpenes. *Contact Dermatitis*, **60**, 32–40.
214. Hagvall, L., Skold, M., Brared-Christensson, J. *et al.* (2008) Lavender oil lacks natural protection against autoxidation, forming strong contact allergens on air exposure. *Contact Dermatitis*, **59**, 143–150.
215. Mozelsio, N.B., Harris, K.E., McGrath, K.G. and Grammer, L.C. (2003) Immediate systemic hypersensitivity reaction associated with topical application of Australian tea tree oil. *Allergy Asthma Proc.*, **24**, 73–75.
216. Garcia-Abujeta, J.L., Larramendi, C.H., Berna, J.P. and Palomino, E.M. (2005) Mud bath dermatitis due to cinnamon oil. *Contact Dermatitis*, **52**, 234.
217. Zacher, K.D. and Ippen, H. (1984) [Contact dermatitis caused by bergamot oil]. *Derm. Beruf Umwelt*, **32**, 95–97.
218. Hammer, K.A., Carson, C.F. and Riley, T.V. (1998) In-vitro activity of essential oils, in particular *Melaleuca alternifolia* (tea tree) oil and tea tree oil products, against *Candida* spp. *J. Antimicrob. Chemother.*, **42**, 591–595.
219. Fu, Y.J., Zu, Y.G., Chen, L.Y. *et al.* (2007) Antibacterial activity of clove and rosemary essential oils alone and in combination. *Phytother. Res.*, **21**, 989–994.
220. Carson, C.F., Hammer, K.A. and Riley, T.V. (1995) Broth micro-dilution method for determining the susceptibility of *Escherichia coli* and *Staphylococcus aureus* to the essential oil of *Melaleuca alternifolia* (tea tree oil). *Microbios*, **82**, 181–185.
221. Papadopoulos, C.J., Carson, C.F., Hammer, K.A. and Riley, T.V. (2006) Susceptibility of pseudomonads to *Melaleuca alternifolia* (tea tree) oil and components. *J. Antimicrob. Chemother.*, **58**, 449–451.
222. Kawakami, E., Washizu, M., Hirano, T. *et al.* (2006) Treatment of prostatic abscesses by aspiration of the purulent matter and injection of tea tree oil into the cavities in dogs. *J. Vet. Med. Sci.*, **68**, 1215–1217.
223. Kristinsson, K., Magnusdottir, A., Petersen, H. and Hermansson, A. (2005) Effective treatment of experimental acute otitis media by application of volatile fluids into the ear canal. *J. Infect. Dis.*, **191**, 1876–1880.
224. Tsao, N., Kuo, H.-Y., Lu, S.-L. and Huang, K.-J. (2010) Inhibition of group A streptococcal infection by *Melaleuca alternifolia* (tea tree) oil concentrate in the murine model. *J. Appl. Microbiol.*, **108**, 936–944.
225. Pisseri, F., Bertoli, A., Nardoni, S. *et al.* (2009) Antifungal activity of tea tree oil from *Melaleuca alternifolia* against *Trichophyton equinum*: an in vivo assay. *Phytomedicine*, **16**, 1056–1058.
226. Sokovic, M., Glamoclija, J., Ciric, A. *et al.* (2008) Antifungal activity of the essential oil of *Thymus vulgaris* L. and thymol on experimentally induced dermatomycoses. *Drug Dev. Ind. Pharm.*, **34**, 1388–1393.
227. Tiwari, T., Chansouria, J. and Dubey, N. (2003) Antimycotic potency of some essential oils in the treatment of induced dermatomycosis of an experimental animal. *Pharmaceut. Biol.*, **41**, 351–356.
228. Manohar, V., Ingram, C., Gray, J. *et al.* (2001) Antifungal activities of origanum oil against *Candida albicans*. *Mol. Cell. Biochem.*, **228**, 111–117.

229. Hood, J., Burton, D., Wilkinson, J. and Cavanagh, H. (2010) Antifungal activity of *Leptospermum petersonii* oil volatiles against *Aspergillus* spp. in vitro and in vivo. *J. Antimicrob. Chemother.*, **65**, 285–288.
230. Blackwell, A.L. (1991) Tea tree oil and anaerobic (bacterial) vaginosis. *Lancet*, **337**, 300.
231. Jandourek, A., Vaishampayan, J.K. and Vazquez, J.A. (1998) Efficacy of melaleuca oral solution for the treatment of fluconazole refractory oral candidiasis in AIDS patients. *AIDS*, **12**, 1033–1037.
232. Sherry, E., Reynolds, M., Sivananthan, S. *et al.* (2004) Inhalational phytochemicals as possible treatment for pulmonary tuberculosis: Two case reports. *Am. J. Infect. Control*, **32**, 369–370.
233. Bassett, I.B., Pannowitz, D.L. and Barnetson, R.S. (1990) A comparative study of tea-tree oil versus benzoylperoxide in the treatment of acne. *Med. J. Aust.*, **153**, 455–458.
234. Caelli, M., Porteous, J., Carson, C.F. *et al.* (2000) Tea tree oil as an alternative topical decolonization agent for methicillin-resistant *Staphylococcus aureus*. *J. Hosp. Infect.*, **46**, 236–237.
235. Dryden, M.S., Dailly, S. and Crouch, M. (2004) A randomized, controlled trial of tea tree topical preparations versus a standard topical regimen for the clearance of MRSA colonization. *J. Hosp. Infect.*, **56**, 283–286.
236. Satchell, A.C., Saurajen, A., Bell, C. and Barnetson, R.S. (2002) Treatment of dandruff with 5% tea tree oil shampoo. *J. Am. Acad. Dermatol.*, **47**, 852–855.
237. Vazquez, J.A. and Zawawi, A.A. (2002) Efficacy of alcohol-based and alcohol-free melaleuca oral solution for the treatment of fluconazole-refractory oropharyngeal candidiasis in patients with AIDS. *HIV Clin. Trials*, **3**, 379–385.
238. Tong, M.M., Altman, P.M. and Barnetson, R.S. (1992) Tea tree oil in the treatment of tinea pedis. *Australas. J. Dermatol.*, **33**, 145–149.
239. Satchell, A.C., Saurajen, A., Bell, C. and Barnetson, R.S.C. (2002) Treatment of interdigital tinea pedis with 25% and 50% tea tree oil solution: a randomized, placebo-controlled, blinded study. *Australas. J. Dermatol.*, **43**, 175–178.
240. Opdyke, D.L. (1973) Monographs on fragrance raw materials: coriander oil. *Food Cosmet. Toxicol.*, **11**, 1077–1081.
241. Opdyke, D.L. (1976) Monographs on fragrance raw materials: lavender. *Food Cosmet. Toxicol.*, **14**, 451.
242. Opdyke, D.L. and Letizia, C. (1982) Monographs on fragrance raw materials. *Food Chem. Toxicol.*, **20** (Suppl), 633–852.
243. Opdyke, D.L. (1976) Monographs on fragrance raw materials: marjoram oil, sweet. *Food Cosmet. Toxicol.*, **14**, 469.
244. Ford, R. (1988) Fragrance raw materials monographs: tea tree oil. *Food Chem. Toxicol.*, **26**, 407.
245. Opdyke, D.L. (1973) Monographs on fragrance raw materials: basil oil, sweet. *Food Cosmet. Toxicol.*, **11**, 867–868.
246. Opdyke, D.L. (1973) Monographs on fragrance raw materials: bay oil. *Food Cosmet. Toxicol.*, **11**, 869–870.

Index

References to tables are given in bold type. References to figures are given in italic type.

Lipids and Essential Oils as Antimicrobial Agents   Halldor Thormar
© 2011 John Wiley & Sons, Ltd